Actuating, Sensing, Control, and Instrumentation for Ultra Precision Engineering

Actuating, Sensing, Control, and Instrumentation for Ultra Precision Engineering

Editors

Limin Zhu
Yuen Kuan Yong
Yanling Tian
Yingxiang Liu

MDPI

Basel • Beijing • Wuhan • Barcelona • Belgrade • Novi Sad • Cluj • Manchester

Editors

Limin Zhu
School of Mechanical Engineering
Shanghai Jiao Tong University
Shanghai, China

Yuen Kuan Yong
School of Electrical Engineering and Computer Science
University of Newcastle
Callaghan, NSW, Australia

Yanling Tian
School of Engineering
University of Warwick
Coventry, UK

Yingxiang Liu
State Key Laboratory of Robotics and System
Harbin Institute of Technology
Harbin, China

Editorial Office
MDPI
St. Alban-Anlage 66
4052 Basel, Switzerland

This is a reprint of articles from the Special Issue published online in the open access journal *Actuators* (ISSN 2076-0825) (available at: https://www.mdpi.com/journal/actuators/special_issues/Actuating_Ultra_Precision_Engineering).

For citation purposes, cite each article independently as indicated on the article page online and as indicated below:

Lastname, A.A.; Lastname, B.B. Article Title. *Journal Name* **Year**, *Volume Number*, Page Range.

ISBN 978-3-0365-9526-9 (Hbk)
ISBN 978-3-0365-9527-6 (PDF)
doi.org/10.3390/books978-3-0365-9527-6

Contents

actuators MDPI

Article

Hysteresis Modeling and Compensation of Piezoelectric Actuators Using Gaussian Process with High-Dimensional Input

Yixuan Meng, Xiangyuan Wang, Linlin Li, Weiwei Huang and Limin Zhu *

State Key Laboratory of Mechanical System and Vibration, School of Mechanical Engineering, Shanghai Jiao Tong University, Shanghai 200240, China; mengyixuan@sjtu.edu.cn (Y.M.); wangxiangyuan@sjtu.edu.cn (X.W.); lilinlin321@sjtu.edu.cn (L.L.); 12510102161080@sjtu.edu.cn (W.H.)
* Correspondence: zhulm@sjtu.edu.cn

Abstract: Rate-dependent hysteresis seriously deteriorates the positioning accuracy of the piezoelectric actuators, especially when tracking high-frequency signals. As a widely-used nonparametric Bayesian method, the Gaussian process (GP) has proven its effectiveness in nonlinear hysteresis modeling. In this paper, the dimension of the input to the GP model is extended to consider more dynamic features of the tracking signal so as to improve the rate-dependent hysteresis modeling accuracy. In contrast with the traditional training set containing only the position and speed information, the acceleration and jerk information, as well as their temporal distribution information, is also included in the input of the model. An inverse hysteresis compensator (IHC) is established in the same way, and open-loop and closed-loop controllers are developed by using the IHC. Experimental results on a PEA stage show that with the increase in the input dimension, the hysteresis modeling accuracy improves greatly and, thus, the controllers based on IHC can achieve a better tracking performance.

Keywords: piezoelectric actuators; hysteresis modeling; Gaussian process; training input; tracking control

Citation: Meng, Y.; Wang, X.; Li, L.; Huang, W.; Zhu, L. Hysteresis Modeling and Compensation of Piezoelectric Actuators Using Gaussian Process with High-Dimensional Input. *Actuators* **2022**, *11*, 115. https://doi.org/10.3390/act11050115

Academic Editor: Katsushi Furutani

Received: 29 March 2022
Accepted: 20 April 2022
Published: 20 April 2022

1. Introduction

Piezoelectric-driven nano-positioning stages are widely used in ultra-precision machining and measurement, such as fast tool servo (FTS) [1,2] and atomic force microscope (AFM) [3,4]. However, the rate-dependent hysteresis of the piezoelectric actuators (PEAs) seriously deteriorates the positioning accuracy, resulting in the morphological error of FTS and the image distortion of AFM [5]. In addition, as the input voltage frequency increases, the hysteresis loop becomes larger and rounder, which seriously limits the positioning performance, especially in high-speed positioning applications [6]. Therefore, in order to improve the positioning accuracy of the PEAs, it is important to model and compensate for the rate-dependent hysteresis nonlinearity.

In the past few decades, researchers have conducted extensive works to deal with the challenge of hysteretic nonlinearity modeling [7–9]. The relevant models can be generally divided into two types: the physics-based model and the phenomenological model [10,11]. The physics-based hysteresis model is derived from the basic physical principles of hysteresis materials via the relationship between physical quantities and empirical formulas [12]. Nevertheless, due to the complex physical causes of the actual nonlinear hysteresis system, it is difficult to establish the model based on physical principles. On the other hand, the phenomenological hysteresis model directly uses mathematical models to describe the nonlinear input and output relationship of hysteresis, without considering its inherent physical characteristics. Because of its relatively simple structure, the phenomenological model has become the most widely used one in hysteresis research, which mainly includes the Preisach model [13,14], Prandtl–Ishlinskii (P-I) model [15,16], Bouc–Wen model [17–19],

Dahl model [20], Duhem model [21], etc. Unfortunately, most of these models can only describe the rate-independent hysteresis.

Owning to the fact that hysteretic nonlinearity is generally rate-dependent [6], more rate-dependent models were developed to consider the changing rate of the input voltage, including the rate-dependent Preisach model [22] and rate-dependent PI model [23–26]. Nonetheless, these models usually contain plentiful unknown parameters, which makes the identification more difficult. Therefore, in 2015, Yang et al. [27] introduced a velocity damping mechanism into the traditional PI model, which successfully constructed a modified rate-dependent PI (MPI) model. This model can accurately describe the dynamic hysteresis characteristics with a relatively simple model structure and relatively few unknown parameters. Recently, with the development of machine learning methods, intelligent models are becoming more popular due to their powerful nonlinear approximation capability [28,29]. Researchers have developed various machine learning methods to model the rate-dependent hysteresis, such as neural networks [30], fuzzy system [31], least-squares support vector machines [32], adaptive fuzzy internal model [33] and so on. These intelligent methods have shown great improvements in modeling accuracy, which provide a new way for modeling the hysteresis of the PEAs.

As a widely-used machine learning method, the Gaussian process (GP), which is based on Bayesian probabilistic inference, removes the need for selecting many parameters while retaining the power of nonlinear dynamic description. Therefore, the usage of GP could make the model flexible and accurate without specifying the functional form and the parameters. It is a proper choice for rate-dependent hysteresis modeling. In 2019, Tao et al. [34] firstly applied the GP to the modeling and feedforward compensation of the PEA hysteresis. Since the hysteresis behavior of PEA is rate-dependent, the voltage value and its changing rate were utilized as a two-dimensional input of the training set and the displacement of the PEA was used as the output, which outperforms the aforementioned MPI model and many other classical models. However, for mixed-frequency signals that are widely used in nano-positioning tracking, the model performance is still unsatisfactory. Therefore, it is necessary to develop an effective model for complex tracking signals.

In this work, besides the position and velocity information used in the traditional models, the acceleration and jerk information, as well as their temporal distribution information, is introduced into the input of the training set and the dimension of the input is increased. The effect of input dimension on prediction is studied and the inverse hysteresis model is obtained via exchanging the input and output signals of the experiment. The IHC is established through the inverse hysteresis model, and open-loop and closed-loop controllers are developed by using the inverse hysteresis compensator (IHC). Comparative experimental investigations on a commercial piezo-actuated stage validate the effectiveness of the proposed GP-based model with high-dimensional input and the controllers based on the IHC. The results show that the increase in the input dimension would improve the accuracy of model prediction and the controller, especially for complex tracking signals. Therefore, the tracking performance of the controllers with the proposed model is greatly improved compared to those controllers with the traditional 2-dimensional GP model and the MPI model.

2. Rate-Dependent Hysteresis Nonlinear Modeling of PEA

For piezoelectric actuators, the mapping between the input voltage and the output displacement is nonlinear, rate-dependent and memorable because of hysteresis. Thus, the critical issue is to predict the output in the case of a particular input under the influence of hysteresis. The input and output of the GP training set can be expressed as $\mathbf{D} = \{X, y\} = \{x_i, y_i\}_{i=1}^{n}$, where n is the length of the training input and output vectors and the training output $y_i \in \mathbb{R}$ is the measured displacement of the PEA. Different from the traditional two-dimensional input $x_i = (v_i, \dot{v}_i) \in \mathbb{R}^2$, where v_i is the input voltage value and $\dot{v}_i$ is its relative changing rate, in this work, the dimensions of the input are extended to six, nine and twelve, which will be discussed in detail in Section 2.2. With these training sets, the latent function $f(x)$ between the input and output could be mapped via training the GP model and the output displacement for a new input X^* out of the training set can be predicted.

2.1. Principle of GP-Based Hysteresis Modeling

Essentially, GP is a series of random variables, in which any finite random variables are subject to joint Gaussian distribution over functions, $p(f)$. Therefore, if two or more points are picked in a function, observation of the outputs at these points will follow a joint Gaussian distribution. From a function-space view, a GP is formally specified by a mean function $m(x)$ and a covariance function $k(x, x')$, where:

$$m(x) = \mathbf{E}[f(x)] \tag{1}$$

$$k(x, x') = \mathbf{E}[(f(x) - m(x))(f(x') - m(x'))] \tag{2}$$

Therefore, the latent function of GP can be written as:

$$f(x) \sim \mathrm{GP}(m(x),\, k(x, x')) \tag{3}$$

In order to facilitate derivation, it is necessary to preprocess the data and subtract its mean to obtain the zero-mean distribution. The following inferences assume $m(x) = 0$. The covariance function $k(x, x')$, also called the kernel, plays a pivotal role in the Gaussian regression model.

Suppose that the data noise obtained by the sensor is ε, and assume it is independently and identically distributed and its variance is σ_n^2 [34], then:

$$y = f(x) + \varepsilon \tag{4}$$

and the joint prior distribution is:

$$\begin{bmatrix} y \\ y^* \end{bmatrix} \sim \mathbf{N}\left(0, \begin{bmatrix} K(\mathbf{X},\mathbf{X}) + \sigma_n^2 \mathbf{I} & K(\mathbf{X},\mathbf{X}^*) \\ K(\mathbf{X}^*,\mathbf{X}) & K(\mathbf{X}^*,\mathbf{X}^*) \end{bmatrix}\right) \tag{5}$$

where $K(\mathbf{X},\mathbf{X})$ is the covariance matrix between all the data sets in input $\mathbf{X}$ and it is expressed as:

$$K(\mathbf{X},\mathbf{X}) = \begin{bmatrix} k(x_1, x_1) & \cdots & k(x_1, x_n) \\ \vdots & \ddots & \vdots \\ k(x_n, x_1) & \cdots & k(x_n, x_n) \end{bmatrix} \tag{6}$$

Other entries $K(\mathbf{X},\mathbf{X}^*)$, $K(\mathbf{X}^*,\mathbf{X})$ and $K(\mathbf{X}^*,\mathbf{X}^*)$ have similar definitions. According to the joint Gaussian prior distribution obtained from the observed value, the joint posterior distribution can be calculated as:

$$p(y^* | \mathbf{X}, y, \mathbf{X}^*) \sim \mathbf{N}(\overline{y}^*, cov(y^*)) \tag{7}$$

It is the critical predictive equation for GP regression with:

$$\overline{y}^* \triangleq \mathbf{E}\left(y^* \middle| \mathbf{X}, y, \mathbf{X}^*\right) = K(\mathbf{X}^*,\mathbf{X})\left[K(\mathbf{X},\mathbf{X}) + \sigma_n^2 \mathbf{I}\right]^{-1} y \tag{8}$$

$$cov(y^*) = K(\mathbf{X}^*,\mathbf{X}^*) - K(\mathbf{X}^*,\mathbf{X})\left[K(\mathbf{X},\mathbf{X}) + \sigma_n^2 \mathbf{I}\right]^{-1} K(\mathbf{X},\mathbf{X}^*) \tag{9}$$

Finally, $\overline{y}^*$ will be taken as the prediction of the displacement of the PEA.

As mentioned above, the covariance function $k(x, x')$ is the kernel function of the GP, which defines the similarity between samples. As one of the most popular covariance functions, the squared exponential (SE) function is chosen in this work, which is expressed as:

$$k(x,x') = \sigma_f^2 \exp\left(-\frac{||x-x'||^2}{2l^2}\right) \tag{10}$$

where σ_f^2 is the signal variance and l is the length scale. These parameters contained in the covariance function are called hyperparameters, which are defined as $\boldsymbol{\theta}$. Considering the noise term ε in Equation (4), the hyperparameters are:

$$\boldsymbol{\theta} = \left[l,\ \sigma_f^2,\ \sigma_n^2\right] \tag{11}$$

Hence, the logarithmic marginal likelihood is:

$$logp(\boldsymbol{y}|\boldsymbol{X},\ \boldsymbol{\theta}) = -\frac{1}{2}\boldsymbol{y}^T\left[K(\boldsymbol{X},\boldsymbol{X}) + \sigma_n^2\boldsymbol{I}\right]^{-1}\boldsymbol{y} - \frac{1}{2}\log\left|K(\boldsymbol{X},\boldsymbol{X}) + \sigma_n^2\boldsymbol{I}\right| - \frac{n}{2}log2\pi \tag{12}$$

The above logarithmic marginal likelihood can be maximized iteratively by the numerical optimization methods and the hyperparameters can be obtained using the training input information. For more information about the hyperparametric optimization, readers can refer to [34].

2.2. Extension of the Input Dimensions

As a rate-dependent behavior, the hysteresis nonlinearity is dependent on not only the voltage and its changing rate, but also the changing acceleration and even the jerk. Hence, in order to improve the modeling accuracy, it is necessary to increase the input dimension to include more information into the training set. Inspired by the work of Hu et al. to predict the contouring error dynamics of the multi-axis systems with a deep gated recurrent unit (GRU) neural network [35], the input dimension is extended to six, nine and twelve to improve the modeling performance.

Suppose that the voltage of PEA is:

$$\boldsymbol{v} = \left[v_T,\ v_{2T}, v_{3T}, \ldots, v_{nT}\right]^{\mathrm{T}} \tag{13}$$

where T is the sampling period. By differentiating the voltage signal $\boldsymbol{v}$, the changing rate $\dot{\boldsymbol{v}}$, the changing acceleration $\ddot{\boldsymbol{v}}$ and the changing jerk $\dddot{\boldsymbol{v}}$ can be obtained as:

$$\dot{\boldsymbol{v}} = \left[\dot{v}_T,\ \dot{v}_{2T}, \dot{v}_{3T}, \ldots, \dot{v}_{nT}\right]^{\mathrm{T}} \tag{14}$$

$$\ddot{\boldsymbol{v}} = \left[\ddot{v}_T,\ \ddot{v}_{2T}, \ddot{v}_{3T}, \ldots, \ddot{v}_{nT}\right]^{\mathrm{T}} \tag{15}$$

$$\dddot{\boldsymbol{v}} = \left[\dddot{v}_T,\ \dddot{v}_{2T}, \dddot{v}_{3T}, \ldots, \dddot{v}_{nT}\right]^{\mathrm{T}} \tag{16}$$

Different from the two-dimensional input of the training set, which contains only the voltage and its changing rate information at time t, the six-dimensional input, nine-dimensional input and twelve-dimensional input at time t are constructed by considering not only more temporal derivative information but also the temporal distribution information. They are described as follows:

$$\boldsymbol{x}_{6t} = \left(v_{t-T}, v_t, v_{t+T}, \dot{v}_{t-T}, \dot{v}_t, \dot{v}_{t+T}\right) \tag{17}$$

$$\boldsymbol{x}_{9t} = \left(v_{t-T}, v_t, v_{t+T}, \dot{v}_{t-T}, \dot{v}_t, \dot{v}_{t+T}, \ddot{v}_{t-T}, \ddot{v}_t, \ddot{v}_{t+T}\right) \tag{18}$$

$$\boldsymbol{x}_{12t} = \left(v_{t-T}, v_t, v_{t+T}, \dot{v}_{t-T}, \dot{v}_t, \dot{v}_{t+T}, \ddot{v}_{t-T}, \ddot{v}_t, \ddot{v}_{t+T}, \dddot{v}_{t-T}, \dddot{v}_t, \dddot{v}_{t+T}\right) \tag{19}$$

It is worth noting that the original signal collected by the displacement sensor includes noise. The differential process is that the displacement difference between two adjacent sampling points divides by the sampling period. If there exists displacement noise, the noise will be enlarged greatly since it is divided by a rather small denominator, the sampling period. As the differential order increases, the noise will be consistently enlarged in the same way. Thus, a low-pass finite impulse response (FIR) filter is utilized to eliminate the noise of the training set.

2.3. Experimental Setup

The experiments are conducted on a commercial one-dimension piezoelectric-driven nano-positioning stage (P66X30, Harbin Core Tomorrow Science and Technology Co., Ltd., Harbin, China). Its displacement is realized by the deformation of a flexure hinge guiding mechanism with a travel range from 0~26.52 μm. The stage is driven by a piezo-ceramic-made PEA composed of lead zirconate titanate and an integrated high-resolution strain gauge sensor is used to measure the displacement. A voltage amplifier with an amplification factor of 15 is employed to drive the PEA and a signal conditioner is used to capture the position signal from the strain gauge sensor. The control algorithms are programmed in the environment of MATLAB/Simulink and downloaded to a rapid prototyping system (dSPACE-DS1103) equipped with several 16-bit Analog-to-Digital Converters (ADCs) and the 16-bit Digital to Analog Convertors (DACs) to realize real-time control. The sampling frequency of the experiments is selected as 20 kHz. The whole experimental setup is illustrated in Figure 1a, while its signal flow block diagram is shown in Figure 1b.

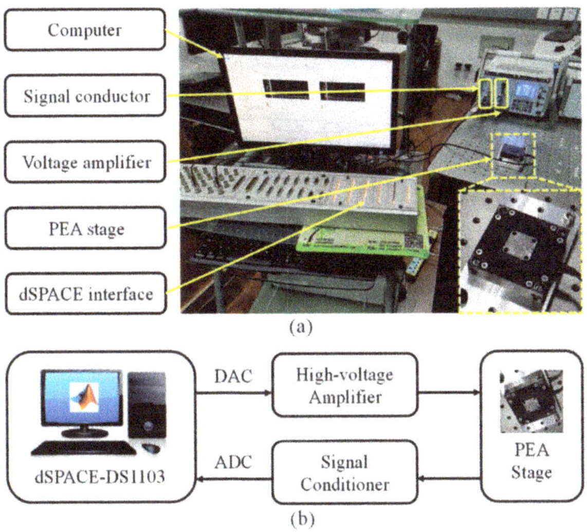

Figure 1. Experimental setup of PEA nano-positioning system: (**a**) experimental platform; (**b**) block diagram.

2.4. Modeling Results with Different Input Dimension

Since the bandwidth of the proportional-integral (PI) controller of the experimental stage is approximately 1700 Hz, 50–1000 Hz is selected as the interesting frequency range. Therefore, to obtain the training set, a chirp signal with a frequency from 25 to 1025 Hz and an amplitude of up to 30 V is selected as the voltage input to drive the PEA, as shown in Figure 2. And the displacement of the PEA stage is captured by the strain gauge sensor, which is used as the output of the training set. Additionally, the MPI rate-dependent model [27], a classical phenomenological model (please refer to the Appendix A for more details), is built to describe the hysteresis nonlinearity via the same dataset for comparison.

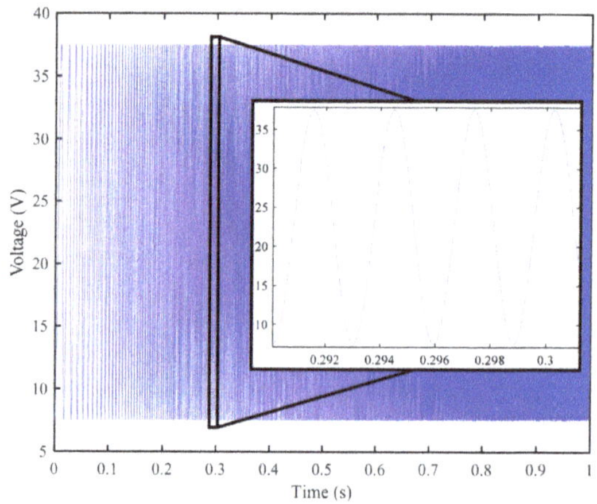

Figure 2. Driven signal for training.

To evaluate the modeling accuracy, the widely-used normalized root mean squared error (NRMSE) and the relative maximum error (RME) are chosen as the evaluation indexes [7,8,27,34], which are defined respectively as:

$$\text{NRMSE} = \frac{\sqrt{\frac{1}{n}\sum_{i=1}^{n}(y_i - \hat{y}_i)^2}}{\max_i(y_i) - \min_i(y_i)} \times 100\% \tag{20}$$

$$\text{RME} = \frac{\max_i(|y_i - \hat{y}_i|)}{\max_i(y_i) - \min_i(y_i)} \times 100\% \tag{21}$$

where y_i and $\hat{y}_i$ are the true displacements and the model predictions of the PEA, respectively.

To evaluate the modeling accuracy with different input dimensions, sinusoidal signals with a frequency of 100 Hz, 200 Hz, 300 Hz, 500 Hz and 1000 Hz and an amplitude of up to 30 V, a two-frequency mixed-signal $x(t) = 7.5(\sin(120 \times 2\pi t) + \sin(180 \times 2\pi t)) + 22.5$, a four-frequency mixed-signal $x(t) = 6\sin(100 \times 2\pi t) + 3\sin(150 \times 2\pi t) + 3\sin(200 \times 2\pi t) + 6\sin(250 \times 2\pi t) + 22.5$ and a triangular wave of 50 Hz and an amplitude of up to 20 V, are chosen as the driven signals, respectively. It is noted that under high frequency inputs, the actual control voltage will be greatly enlarged due to the lightly-damped resonance of the nanopositioning stage, especially for the closed-loop controllers. To avoid the control saturation, this study is dedicated to the range from 0–10 μm as a reference for any range study. The hyperparameters are obtained by a typical GP algorithm implemented in our MATLAB environment, converged after some 200 iterations on the measured data. No specific parameter optimization is applied at this stage of the investigation that is focused on the adequate performance of the GP method. The results obtained with the MPI model and the GP-based model with different input dimensions are listed in Table 1, and a comparison results with different methods is shown in Figure 3.

From Table 1, it can be observed that for the sinusoidal signal inputs, the GP-based models with different input dimensions all outperform the MPI model. With the increase in the input dimension, the prediction accuracy is slightly enhanced. With the increase in the signal frequency, the predictive capability of the MPI model deteriorates rapidly. However, for the GP-based model, the prediction accuracy retains satisfactory even when the input frequency is very high. For the 1000 Hz sinusoidal signal, the NRMSE for the 12-dimensional input GP-based model is just 1.2855%, much smaller than that of the MPI model. For the mixed-frequency signals, the NRMSEs for the GP-based models with 2-dimensional and 6-dimensional inputs are similar to those for the MPI model. With the increase in the input dimension, the prediction accuracy can be greatly improved. The NRMSE for the GP-based model with the 12-dimensional input is only 1.0759% for the four-frequency mixed-signal, which is just 37.26% of that for the MPI model and 35.62% of that for the GP-based model with the 2-dimensional input. For the triangular wave, the GP-based model yields the similar prediction accuracy as compared with the MPI model. The above testing results verify the effectiveness of GP-based rate-dependent hysteresis modeling with high-dimensional input and show that the modeling accuracy is greatly improved with the enhancement of the input dimension. The proposed high-dimensional GP-based model outperforms the classical phenomenological MPI model and the traditional 2-dimensional GP-based model for all testing signals. Compared with Refs. [7–9], this model method is capable of realizing more accurate prediction results with fewer hyperparameters. It has been proven a great success in modeling complex signals, including mixed-frequency signal and triangular wave. Nonetheless, the effectiveness of modeling these complex signals were not validated in these references.

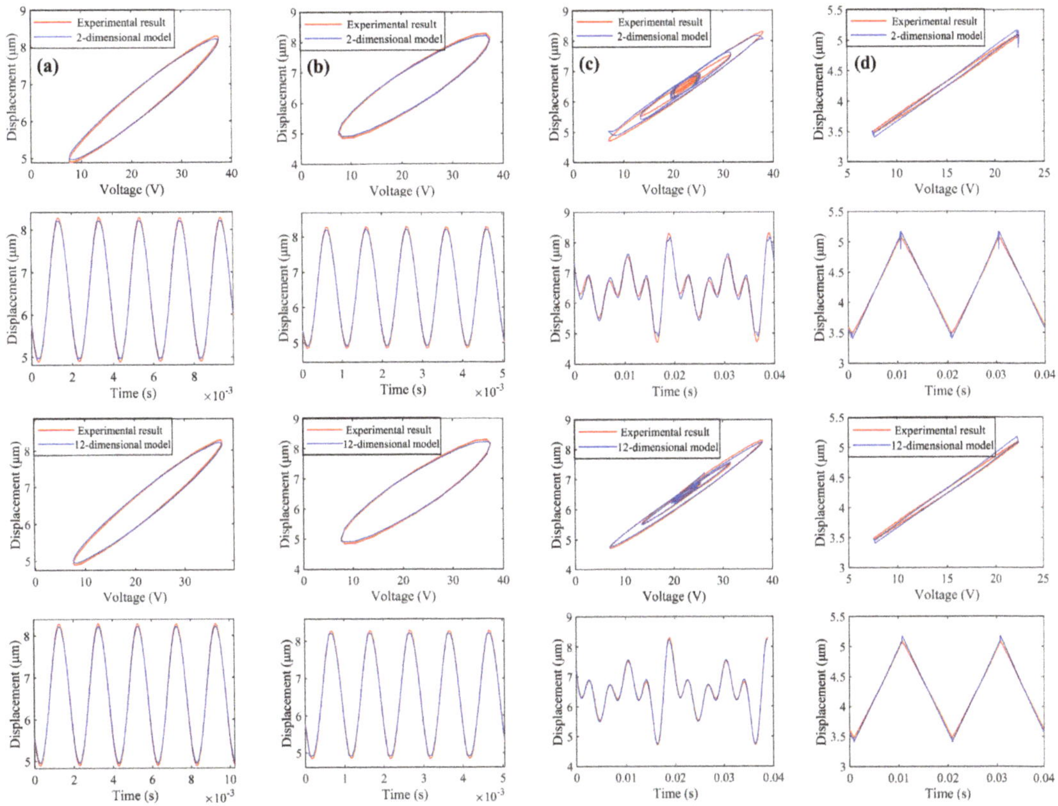

Figure 3. Experimental measured results of the PEA (red lines) as function of voltage and time for 500 Hz sinusoidal signal (column **a** panels); 1000 Hz sinusoidal signal (column **b** panels); four−frequency mixed−signal (column **c** panels); 50 Hz triangular wave (column **d** panels). The prediction results for 2−dimensional and 12−dimensional GP models are also plotted (blue lines) in the upper two and lower two panels, respectively.

Table 1. Experimental results for the MPI model and GP-based models with different dimensions.

The Type of Input Signals		MPI Model (NRMSE/RME, %)	GP-Based Model (NRMSE/RME, %) 2-Dimension	6-Dimension	9-Dimension	12-Dimension
Sinusoid signal	100 Hz	0.6497/2.01	0.4034/0.95	0.1908/0.75	0.1812/0.66	0.1766/0.69
	200 Hz	1.1673/2.93	0.2165/0.96	0.1712/0.56	0.1691/0.55	0.1673/0.56
	300 Hz	2.9334/5.68	0.9519/1.93	0.9223/1.65	0.9187/1.65	0.9196/1.58
	500 Hz	3.2692/6.33	1.3732/2.74	1.2974/2.31	1.3116/2.40	1.2749/2.29
	1000 Hz	4.9547/8.96	1.3183/2.77	1.2875/2.32	1.2874/2.32	1.2855/2.31
Mixed-frequency signal	120 + 180 Hz	2.1251/4.97	2.1670/4.65	2.1462/4.28	1.9929/3.83	1.6752/3.62
	100 + 150 + 200 + 250 Hz	2.8872/6.69	3.0209/7.28	2.7079/6.19	1.6165/4.03	1.0759/2.76
Triangular wave	50 Hz	2.1725/7.29	2.2333/13.8	2.1070/7.18	2.0444/7.41	2.0192/6.24

3. Controller Design Based on Hysteresis Compensation

As discussed above, the high-dimensional GP model can effectively predict the rate-dependent hysteresis. It can be used to correct the control of the PEAs via the IHC designed with the inverse GP model. Based on the IHC, open-loop and closed-loop controllers are constructed and tested to validate the effectiveness of the proposed model in this section.

3.1. *The Open-Loop Controller Based on IHC*

Owing to the fact that the inversion of hysteresis effect is by nature hysteresis loops, the direct inverse hysteresis compensation concept is widely used in the literature to help eliminate the hysteresis effect [36], which is shown in Figure 4. The compensation process can be written as:

$$y_{out}(t) = H\left[H^{-1}[y_r]\right](t) \tag{22}$$

where $H[\cdot]$ denotes the rate-dependent hysteresis model, $H^{-1}[\cdot]$ denotes its inverse model, $y_r(t)$ is the reference trajectory and $y_{out}(t)$ is the real output trajectory. The input voltage of the PEA, $v_{ff}(t)$, is generated from the IHC and the IHC can be directly modeled by GP with different input dimensions via interchanging the voltage and displacement as the input and output of the aforementioned GP model since the inverse of the hysteresis model is still a hysteresis model. The hyperparameters selected are the same as for Equation (11). For more details about the GP-based IHC modeling, readers can refer to [34].

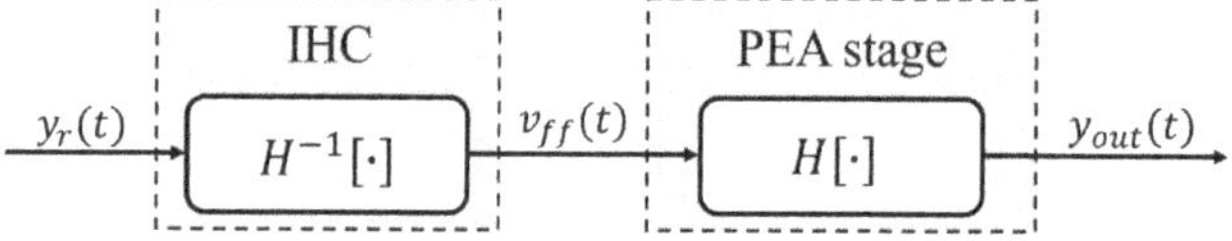

Figure 4. Block diagram of the open-loop controller based on the IHC.

To model the IHC so that $v_{ff}(t)$ can be obtained, the training data is chosen as $\tilde{\mathbf{D}} = \left\{\tilde{\mathbf{X}}, v\right\} = \{\tilde{x}_i, v_i\}_{i=1}^{n}$, where $\tilde{x}_i$ is a two, six, nine or twelve-dimensional vector containing the PEA output displacement y_i and its temporal derivative and temporal distribution information. For example, the 12-dimensional input $\tilde{x}$ at time t is expressed as:

$$\tilde{x}_{12t} = \left(y_{t-T}, y_t, y_{t+T}, \dot{y}_{t-T}, \dot{y}_t, \dot{y}_{t+T}, \ddot{y}_{t-T}, \ddot{y}_t, \ddot{y}_{t+T}, \dddot{y}_{t-T}, \dddot{y}_t, \dddot{y}_{t+T}\right) \tag{23}$$

where y, $\dot{y}$, $\ddot{y}$ and $\dddot{y}$ are the displacement, velocity, acceleration and jerk of the PEA, respectively. For other dimensional input, $\tilde{x}$ can be constructed in a similar way.

For comparison, the inverse MPI model is built using the same training data as that used for the two-dimensional GP-based inverse hysteresis model. Sinusoidal signals with different frequencies from 100 Hz to 1000 Hz, a two-frequency mixed-signal $x(t) = 7.5(\sin(120 \times 2\pi t) + \sin(180 \times 2\pi t)) + 22.5$, a four-frequency mixed-signal $x(t) = 6\sin(100 \times 2\pi t) + 3\sin(150 \times 2\pi t) + 3\sin(200 \times 2\pi t) + 6\sin(250 \times 2\pi t) + 22.5$ and a triangular wave of 50 Hz are chosen as the reference trajectories, respectively. The testing results of the open-loop controllers with different compensators are listed in Table 2.

Table 2. Tracking results of the open-loop controllers.

The Type of Reference Trajectories		MPI-Based Compensator (NRMSE/RME, %)	GP-Based Compensator (NRMSE/RME, %)			
			2-Dimension	6-Dimension	9-Dimension	12-Dimension
Sinusoid signal	100 Hz	0.8204/1.80	0.5118/1.44	0.3565/1.01	0.5098/1.33	0.3178/0.96
	200 Hz	1.5220/2.61	0.5622/1.50	0.5177/1.39	0.4199/1.25	0.5028/1.29
	300 Hz	2.8829/4.67	1.7609/5.02	1.7623/3.60	1.3508/4.51	0.9646/2.05
	400 Hz	3.1589/7.46	2.8196/8.65	1.9965/4.07	1.7018/3.84	1.0976/2.50
	500 Hz	3.5382/12.1	2.7784/6.83	2.6411/5.74	2.0716/4.66	1.9783/4.85
	600 Hz	5.4530/13.2	3.1666/7.58	2.8752/7.02	2.8545/6.14	1.4955/4.98
	700 Hz	6.8551/19.0	3.5210/7.36	2.8407/8.30	2.5233/7.33	1.6366/5.48
	800 Hz	7.9379/19.9	4.4801/10.0	3.5891/8.19	3.1453/7.77	2.7030/6.87
	900 Hz	6.0733/16.1	5.5373/12.7	3.6697/8.93	3.7629/9.53	2.6164/6.67
	1000 Hz	9.3307/19.6	4.6383/10.5	4.1660/9.26	4.0557/10.5	3.2564/8.85
Mixed-frequency signal	120 + 180 Hz	1.6816/3.85	2.2292/7.72	1.2907/3.30	1.1217/2.83	1.1096/2.51
	100 + 150 + 200 + 250 Hz	2.5270/8.65	3.5872/9.85	2.3128/5.79	1.9443/5.19	1.8159/4.86
Triangular wave	50 Hz	2.0806/4.21	3.8415/9.04	1.0245/4.74	0.9673/2.78	0.8783/2.27

From Table 2, it can be observed that the GP-based compensators can provide better tracking accuracy for the sinusoidal signals as compared with the MPI compensator, especially when the tracking frequency becomes high. The 12-dimensional and 9-dimensional GP-based compensators outperform the 6-dimensional and 2-dimensional GP-based compensators. For the mixed-frequency signals and triangular wave, the 12-dimensional GP-based compensator and 9-dimensional GP-based compensator show better tracking performance as well. For the 1000 Hz sinusoidal signal, the NRMSE for the 12-dimensional GP-based compensator is only 34.90% of that for the MPI compensator and 70.21% of that for the 2-dimensional GP-based compensator. For the 120 and 180 Hz mixed-frequency signal, the NRMSE for the 12-dimensional GP-based compensator is just 65.98% of that for the MPI compensator and 49.78% of that for the 2-dimensional GP-based compensator. The evaluation index values summarized in Table 2 strongly validate the effectiveness of the proposed high-dimensional GP-based compensator.

Comparisons of tracking accuracy for some representative signals with different compensators are shown in Figure 5. It can be observed that, even with complex signals, the proposed compensator still retains an excellent tracking accuracy as more information, including more temporal derivation as well as the temporal distribution, is introduced into the training sets compared to the normal 2-dimensional GP-based compensator and the inverse MPI based compensator. The superiority of this method is demonstrated clearly via comparing these tracking curves.

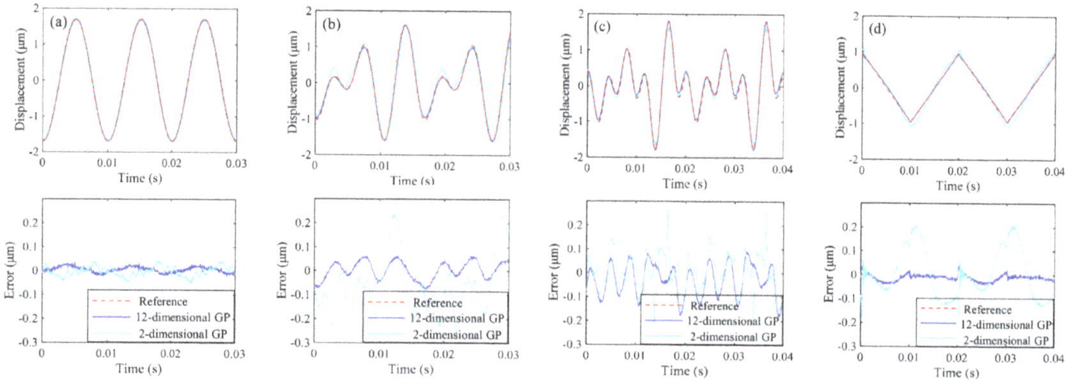

Figure 5. Tracking results of the open–loop controller under different reference signals: (**a**) 100 Hz sinusoidal signal; (**b**) dual–frequency mixed–signal; (**c**) four–frequency mixed–signal; (**d**) 50 Hz triangular wave.

3.2. Closed-Loop Controller

Although the GP-based IHC open-loop controllers perform well for the reference trajectories, especially when the dimensions of the input are high, it is difficult for them to cope with the creep of PEA and other kinds of external disturbances. Therefore, in order to improve the stability of the control system and the ability of the disturbance rejection, a feedforward/feedback closed-loop controller is designed based on the IHC, whose block diagram is shown in Figure 6. Here, *P*(s) denotes the PEA system, and a feedback loop is used to eliminate the influence of the modeling error and various kinds of external disturbances. A widely used PI controller is utilized as the tracking controller *C*(s). It can be written as:

$$v_{fb}(t) = K_p e(t) + K_i \int_0^t e(\tau) d\tau \tag{24}$$

where $e(t)$ is the tracking error, K_p is the proportional gain and K_i is the integral gain of the PI controller. It is known that the tracking performance of the PI controller becomes better with an increase in the control gain. However, overlarge control gains may cause a low relative stability margin. Hence, the specific parameters of K_p and K_i are finally maximized with the trial-and-error method in the step response experiments before the unstable vibration occurs.

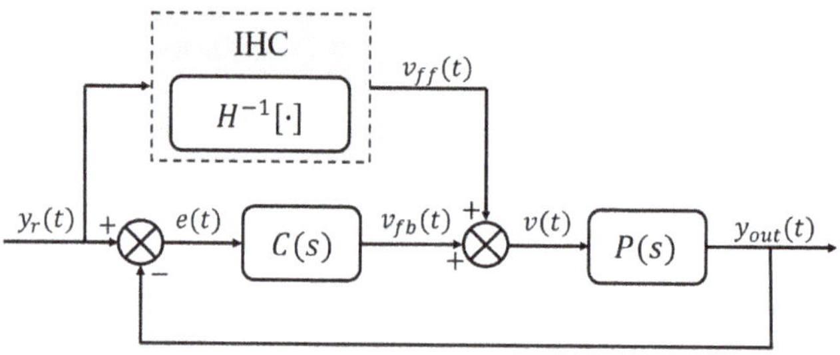

Figure 6. Block diagram of close-loop controller based on IHC.

Hence, the input voltage $v(t)$ of PEA is composed of two parts:

$$v(t) = v_{ff}(t) + v_{fb}(t) \tag{25}$$

where $v_{ff}(t)$ is the voltage generated from the IHC and $v_{fb}(t)$ is the voltage generated from the PI controller C(s).

To evaluate the performance of the feedforward/feedback closed-loop controller, the reference trajectories are selected the same as those used in the open-loop tracking tests. The pure PI controller without the feedforward compensator is also tested for comparisons. In the experiments, $K_p = 0.3$ and $K_i = 20,000$ for all controllers. The testing results are summarized in Table 3.

Table 3. Tracking results of the closed-loop controllers.

The Type of Reference Trajectories		Pure PI Controller (NRMSE/RME, %)	MPI-Based Method (NRMSE/RME, %)	GP-Based Method (NRMSE/RME, %)			
				2-Dimension	6-Dimension	9-Dimension	12-Dimension
Sinusoid signal	100 Hz	1.1033/2.30	0.8094/2.15	0.7000/1.48	0.6940/1.62	0.6672/1.42	0.6407/1.38
	200 Hz	2.2465/3.74	1.4492/2.73	1.4192/2.38	1.4005/2.43	1.3979/2.53	1.3883/2.48
	300 Hz	3.1921/5.34	2.5918/6.57	3.0714/7.57	2.2458/4.17	2.1828/3.82	2.1721/3.86
	400 Hz	4.2390/6.53	4.3373/12.4	4.2570/12.0	3.3804/5.70	3.2419/5.57	3.2052/5.44
	500 Hz	5.3924/8.86	7.5462/17.6	5.1274/11.0	4.7590/8.32	4.2756/6.76	3.5501/6.86
	600 Hz	6.5219/10.6	10.362/24.4	6.5440/12.4	5.8782/12.0	5.0000/9.19	4.9335/9.40
	700 Hz	7.7353/12.3	10.020/23.3	7.2384/18.5	6.9444/14.1	6.6222/12.7	6.2461/10.8
	800 Hz	8.9472/13.9	11.016/30.3	8.3979/19.4	7.5350/15.8	6.9765/12.7	6.6185/12.8
	900 Hz	10.1254/16.2	11.556/32.6	8.8955/23.1	9.2630/19.1	7.7958/13.3	7.3941/11.9
	1000 Hz	11.1463/18.2	14.633/31.5	9.4682/17.1	10.4491/20.4	9.05513/16.3	8.3210/14.6
Mixed-frequency signal	120 + 180 Hz	1.3174/2.91	0.8590/2.09	0.8139/2.03	0.8012/2.05	0.7527/2.00	0.7497/1.84
	100 + 150 + 200 + 250 Hz	2.2445/3.90	0.9088/2.87	1.1093/6.28	0.8894/2.64	0.8787/2.69	0.8065/2.43
Triangular wave	50 Hz	1.3380/3.19	0.8782/2.84	0.9745/4.42	0.5786/3.22	0.4682/2.55	0.4405/2.41

From the table, it can be observed that the GP-based closed-loop controllers outperform the pure PI controller and the MPI-based closed-loop controller. The MPI-based controller performs better than the pure PI controller when tracking low-frequency sinusoidal trajectories, mixed-frequency trajectories and triangular trajectories since hysteresis is compensated to some extent. Nonetheless, when the tracking frequency becomes high, the tracking accuracy of this controller decreases rapidly and is even worse than that of the pure PI controller since it suffers from the rapid deterioration of its modeling accuracy. For the GP-based method, however, the better tracking accuracy can still be maintained at high frequencies due to its smaller modeling errors, especially for the models with 12-dimensional and 9-dimensional inputs. It is worth mentioning that for the mixed-frequency signal and triangular wave, with the increase in the input dimension of the model, the tracking errors decrease greatly. For the four-frequency mixed reference trajectory, the NRMSE for the 12-dimensional GP-based method is just 35.93% and 72.70% of those for the pure PI method and the 2-dimensional GP-based method, respectively. For the triangular reference trajectory, the NRMSE for the 12-dimensional GP-based method is just 32.92% and 50.16% of those for the pure PI method and 2-dimensional GP-based method, respectively. In order to better illustrate the performance of the proposed high-dimensional GP-based closed-loop controller, comparisons of tracking performance with different methods and input dimensions are plotted in Figure 7. From this figure, it can be observed that the results with the proposed controller outperform those obtained with the traditional 2-dimensional GP-based controller and the pure PI controller, which demonstrates its priority. In conclusion, the experimental results shown in the table and the figure validate the effectiveness of the high-dimensional GP-based method.

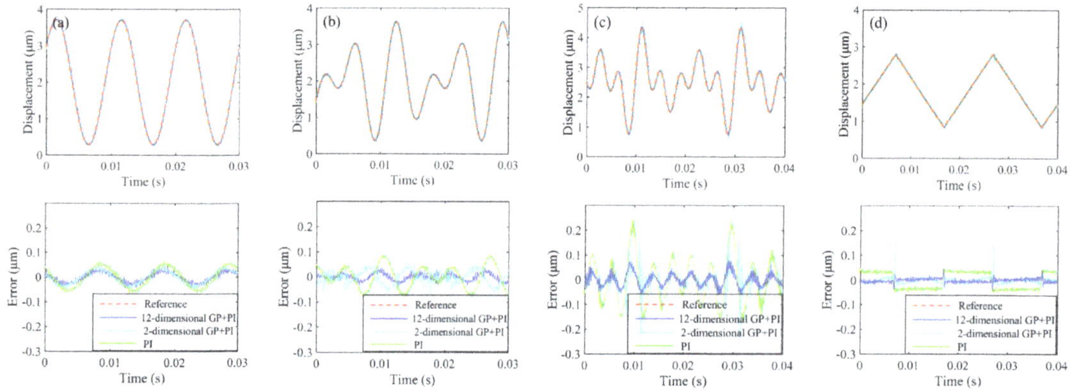

Figure 7. Tracking results of the closed−loop controller under different reference signals: (**a**) 100 Hz sinusoidal signal; (**b**) dual−frequency mixed−signal; (**c**) four−frequency mixed−signal; (**d**) 50 Hz triangular wave.

4. Conclusions

GP regression is a promising method to model rate-dependent hysteretic nonlinearity. However, when tracking high-frequency (i.e., higher than 500 Hz) or complex (i.e., mixed-frequency and triangular) signals, the accuracy of the traditional GP-based model with 2-dimensional input will deteriorate significantly. In this paper, more temporal derivative information, as well as the temporal distribution information, is introduced into the training set by increasing the input dimension of the GP model to improve its modeling accuracy. Experimental results show that with the increase in the input dimension, the prediction accuracy of the model improves greatly, especially for the complex signals, such as the mixed-frequency signal and the triangular wave. For the four-frequency mixed reference trajectory, the NRMSE for the 12-dimensional GP-based method is just 35.93% and 72.70% of those for the pure PI method and the 2-dimensional GP-based method, respectively. As these two types of trajectories are widely used in PEA applications, such as AFM and FTS, the improvement of modeling accuracy is of great importance. GP-based IHCs with different input dimensions are also constructed, and the open-loop and closed-loop controllers based on these IHCs are designed. Testing results show that the 12-dimensional GP-based open-loop and closed-loop controllers both exhibit the best performance among the similar controllers and the tracking performance is much better than that of the classical MPI based controllers, which validates the effectiveness of modeling the rate-dependent hysteresis of PEAs with high-dimensional input GPs.

Author Contributions: Y.M.: Methodology, Formal analysis, Investigation, Writing—Original Draft. X.W.: Formal analysis, Writing—Review and Editing. L.L.: Data curation. W.H.: Software. L.Z.: Conceptualization, Methodology, Project administration. All authors have read and agreed to the published version of the manuscript.

Funding: This paper was funded by the National Natural Science Foundation of China under Grant Nos. U2013211, 51975375 and 52105581, and the China Postdoctoral Science Foundation (No. 2021M692065).

Institutional Review Board Statement: Not applicable.

Informed Consent Statement: Not applicable.

Data Availability Statement: Not applicable.

Acknowledgments: This work was partially supported by the National Natural Science Foundation of China under Grant Nos. U2013211, 51975375 and 52105581, and the China Postdoctoral Science Foundation (No. 2021M692065).

Conflicts of Interest: The authors declare no conflict of interest.

Appendix A. MPI Model for Rate-Dependent Hysteresis

The play operator-based MPI model can be expressed as [27]:

$$y(t) = H[v](t) = g(v(t)) + \sum_{i=1}^{N} q_i F_{ori}^{h}[v](t) \tag{A1}$$

where $y(t)$ denotes the output, N is the number of the play operators, $F_{ori}^{h}(t)$ represents the play operator which has taken the rate term $\dot{v}(t)$ into account, q_i denotes the corresponding weight of the ith play operator and $g(v(t)) = a_1 v^3(t) + a_2 v(t) + a_3$ is the modified term in order to describe the asymmetric hysteresis of the PEA with constant parameters a_1, a_2 and a_3.

In practical applications, N is often set to be 10 and there are also two parameters in the play operator [27]. Therefore, altogether there are 15 parameters that need to be identified in the MPI rate-dependent model, which are obtained by particle swarm optimization. For more information about the MPI model, readers may refer to [27].

References

1. Zhu, Z.; To, S.; Ehmann, K.F.; Zhou, X. Design, analysis, and realization of a novel piezoelectrically actuated rotary spatial vibration system for micro-/nano-machining. *IEEE/ASME Trans. Mechatron.* **2017**, *22*, 1227–1237. [CrossRef]
2. Vagia, M.; Eielsen, A.A.; Gravdahl, J.T.; Pettersen, K.Y. Design of a nonlinear damping control scheme for nanopositioning. In Proceedings of the 2013 IEEE/ASME International Conference on Advanced Intelligent Mechatronics, Wollongong, Australia, 9–12 July 2013; pp. 94–99.
3. Schitter, G.; Astrom, K.J.; DeMartini, B.E.; Thurner, P.J.; Turner, K.L.; Hansma, P.K. Design and modeling of a high-speed AFM-scanner. *IEEE Trans. Control. Syst. Technol.* **2007**, *15*, 906–915. [CrossRef]
4. Devasia, S.; Eleftheriou, E.; Moheimani, S.O.R. A survey of control issues in Nanopositioning. *IEEE Trans. Control Syst. Technol.* **2007**, *15*, 802–823. [CrossRef]
5. Gu, G.; Zhu, L.; Su, C.; Ding, H.; Fatikow, S. Modeling and Control of Piezo-Actuated Nanopositioning Stages: A Survey. *IEEE Trans. Autom. Sci. Eng.* **2016**, *13*, 313–332. [CrossRef]
6. Shan, Y.; Leang, K. Accounting for hysteresis in repetitive control design: Nanopositioning example. *Jpn. J. Appl. Phys.* **2002**, *41*, 4851–4856. [CrossRef]
7. Yu, Z.; Wu, Y.; Fang, Z.; Sun, H. Modeling and compensation of hysteresis in piezoelectric actuators. *Heliyon* **2020**, *6*, e03999. [CrossRef]
8. Seki, K.; Ruderman, M.; Iwasaki, M. Modeling and compensation for hysteresis properties in piezoelectric actuators. In Proceedings of the 2014 IEEE 13th International Workshop on Advanced Motion Control (AMC), Yokohama, Japan, 14–16 March 2014; pp. 687–692.
9. Li, K.; Yang, Z.; Lallart, M.; Zhou, S.; Chen, Y.; Liu, H. Hybrid hysteresis modeling and inverse model compensation of piezoelectric actuators. *Smart Mater. Struct.* **2019**, *28*, 115038. [CrossRef]
10. Richter, H.; Misawa, E.; Lucca, D.; Lu, H. Modeling nonlinear behavior in a piezoelectric actuator. *Precis. Eng.* **2001**, *25*, 128–137. [CrossRef]
11. Gu, G.; Zhu, L. Modeling of rate-dependent hysteresis in piezoelectric actuators using a family of ellipses. *Sens. Actuators Phys.* **2011**, *165*, 303–309. [CrossRef]
12. Gawthrop, P.; Bhikkaji, B.; Moheimani, S. Physical-model-based control of a piezoelectric tube for nano-scale positioning applications. *Mechatronics* **2010**, *20*, 74–84. [CrossRef]
13. Zhao, X.; Tan, Y. Neural network based identification of Preisach-type hysteresis in piezoelectric actuator using hysteretic operator. *Sens. Actuators Phys.* **2006**, *126*, 306–311. [CrossRef]
14. Hu, H.; Mrad, R. On the classical Preisach model for hysteresis in piezoceramic actuators. *Mechatronics* **2002**, *13*, 85–94. [CrossRef]
15. Kuhnen, K. Modeling, identifification and compensation of complex hysteretic nonlinearities: A modified Prandtl-Ishlinskii approach. *Eur. J. Control* **2003**, *9*, 407–418. [CrossRef]
16. Janocha, H.; Kuhnen, K. Real-time compensation of hysteresis and creep in piezoelectric actuators. *Sens. Actuators Phys.* **2000**, *79*, 83–89. [CrossRef]
17. Iamail, M.; Ikhouane, F.; Rodellar, J. The hysteresis Bouc-Wen model, a survey. *Arch. Comput. Methods Eng.* **2009**, *16*, 161–188.

18. Charalampakis, A.; Dimou, C. Identifification of Bouc-Wen hysteretic systems using particle swarm optimization. *Comput. Struct.* **2010**, *88*, 1197–1205. [CrossRef]
19. Kwok, N.; Ha, Q.; Nguyen, M.; Li, J.; Samali, B. Bouc-Wen model parameter identifification for a MR flfluid damper using computationally effificient GA. *ISA Trans.* **2007**, *46*, 167–179. [CrossRef]
20. Xu, Q.; Li, Y. Dahl model-based hysteresis compensation and precise positioning control of an XY parallel micromanipulator with piezoelectric actuation. *ASME Trans. J. Dyn. Syst. Meas. Control* **2010**, *132*, 041011. [CrossRef]
21. Lin, C.; Lin, P. Tracking control of a biaxial piezo-actuated positioning stage using generalized Duhem model. *Comput. Math. Appl.* **2012**, *64*, 766–787. [CrossRef]
22. Mayergoyz, I.D. Dynamic Preisach models of hysteresis. *IEEE Trans. Magn.* **1988**, *24*, 2925–2927. [CrossRef]
23. Tan, U.X.; Win, T.L.; Ang, W.T. Modeling piezoelectric actuator hysteresis with singularity Free Prandtl-Ishlinskii model. In Proceedings of the 2006 IEEE International Conference on Robotics and Biomimetics, Kunming, China, 17–20 December 2006; pp. 251–256.
24. Janaideh, M.A.; Rakheja, S.; Su, C.Y. Experimental characterization and modeling of rate-dependent hysteresis of a piezoceramic actuator. *Mechatronics* **2009**, *19*, 656–670. [CrossRef]
25. Qin, Y.; Shirinzadeh, B.; Tian, Y.; Zhang, D. Design issues in a decoupled XY stage: Static and dynamics modeling, hysteresis compensation, and tracking control. *Sens. Actuators Phys.* **2013**, *194*, 95–105. [CrossRef]
26. Al Janaideh, M.; Krejci, P. Inverse rate-dependent prandtl–Ishlinskii model for feedforward compensation of hysteresis in a piezomicropositioning actuator. *IEEE/ASME Trans. Mechatron.* **2013**, *18*, 1498–1507. [CrossRef]
27. Yang, M.J.; Li, C.X.; Gu, G.Y.; Zhu, L.M. Modeling and compensating the dynamic hysteresis of piezoelectric actuators via a modified rate-dependent Prandtl–Ishlinskii model. *Smart Mater. Struct.* **2015**, *24*, 125006. [CrossRef]
28. Raj, R.A.; Samikannu, R.; Yahya, A.; Mosalaosi, M. Performance evaluation of natural esters and dielectric correlation assessment using artificial neural network (ANN). *J. Adv. Dielectr.* **2020**, *10*, 2050025. [CrossRef]
29. Haggag, S.; Nasrat, L.; Ismail, H. ANN approaches to determine the dielectric strength improvement of MgO based low density polyethylene nanocomposite. *J. Adv. Dielectr.* **2021**, *11*, 2150016-48727. [CrossRef]
30. Dong, R.; Tan, Y.; Chen, H.; Xie, Y. A neural networks based model for rate-dependent hysteresis for piezoceramic actuators. *Sens. Actuators Phys.* **2008**, *143*, 370–376. [CrossRef]
31. Li, P.; Yan, F.; Ge, C.; Wang, X.; Xu, L.; Guo, J.; Li, P. A simple fuzzy system for modelling of both rate-independent and rate-dependent hysteresis in piezoelectric actuators. *Mech. Syst. Signal Process.* **2013**, *36*, 182–192. [CrossRef]
32. Mao, X.; Wang, Y.; Liu, X.; Guo, Y. A hybrid feedforward-feedback hysteresis compensator in piezoelectric actuators based on least-squares support vector machine. *IEEE Trans. Ind. Electron.* **2018**, *65*, 5704–5711. [CrossRef]
33. Li, P.; Li, P.; Sui, Y. Adaptive Fuzzy Hysteresis Internal Model Tracking Control of Piezoelectric Actuators with Nanoscale Application. *IEEE Trans. Fuzzy Syst.* **2016**, *24*, 1246–1254. [CrossRef]
34. Tao, Y.D.; Li, H.X.; Zhu, L.M. Rate-dependent hysteresis modeling and compensation of piezoelectric actuators using Gaussian process. *Sens. Actuators Phys.* **2019**, *295*, 357–365. [CrossRef]
35. Hu, C.X.; Ou, T.S.; Zhu, Y.; Zhu, L.M. GRU-Type LARC Strategy for Precision Motion Control with Accurate Tracking Error Prediction. *IEEE Trans. Ind. Electron.* **2021**, *68*, 812–820. [CrossRef]
36. Sun, Z.; Song, B.; Xi, N.; Yang, R.; Hao, L.; Yang, Y.; Chen, L. Asymmetric hysteresis modeling and compensation approach for nanomanipulation system motion control considering working-range effect. *IEEE Trans. Ind. Electron.* **2017**, *64*, 5513–5523. [CrossRef]

actuators

MDPI

Article

Development of a Parallel Dual-Stage Compliant Nanopositioning System

Xu Yang [1], Lichao Ji [1], Ying Shang [2], Wule Zhu [3] and Shizhen Li [1,*]

1 Institute of Marine Science and Technology, Shandong University, Qingdao 266237, China; xu_y@sdu.edu.cn (X.Y.); 202016936@mail.sdu.edu.cn (L.J.)
2 Laser Institute, Qilu University of Technology, Jinan 250353, China; shangying@sdlaser.cn
3 State Key Laboratory of Fluid Power and Mechatronic Systems, Zhejiang University, Hangzhou 310027, China; wulezhu@zju.edu.cn
* Correspondence: lishizhen@sdu.edu.cn

Abstract: This paper presents a novel parallel dual-stage compliant nanopositioning system (PDCNS), aimed at nanoscale positioning for microscale manipulation. In the developed PDCNS, the coarse stage actuated by the voice coil motor and the fine stage driven by the piezoelectric actuator are integrated in a parallel manner by a specially devised A-shaped compliant mechanism, which leads to many excellent performances, such as good resolution and large stroke and broadband. To enhance the closed-loop-positioning capability of the proposed PDCNS, a double-servo cooperative control (DSCC) strategy is specially constructed. The performance of the proposed PDCNS is evaluated by analytical model, finite element analysis, and experimental research. Results show that the first-order resonance frequency of the designed A-shaped compliant mechanism can reach 99.7 Hz. Combined with the designed DSCC, the developed PDCNS prototype is demonstrated to provide a stroke of 1.49 mm and a positioning resolution of $\leq$50 nm.

Keywords: nanopositioning system; microscale manipulation; compliant mechanism; double-servo cooperative control

Citation: Yang, X.; Ji, L.; Shang, Y.; Zhu, W.; Li, S. Development of a Parallel Dual-Stage Compliant Nanopositioning System. *Actuators* **2022**, *11*, 136. https://doi.org/10.3390/act11050136

Academic Editors: Nicola Pio Belfiore and Jose Luis Sanchez-Rojas

Received: 19 March 2022
Accepted: 9 May 2022
Published: 13 May 2022

1. Introduction

Precision-positioning platforms with nanometer resolution play more and more important roles in many scientific and industrial applications [1–6]. In previous research, piezoelectric actuators were often adopted to configure these nanopositioning stages because of their fast speed and nanometer-level resolution [7–10]. Nevertheless, the maximum stroke of the piezoelectric actuator is only about 0.1% of its length [11]. Piezoelectric motors which can realize continuous nanoscale movement without stroke limit have been applied in nanopositioning systems as well [12–15]. However, the piezoelectric motors encounter severe nonlinear friction. To overcome these drawbacks, the dual-stage nanopositioning systems are employed to deliver high-resolution motion over a long range.

In the literature, some types of dual-stage positioning systems were proposed. To improve the areal density of the magnetic hard disk, a dual-stage positioning mechanism was designed to regulate the position of read–write head, in which an additional piezoelectric actuator was mounted on the primary stage for fast and fine positioning [16]. Fast tool servo units based on the piezoelectric actuator or electromagnetic actuator were used to cooperate with the feed shaft of machine tools for high-speed and high-accuracy nanocutting [17–19]. A piezo-driven compliant injector was configured at the terminal of a traditional XYZ-positioning stage to accomplish the automatic injection and batch micromanipulation of cells [20,21]. However, the previous design concepts of the dual-stage system were limited to stacking a fine stage on top of a coarse stage in a serial manner, which encountered many technical bottlenecks, such as large moving mass, assembly error, and uncertain disturbance from the cable. Therefore, new configurations of dual-stage positioning systems are still required.

Apart from the structure design, the control strategy also significantly influences the performance of the dual-stage positioning system. Extensive research on the closed-loop control of the dual-stage positioning system was carried out in the past few years. For example, Xu et al. [22] and Michellod et al. [23] designed single-feedback control schemes for the dual-stage nanopositioning system that determines the relative position between two stages using observers. By employing the position feedbacks from coarse and fine stages, Dong et al. [24] and Zhu et al. [25] proposed multiple-feedback control systems for the dual-stage nanopositioning system. Despite this, it is still a challenging job to control a dual-stage positioning system.

In this paper, a novel parallel dual-stage compliant nanopositioning system (PDCNS) with a corresponding double-servo cooperative control (DSCC) strategy is developed. The PDCNS employs a new A-shaped compliant parallel mechanism to combine the coarse motion from the voice coil motor and the fine motion from the piezoelectric actuator. Considering the uncertain disturbances and the double-input-single-output configuration, a DSCC is designed in this paper to enhance the motion capability of the proposed PDCNS. The static and dynamic performances of the PDCNS are investigated by the analytical-model-based calculation and the finite element analysis. Moreover, a series of experimental studies are carried out to demonstrate the kinematic properties and closed-loop-positioning performance of the PDCNS.

The rest of this paper is organized as follows: Section 2 presents the mechanical design of the PDCNS. An analytical model of the PDCNS is established in Section 3. With the analytical model and finite element analysis, the static and dynamic performances of the proposed PDCNS are analyzed in Section 4. Moreover, experiments are conducted to evaluate the static and dynamic performances of the proposed PDCNS in Section 5. Finally, conclusions are given in Section 6.

2. Design of PDCNS

The PDCNS consists of an A-shaped compliant parallel mechanism (ACPM), a voice coil motor and a piezoelectric actuator, as shown in Figure 1. The ACPM can transmit the motion of the voice coil motor (VCM) to the end effector through two parallel limbs. At the same time, the piezoelectric actuator (PEA) contributes to the motion of the end effector via a half bridge-type compliant chain. For compactness, the upper part of the ACPM and the half bridge-type chain share the same flexural structure. To eliminate parasitic motions, two guiding flexible beams are designed for the end effector which help to improve the resonant frequency as well. In addition, two input decoupling compliant mechanisms are devised to protect the piezoelectric actuator from undesired lateral forces. A support bearing is adopted to guide the actuation motion of the voice coil motor.

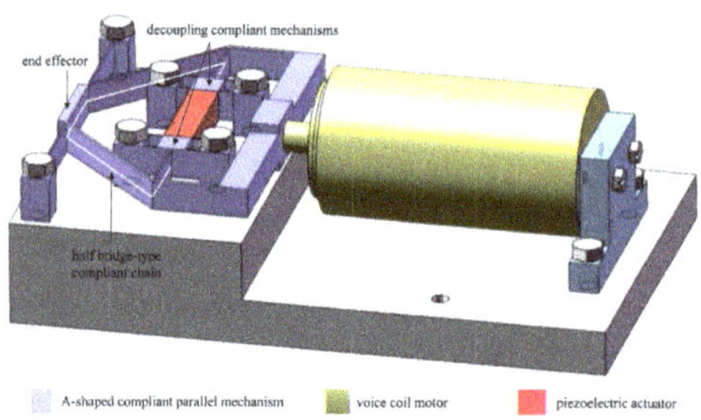

Figure 1. Structure of the PDCNS.

With the above designs, both the voice coil motor and the piezoelectric actuator can contribute to the output motion. As the voice coil motor has a large-range motion capability and the piezoelectric actuator features a high-resolution motion characteristic, the proposed PDCNS is expected to position in the millimeter range with nanometer resolution. Moreover, as the bases of the voice coil motor and the piezoelectric actuator remain static to the ground, low-moving mass and high dynamics can be realized.

3. Analytical Modelling

In this section, the kinematic and stiffness characteristics of PDCNS are investigated. Based on the finite element theory [26,27], an analytical model is built for the ACPM, as shown in Figure 2. We can see from the figure, rigid nodes with three degrees of freedom are defined at the two ends of the beams to describe the deformation of the flexible beam. A global coordinate system O-xy is defined to analyze the overall deformation of ACPM.

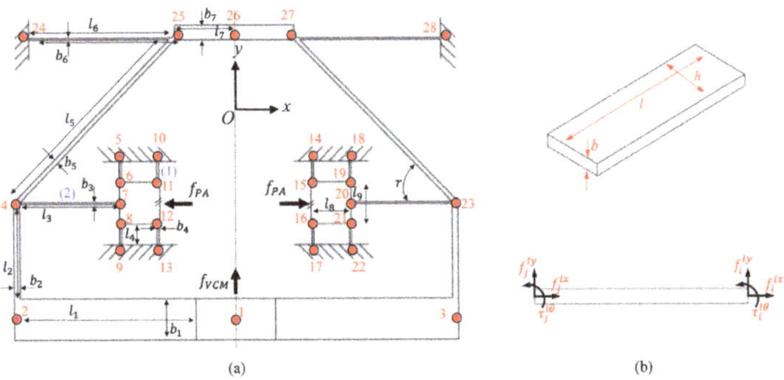

Figure 2. (**a**) Analytical model of ACPM, (**b**) schematic of flexible beam.

According to the structural mechanics theory [26,28], in the local coordinate system of the flexible beam, the relationship between the applied force and the induced displacement of the flexible beam can be expressed as

$$\begin{bmatrix} f_i^{lx} \\ f_i^{ly} \\ \tau_i^{l\theta} \\ f_j^{lx} \\ f_j^{ly} \\ \tau_j^{l\theta} \end{bmatrix} = \begin{bmatrix} \frac{Ebh}{l} & 0 & 0 & -\frac{Ebh}{l} & 0 & 0 \\ 0 & \frac{12EI}{l^3} & \frac{6EI}{l^2} & 0 & -\frac{12EI}{l^3} & \frac{6EI}{l^2} \\ 0 & \frac{6EI}{l^2} & \frac{4EI}{l} & 0 & -\frac{6EI}{l^2} & \frac{2EI}{l} \\ -\frac{Ebh}{l} & 0 & 0 & \frac{Ebh}{l} & 0 & 0 \\ 0 & -\frac{12EI}{l^3} & -\frac{6EI}{l^2} & 0 & \frac{12EI}{l^3} & -\frac{6EI}{l^2} \\ 0 & \frac{6EI}{l^2} & \frac{2EI}{l} & 0 & -\frac{6EI}{l^2} & \frac{4EI}{l} \end{bmatrix} \begin{bmatrix} \delta_i^{lx} \\ \delta_i^{ly} \\ \phi_i^{l\theta} \\ \delta_j^{lx} \\ \delta_j^{ly} \\ \phi_j^{l\theta} \end{bmatrix} = k_{loc} \begin{bmatrix} \delta_i^{lx} \\ \delta_i^{ly} \\ \phi_i^{l\theta} \\ \delta_j^{lx} \\ \delta_j^{ly} \\ \phi_j^{l\theta} \end{bmatrix} \tag{1}$$

where $f_i^{lx}, f_i^{ly}, \tau_i^{l\theta}$ are the x-, y-directional forces and z-directional torque applied on node i in the local coordinate system, $f_j^{lx}, f_j^{ly}, \tau_j^{l\theta}$ are the x-, y-directional forces and z-directional torque applied on node j in the local coordinate system, $\delta_i^{lx}, \delta_i^{ly}, \phi_i^{l\theta}$ are the x-, y-directional translations and z-directional angular displacement of node i in the local coordinate system, $\delta_j^{lx}, \delta_j^{ly}, \phi_j^{l\theta}$ are the x-, y-directional translations and z-directional angular displacement of node j in the local coordinate system, E is Young's modulus, b, h are the section sizes of the flexible beam, l is the length of the flexible beam, and I is the moment of inertia of the flexible beam which can be calculated by $I = hb^3/12$, k_{loc} is the local stiffness matrix of the flexible beam.

With Equation (1), the loads applied on the two end ports of the flexible beam can be calculated according to the displacements of the corresponding two rigid nodes. Nevertheless, the stiffness matrix in Equation (1) is expressed in the local coordinate system, which relates the deformations and loads of the flexible beam in the local coordinate system. To combine the deformations and loads of all the flexible beams, a rotation transformation is employed to transfer the local stiffness matrix, $\boldsymbol{k}_{loc}$, to be a global stiffness matrix, $\boldsymbol{k}_{glo}$, as shown in Equation (3) [26,29].

$$\boldsymbol{R}(\boldsymbol{\theta}) = \begin{bmatrix} \cos(\theta) & \sin(\theta) & 0 & & & \\ -\sin(\theta) & \cos(\theta) & 0 & & \mathbf{0} & \\ 0 & 0 & 1 & & & \\ & & & \cos(\theta) & \sin(\theta) & 0 \\ & \mathbf{0} & & -\sin(\theta) & \cos(\theta) & 0 \\ & & & 0 & 0 & 1 \end{bmatrix} \quad (2)$$

$$\boldsymbol{k}_{glo} = \boldsymbol{R}(\boldsymbol{\theta}) \cdot \boldsymbol{k}_{loc} \cdot \boldsymbol{R}(\boldsymbol{\theta})^T \quad (3)$$

where $\boldsymbol{R}(\boldsymbol{\theta})$ is the rotation matrix, θ is the angle between the local and global coordinate systems (defined as the angle from the x axis of the local coordinate system to the x axis of the global coordinate system around the o point of the local coordinate system).

With the global stiffness matrix in (3) and the analytical model in Figure 2a, an overall stiffness matrix, $\boldsymbol{K}$, can be established for the ACPM with the finite element theory [26,30], which is able to depict the relationship between the external forces and induced deformations as follows.

$$\begin{bmatrix} f_1^x & f_1^y & \tau_1^\theta & \cdots & f_{28}^x & f_{28}^y & \tau_{28}^\theta \end{bmatrix}^T = \boldsymbol{K} \cdot \begin{bmatrix} \delta_1^x & \delta_1^y & \phi_1^\theta & \cdots & \delta_{28}^x & \delta_{28}^y & \phi_{28}^\theta \end{bmatrix}^T \quad (4)$$

where f_i^x and f_i^y are the x- and y-directional external forces applied on node i, τ_i^θ is the external torque around the z-axis applied on node i, δ_i^x, and δ_i^y are the x- and y-directional translations of node i, and ϕ_i^θ is the rotational angle (around z-axis) of node i.

While two input forces, f_{PA}, are applied on nodes 7 and 20, the resulting input displacements, δ_1^y, from static equilibrium Equation (4) can be set to zero (unchanged with respect to the ground) by searching for a suitable actuation force, f_{VCM}, for node 1. Under the above conditions, the input displacements, δ_7^x and δ_{20}^x, as well as the output displacement, δ_{26}^y, can be calculated.

(1) The input stiffness for the *PEA* can be derived as:

$$k_{in}^{PEA} = \frac{f_{PA}}{|\delta_7^x| + |\delta_{20}^x|} \quad (5)$$

(2) The displacement amplification ratio for the PEA can be derived as:

$$\lambda_{amp}^{PEA} = \frac{|\delta_{26}^y|}{|\delta_7^x| + |\delta_{20}^x|} \quad (6)$$

(3) The maximum stress coefficient of the beam 1 contributed by the input displacement of PEA can be calculated by [30,31].

$$\xi_{11} = \frac{3 \cdot k_c \cdot E \cdot b_4 \cdot |\delta_{11}^x|}{2 \cdot l_4{}^2 \cdot (|\delta_7^x| + |\delta_{20}^x|)} \quad (7)$$

where k_c is the stress concentration factor which is set to 2 in this paper.

By applying one input force, f_{VCM}, on node 1 and in the meantime imposing a suitable actuation force, f_{PA}, on nodes 7 and 20, the resulting input displacement $|\delta_7^x| + |\delta_{20}^x|$ from static equilibrium Equation (4) can be set to zero. Under the above conditions, the induced input displacement, δ_1^y, and output displacement, δ_{26}^y, can be calculated.

(4) The input stiffness for the *VCM* can be derived as:

$$k_{in}^{VCM} = \frac{f_{VCM}}{|\delta_1^y|} \quad (8)$$

(5) The displacement amplification ratio for the VCM can be derived as:

$$\lambda_{amp}^{VCM} = \frac{\left|\delta_{26}^{y}\right|}{\left|\delta_{1}^{y}\right|} \tag{9}$$

(6) The maximum stress coefficient of beam 2 contributed by the input displacement of the VCM can be calculated by [30,31].

$$\xi_{22} = \frac{3 \cdot k_c \cdot E \cdot b_3 \cdot \left|\delta_{4}^{y}\right|}{2 \cdot l_3{}^2 \cdot \left|\delta_{1}^{y}\right|} \tag{10}$$

(7) For the PDCNS, the maximum stress usually occurs in beam 1 or beam 2. Hence, a maximum stress matrix is established for the PDCNS as follows:

$$\begin{bmatrix} \sigma_{max}^{1} \\ \sigma_{max}^{2} \end{bmatrix} = \begin{bmatrix} \xi_{11} & 0 \\ 0 & \xi_{22} \end{bmatrix} \begin{bmatrix} \left|\delta_{7}^{x}\right| + \left|\delta_{20}^{x}\right| \\ \left|\delta_{1}^{y}\right| \end{bmatrix} \tag{11}$$

where σ_{max}^{1} and σ_{max}^{2} are the maximum stresses of beam 1 and beam 2, respectively. According to (11), the compressive stress of beam 2 induced by the PEA is ignored since it is small compared to the bending stress induced by the maximum displacement of the VCM. Due to the good input decoupling, the bending stress of beam 1 induced by the VCM is ignored as well.

4. Simulation Analysis

Based on the working principles of the PDCNS, a 3D virtual prototype is built by using the SOLIDWORKS software. The kinematic and stiffness performances of the PDCNS virtual prototype are evaluated by the established analytical model. The key parameters of the analytical model are listed in Table 1. The dimensions of the PDCNS virtual prototype are indicated in Figure 2. The material utilized for analysis is TC4 titanium alloy with Young's modulus E, Poisson's ratio μ, and mass density ρ [32,33]. To validate the static and dynamic performances of the PDCNS virtual prototype, ANSYS Workbench-based FEA simulation is carried out as well in this section. The computational models for FEA analysis are configured as follows: mesh physics preference = mechanical, mesh size function = proximity and curvature, mesh relevance = −75, inner surfaces of six bolt holes is fixed, solver type = program controlled.

Table 1. Key parameters for simulation.

l_1 mm	b_1 mm	l_2 mm	b_2 mm	l_3 mm	b_3 mm	l_4 mm	b_4 mm	l_5 mm	b_5 mm	l_6 mm
43.98	10	22.75	1.5	25	0.5	5	0.5	55.13	1	35
b_6 mm	l_7 mm	b_7 mm	l_8 mm	l_9 mm	r °	$w_i(i-1\sim7)$ mm		E $\times 10^{11}$ Pa	μ	ρ kg/m^3
0.5	15	3.5	10	10	45°	10		1.167	0.36	4414

First, by applying a driving force (38.36 N) to the input end of the VCM and fixing the input end of the PEA in the ANSYS Workbench, the resulting overall deformation of the proposed ACPM can be obtained as shown in Figure 3a. The slight displacement difference between the two input ends of PEA can be attributed to asymmetric probe location, asymmetric mesh generation, computational convergence error, etc. The displacement amplification ratio for the VCM can be derived by relating the displacements of the VCM and the end effector. The input stiffness for VCM can be revealed by identifying the slope of the applied force over the induced displacement. As seen in Table 2, the amplification ratio for the VCM is approximately 1, which means that the motion of the VCM is transmitted to the end effector almost directly. The input stiffness for the VCM is revealed to be only 0.026 N/μm, which helps to achieve a large stroke. Good agreement between the results of the analytical model and the ANSYS Workbench are obtained, confirming the performance of the proposed ACPM.

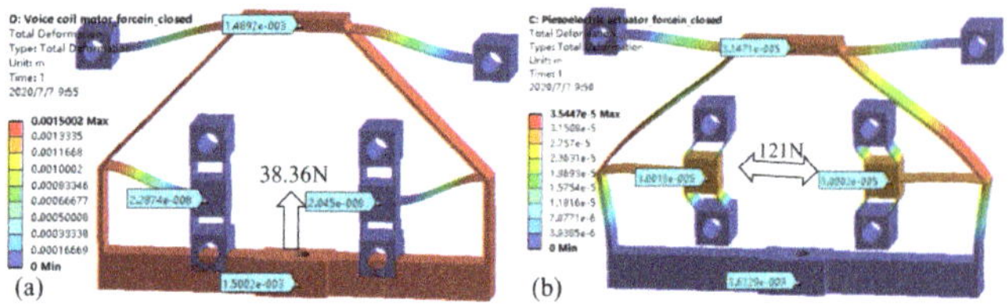

Figure 3. Displacement distribution of ACPM while (**a**) actuating the input end of VCM with 38.36 N and fixing the input end of PEA, (**b**) driving the input end of PEA with 121 N and fixing the input end of VCM.

Table 2. Performance of the ACPM virtual prototype.

	Amplification Ratio		Input Stiffness (N/μm)		Stress Coefficient (Mpa/μm)	
	λ_{amp}^{PEA}	λ_{amp}^{VCM}	k_{in}^{PEA}	k_{in}^{VCM}	ξ_{11}	ξ_{22}
anal.	0.514	0.997	2.41	0.0249	3.49	0.279
ANSYS	0.524	0.993	2.02	0.0256	3.11	0.287

Similarly, by applying a driving force of 121 N to the input ends of the PEA and fixing the input end of the VCM in the ANSYS Workbench, the resulting deformation of the proposed ACPM is shown in Figure 3b. The amplification ratio and input stiffness for the PEA are revealed to be 0.52 and 2.02 N/μm, respectively, which are highly consistent with the analytical predictions, as shown in Table 2.

Corresponding to the deformations shown in Figure 3, the stress distributions of ACPM are also resolved in the ANSYS Workbench, as illustrated in Figure 4. It can be observed that the beam between nodes 20 and 23 (which has the same stress state as beam 2) is revealed to have the maximum stress, while the VCM is working at its maximum stroke of 1.5 mm. The beam between nodes 16 and 17 (which has the same stress state as beam 1) tends to show the maximum stress, while the PEA is working at the displacement of 60 μm. The obtained maximum stress coefficients are tabulated in Table 2, which agree well with the analytical results.

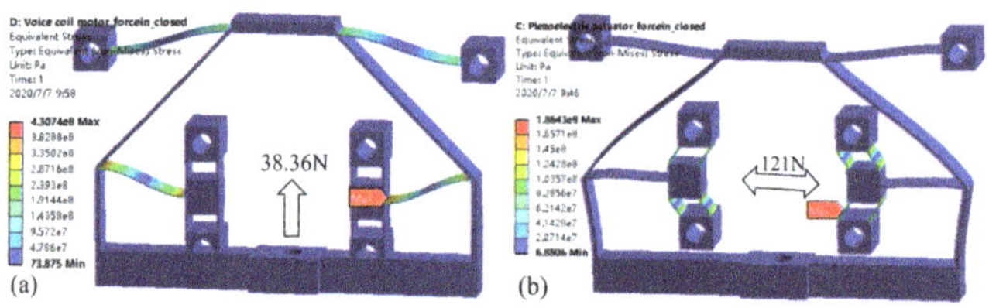

Figure 4. Stress distribution of ACPM while (**a**) actuating the input end of VCM with 38.36 N and fixing the input end of PEA, (**b**) driving the input end of PEA with 121 N and fixing the input end of VCM.

To reveal the maximum stress while the proposed ACPM is actuated by the VCM and PEA simultaneously, two actuation displacements of 1.5 mm ($\left|\delta_1^y\right|$) and 60 μm ($\left|\delta_7^x\right| + \left|\delta_{20}^x\right|$) are applied on the ACPM, as shown in Figure 5. It can be observed that beam 2 shows the maximum stress with a value close to the maximum stress shown in Figure 4a. Furthermore, the maximum stresses of the node16-node17 beams (beam between nodes 16 and 17) in Figures 4b and 5 are close to each other. Therefore, for the proposed ACPM, the maximum stress of beam 2 is mainly contributed by the VCM. Because of the input-decoupling compliant mechanisms, the VCM has limited influence on the stress of beam 1.

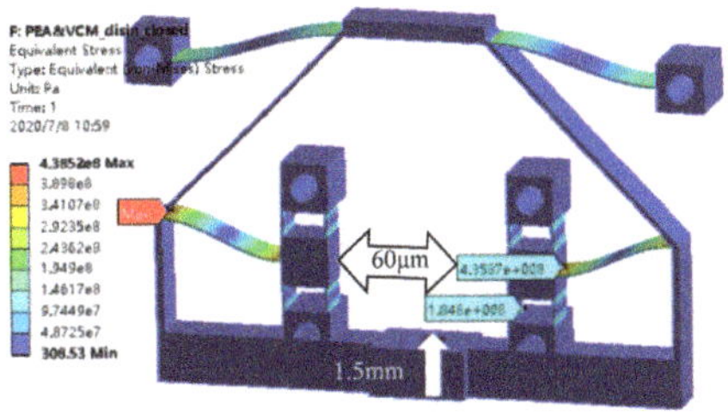

Figure 5. Stress distribution of ACPM while actuating VCM and PEA simultaneously.

Finally, modal analysis of the proposed ACPM is carried out in the ANSYS Workbench software. Considering the guiding function of the bearing of the VCM, a cylindrical support with the Y-translational freedom is applied on the input end of the VCM. Undesired out-of-plane resonances can be attenuated by the VCM bearing. Figure 6 shows the revealed first-order modal shape of the proposed ACPM. It can be seen that the first resonant resonance occurs along the actuation direction of the VCM with a frequency of 99.7 Hz. Therefore, while in open-loop control mode, the VCM actuation should be operated below 99.7 Hz to avoid mechanical resonance.

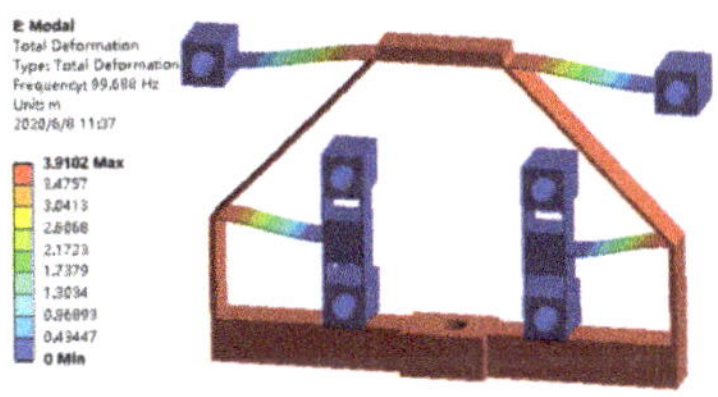

Figure 6. First modal shape of ACPM.

5. Experimental Analysis

A PDCNS prototype is fabricated to verify its kinematic performance as well as study its closed-loop motion capability, as shown in Figure 7. The ACPM with the dimensions shown in Table 1 is monolithically fabricated from TC4 titanium alloy. A commercial VCM (TM motion, model: TMEH0250, with a support bearing) and a commercial PEA (model: NAC2015-H38, maximum displacement: 59.4 μm, maximum thrust: 4200 N) are adopted to actuate the ACPM. To evaluate the motion of the ACPM, two laser displacement sensors (KEYENCE, model: LK-G10, max. measurement range: ±1) mm are used to measure the displacements of the VCM and the end effector, and two capacitance displacement sensors (MTI Instruments, model: ASP-125M, max. measurement range: 125 μm) are employed to detect the input displacement of the PEA. The following experiments are carried out on a PC with a PCI-6259 data acquisition card.

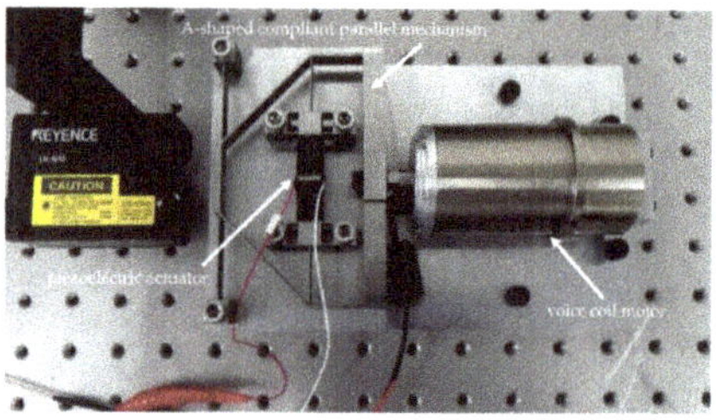

Figure 7. Experimental setup of PDCNS prototype.

5.1. Kinematic Investigation

The displacement amplification ratio for the VCM is identified by maintaining the position of the PEA with a closed-loop control and increasing the control signal of the VCM at the same time. Figure 8a shows the obtained displacement relationship between the end effector and the VCM while the displacement of the PEA is maintained at 0 μm. The displacement amplification ratio for the VCM is calculated to be 0.992. The small difference between the measured and calculated displacement amplification ratios of VCM confirms the effectiveness of the established analytical model.

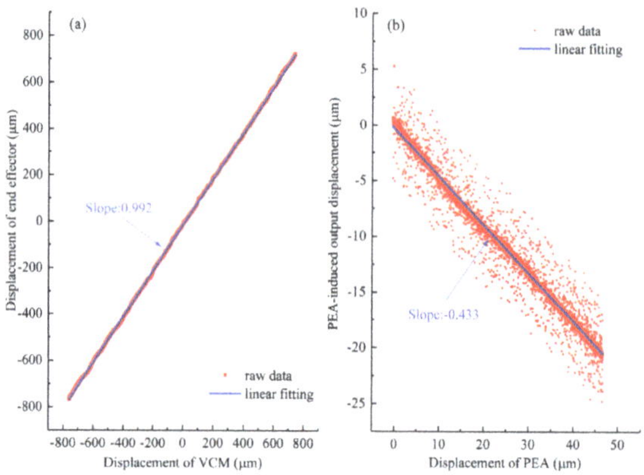

Figure 8. Kinematic test: (**a**) the displacement relationship between VCM and end effector while the displacement of the PEA is maintained at 0 μm, and (**b**) the relationship between the PEA displacement and the PEA-induced output displacement.

By applying a triangular-waveform signal with a frequency of 0.1 Hz and amplitude of 5 V to the PEA (maintaining the control voltage of the VCM at 0 V), the induced displacement of the end effector and the coupling displacement of the VCM are measured and recorded. By calculating the output displacement induced by the VCM (product of the identified amplification ratio and measured displacement of the VCM) and subtracting it from the measured displacement of the end effector, the PEA-induced output displacement can be obtained. Figure 8b shows the measured relationship between the displacement of PEA and the output displacement induced by PEA. Because of measurement noise, the measured raw data are distributed around the linear fitting line. Despite this, the kinematic of the PEA actuation can be identified according to the slope ratio of fitted linear line. The displacement amplification ratio for PEA is obtained to be 0.433. According to the established analytical model, the amplification ratio for the PEA is 0.514. By means of finite element analysis (based on ANSYS Workbench), the amplification ratio for the PEA is evaluated to be 0.524. The deviation between the proposed method and the finite element analysis is only 2.0%. Nevertheless, the deviation between the proposed method and the experiments is 15.7%. This obvious deviation of the PEA displacement amplification ratio can be attributed to manufacturing error, assembly error, mismatched material parameters, etc.

5.2. Stroke and Hysteresis Investigation

The stroke and hysteresis characteristics of the proposed PDCNS are further investigated. First, two 0.1 Hz triangular control signals with amplitudes of 10 V and 5 V are separately applied to the VCM. The displacement responses of the PDCNS are obtained, as in Figure 9a. It can be seen that the maximum stroke of the PDCNS can reach 1.49 mm. While the output displacement of PDCNS is calibrated to be 1.49 mm, the maximum stress of the ACPM is simulated to be 431 Mpa, which is below the yield stress of TC4 titanium alloy (>790 Mpa [32]), as shown in Figures 3a and 4a. Hence, the stroke of the PDCNS prototype is limited by the maximum driving force of the voice coil motor. Besides, due to the friction of the VCM's bearing, a significant dead zone exists in the response curve of PDCNS. Then, another two triangular control signals are applied to the PEA, as illustrated in Figure 9b. Compared to the VCM-actuated PDCNS, the PEA-actuated PDCNS is free of the friction-induced dead zone. In the meantime, the PEA-actuated PDCNS displays a reduced hysteresis behavior.

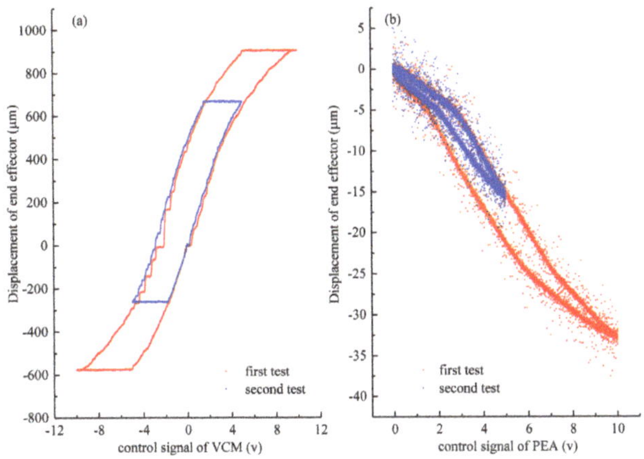

Figure 9. Stroke and hysteresis tests: (**a**) relationship between the displacement of the end effector and the control voltage of the VCM, (**b**) relationship between the displacement of the end effector and the control voltage of the PEA.

5.3. Motion Resolution Investigation

Consecutive-step-positioning experiments are carried out to investigate the motion resolution of the proposed PDCNS, as shown in Figure 10. A clear motion resolution with a step size of 0.74 μm can be achieved while the PDCNS is actuated by PEA, which is mainly limited by the resolution of the displacement sensor. Compared to the PEA-actuated PDCNS, the VCM-actuated PDCNS exhibits a much worse resolution with an average step size of 9.04 μm. The unstable response of the VCM-actuated PDCNS can be attributed to the nonlinearity and friction effects. Hence, the PEA-actuated PDCNS possesses an obvious advantage in high-resolution motion.

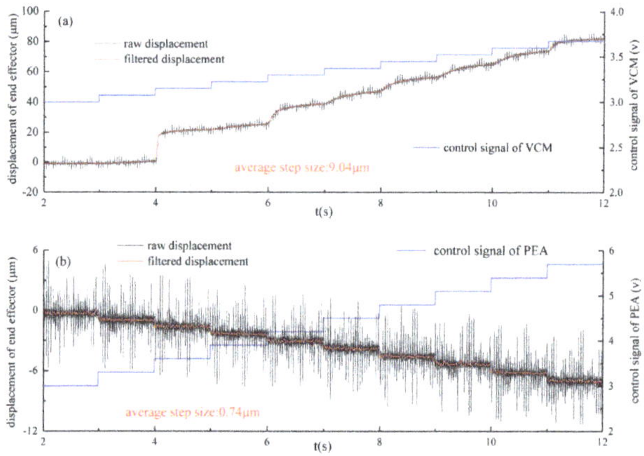

Figure 10. Motion responses with consecutive step control signals applied on (**a**) VCM, (**b**) PEA.

5.4. Closed-Loop Motion Investigation

(1) DSCC Design. Considering that the proposed PDCNS is a two-inputs-one-output system with severe hysteresis nonlinearity, a double-servo cooperative control (DSCC) strategy is designed in this section. Figure 11 depicts the block diagram of the proposed DSCC. The DSCC consists of three main parts: position observer, VCM servo control, and PEA servo control. The position observer is designed to estimate the virtual center of the end effector, x^{vir}, (the position of the end effector if the control voltage of PEA is zero) according to the displacement feedback of end effector as follows: the output displacement induced by PEA is calculated by multiplying the control voltage, V_{PEA}, and the voltage-displacement coefficient, γ_{PEA}; by removing the PEA-induced displacement from the measured position (with the laser displacement sensor) of the end effector, the virtual center of the end effector can be obtained. Using the virtual center of the end effector as feedback, a PID-based servo control strategy with adaptive parameters is designed to control the VCM according to the reference position, x^{ref}. The VCM servo control can not only converge the end effector to the reference position with coarse resolution but also ensure that the PEA is working within its rated range. To cooperate with the VCM actuation, a PID-based PEA servo control method is employed to achieve the fine motion and assure the position accuracy. Additionally, an initial 5 V is added to the control voltage of PEA, V_{PEA}, to ensure that the output signal of PEA servo control, V_{PEA}, can be regulated within −5 V~+5 V (the control voltage range of PEA driver is 0~10 V). With the specially designed DSCC, the PDCNS is expected to realize nm-level positioning over mm-level range.

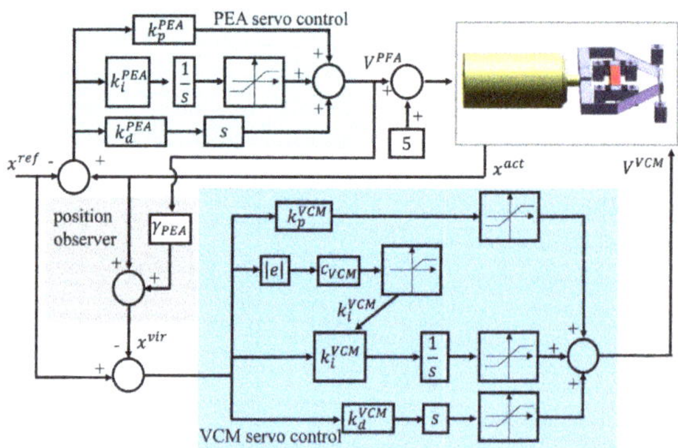

Figure 11. Block diagram of the double-servo cooperative control (DSCC) strategy.

(2) Positioning in mm-level range. A consecutive-step reference trajectory is adopted to guide the PDCNS (with DSCC) from −0.7 mm to 0.65 mm. For comparison, the VCM (with traditional PID) positioning experiments are also conducted based on the PDCNS prototype (while the control voltage of PEA is maintained at zero). The parameters of the PDCNS positioning (cooperation of voice coil motor and piezoelectric actuator) are tuned to be: VCM proportional coefficient = 0.03, VCM integral coefficient is adaptive, VCM derivative coefficient = 1×10^{-4}, voltage-displacement coefficient = 2.5, PEA proportional coefficient = 0.2, PEA integral coefficient = 150, and PEA derivative coefficient = 1×10^{-5}. In contrast, the parameters of the VCM positioning (only using the voice coil motor) are tuned to be: VCM proportional coefficient = 0.03, VCM integral coefficient = 0.3, and VCM derivative coefficient = 1×10^{-4}. The trajectory control results of the PDCNS (with DSCC) and the VCM (with PID) are compared in Figure 12. It can be observed that both PDCNS and VCM are effective for the mm-level positioning. Compared to the VCM, the PDCNS can realize a faster response and eliminate the steady-state error. Due to the fast response, PDCNS also induces an obvious overshoot. Figure 13 depicts the motion-tracking experiments of VCM and PDCNS while facing a consecutive-step trajectory from 0.7 mm to −0.65 mm, which confirms the global-positioning capability of the PDCNS.

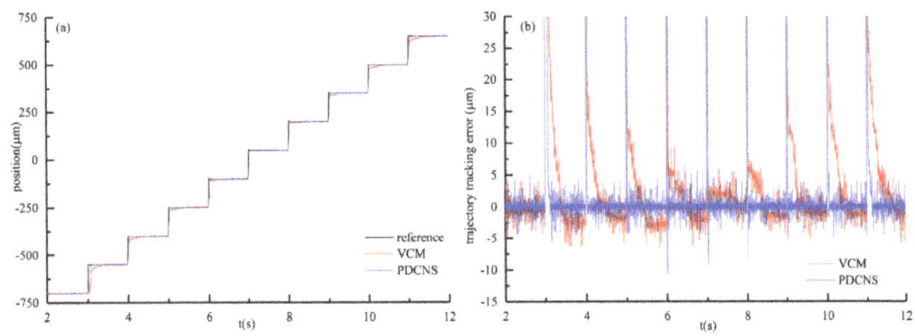

Figure 12. Consecutive-step trajectory (from −0.7 mm to 0.65 mm with a step size of 0.15 mm) control tests (**a**) tracking results and (**b**) tracking errors.

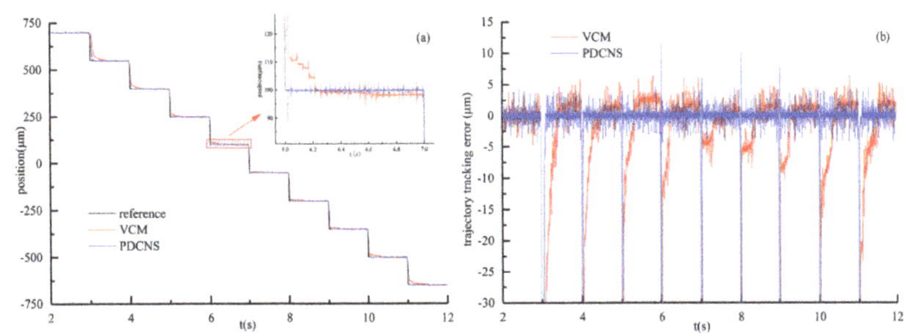

Figure 13. Consecutive-step trajectory (from 0.7 mm to −0.65 mm with a step size of −0.15 mm) control tests (**a**) tracking results and (**b**) tracking errors.

(3) Positioning with μm-level steps. To evaluate the precision positioning capability of the proposed PDCNS, consecutive-step-positioning tests with step sizes of ±1 μm are performed, as shown in Figures 14 and 15. Compared to the VCM (with PID), faster setting and smaller error can be achieved by the PDCNS (with DSCC). Because severe hysteresis exists in the VCM, it is time consuming for the PID to provide a control signal that can start the motion of the VCM. Due to the existence of friction, it is also a challenging job for the VCM to converge the end effector towards the reference position steadily. Different from the VCM, the PDCNS guarantees the convergence of the tracking error by coordinating the PEA and VCM. The VCM is responsible for driving the end effector close to the reference

position as well as decreasing the control effort of the PEA in a slow manner. Meanwhile, the PEA is in charge of eliminating the tracking error in a fast manner. Therefore, the proposed PDCNS can realize large-range, high-speed, and high-resolution positioning.

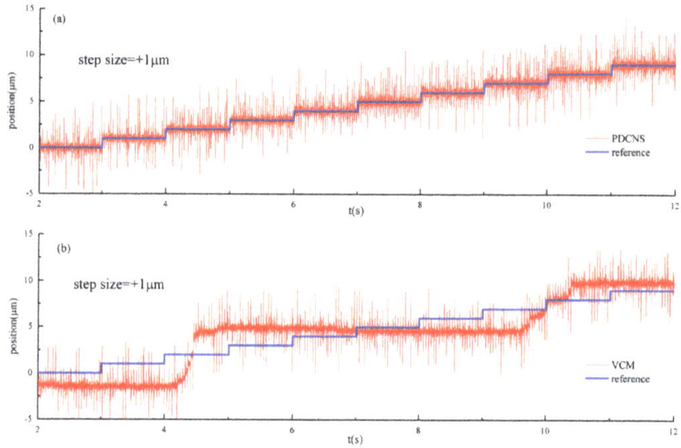

Figure 14. Multistep (step size of +1 μm) response of (**a**) PDCNS and (**b**) VCM.

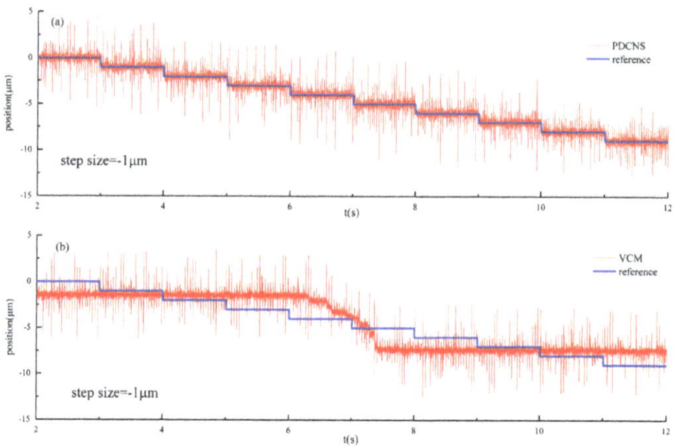

Figure 15. Multistep (step size of −1 μm) response of (**a**) PDCNS and (**b**) VCM.

(4) Positioning with nm-level steps. Finally, the nanometer-level positioning capability of the proposed PDCNS (with DSCC) is investigated by two consecutive-step reference signals with step sizes of ±50 nm. In this subsection, to improve the signal-to-noise ratio, the laser displacement sensor is configured to have a measurement range of −10 μm to 10 μm (corresponding to an output voltage of −10 V to 10 V). A 50 nm height staircase trajectory response is clearly obtained in Figure 16a. Moreover, a continuous stepping movement with amplitude of −50 nm can be observed in Figure 16b. Therefore, the closed-loop-positioning resolution of the proposed PDCNS is better than 50 nm.

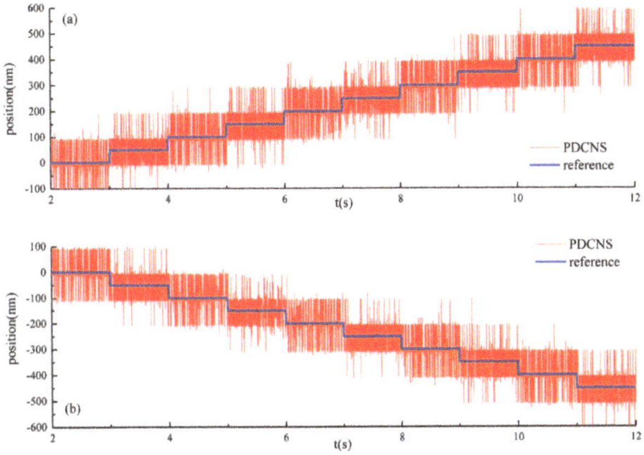

Figure 16. Closed-loop-positioning tests of PDCNS, with step sizes of (**a**) +50 nm (**b**) −50 nm.

6. Conclusions

In this paper, the mechanical design, analytical modeling, and closed-loop control of a novel PDCNS are conducted. The designed PDCNS combines the coarse motion from a voice coil motor and the fine motion from a piezoelectric actuator through a new ACPM, which breaks through many barriers encountered in traditional serial dual-stage systems. To resolve the kinematic characteristics of the proposed ACPM, an analytical model is specially established. Furthermore, a DSCC strategy with good robustness to nonlinearities is devised to enhance the positioning capability of the PDCNS.

Simulations and experiments are conducted to demonstrate the static and dynamic performance of the PDCNS. The results show that the maximum stroke of the PDCNS can reach 1.49 mm, and the positioning resolution can be better than 50 nm. Moreover, the designed DSCC can eliminate the steady-state position error of PDCNS, despite many nonlinear factors.

Author Contributions: Conceptualization, X.Y., L.J.; methodology, X.Y., L.J.; validation, X.Y., L.J.; writing—X.Y., Y.S., W.Z., S.L. All authors have read and agreed to the published version of the manuscript.

Funding: This research was funded by the National Key R&D Program of China under Grant 2019YFB2005304, the National Natural Science Foundation of China under Grant 52175018 & 51875505, the Natural Science Foundation of Shandong Province under Grant ZR2020YQ37 & ZR2019QEE023, the Shandong Provincial Key Research and Development Program Major Scientific and Technological Innovation Project: 2019JZZY010802.

Institutional Review Board Statement: Not applicable.

Informed Consent Statement: Not applicable.

Data Availability Statement: Not applicable.

Conflicts of Interest: The authors declare no conflict of interest.

References

1. Nikooienejad, N.; Maroufi, M.; Moheimani, S.O.R. Iterative Learning Control for Video-Rate Atomic Force Microscopy. *IEEE/ASME Trans. Mechatron.* **2021**, *26*, 2127–2138. [CrossRef]
2. Zhu, W.; Yang, X.; Duan, F.; Zhu, Z.; Ju, B. Design and Adaptive Terminal Sliding Mode Control of a Fast Tool Servo System for Diamond Machining of Freeform Surfaces. *IEEE Trans. Ind. Electron.* **2019**, *66*, 4912–4922. [CrossRef]
3. Wan, N.; Wen, J.; Hu, Y.; Kan, J.; Li, J. A Parasitic Type Piezoelectric Actuator with an Asymmetrical Flexure Hinge Mechanism. *Microsyst. Technol.* **2020**, *26*, 917–924. [CrossRef]
4. Wei, Y.; Xu, Q. A Survey of Force-Assisted Robotic Cell Microinjection Technologies. *IEEE Trans. Automat. Sci. Eng.* **2019**, *16*, 931–945. [CrossRef]

5. Chen, X.; Deng, Z.; Hu, S.; Gao, J.; Gao, X. Designing a Novel Model of 2-DOF Large Displacement with a Stepwise Piezoelectric-Actuated Microgripper. *Microsyst. Technol.* **2020**, *26*, 2809–2816. [CrossRef]
6. Eleftheriou, E. Nanopositioning for Storage Applications. *Annu. Rev. Control* **2012**, *36*, 244–254. [CrossRef]
7. Yang, X.; Zhu, W.; Zhu, Z.; Zhu, L. Design, Assessment, and Trajectory Control of a Novel Decoupled Robotic Nanomanipulator. *IEEE/ASME Trans. Mechatron.* **2022**, 1–12. [CrossRef]
8. Yang, X.; Zhu, W.; Zhu, Z.; Zhu, L. Development of a New Compliant Active-Force Support System. *IEEE/ASME Trans. Mechatron.* **2022**, *27*, 372–382. [CrossRef]
9. Tao, Y.; Li, H.; Zhu, L. Time/Space-Separation-Based Gaussian Process Modeling for the Cross-Coupling Effect of a 2-DOF Nanopositioning Stage. *IEEE/ASME Trans. Mechatron.* **2021**, *26*, 2186–2194. [CrossRef]
10. Fleming, A.J.; Aphale, S.; Moheimani, S.O.R. A New Method for Robust Damping and Tracking Control of Scanning Probe Microscope Positioning Stages. *IEEE Trans. Nanotechnol.* **2010**, *9*, 438–448. [CrossRef]
11. Yong, Y.K.; Moheimani, S.O.R.; Kenton, B.J.; Leang, K.K. Invited review article: High-Speed Flexure-Guided Nanopositioning: Mechanical Design and Control Issues. *Rev. Sci. Instrum.* **2012**, *83*, 121101. [CrossRef] [PubMed]
12. Tan, K.K.; Lee, T.H.; Zhou, H.X. Micro-Positioning of Linear-Piezoelectric Motors Based on a Learning Nonlinear PID Controller. *IEEE/ASME Trans. Mechatron.* **2001**, *6*, 428–436. [CrossRef]
13. Ho, S.T.; Jan, S.J. A Piezoelectric Motor for Precision Positioning Applications. *Precis. Eng.* **2016**, *43*, 285–293. [CrossRef]
14. Li, J.P.; Huang, H.; Morita, T. Stepping Piezoelectric Actuators with Large Working Stroke for Nano-positioning Systems: A Review. *Sens. Actuators A Phys.* **2019**, *292*, 39–51. [CrossRef]
15. Delibas, B.; Koc, B. A Method to Realize Low Velocity Movability and Eliminate Friction Induced Noise in Piezoelectric Ultrasonic Motors. *IEEE/ASME Trans. Mechatron.* **2020**, *25*, 2677–2687. [CrossRef]
16. Mamun, A.A.; Mareels, I.; Lee, T.H.; Tay, A. Dual Stage Actuator Control in Hard Disk Drive—A review. In Proceedings of the Industrial Electronics Conference, Roanoke, VA, USA, 2–6 November 2003.
17. Zhu, Z.; Du, H.; Zhou, R.; Huang, P.; Zhu, W.; Guo, P. Design and Trajectory Tracking of a Nanometric Ultra-Fast Tool Servo. *IEEE Trans. Ind. Electron.* **2020**, *67*, 432–441. [CrossRef]
18. Gutierrez, H.M.; Ro, P.I. Sliding-Mode Control of a Nonlinear-Input System: Application to a Magnetically Levitated Fast-Tool Servo. *IEEE Trans. Ind. Electron.* **1998**, *45*, 921–927. [CrossRef]
19. Yang, X.; Zhu, W. Design, Analysis and Test of a Novel Self-Sensing Fast Tool Servo. *IEEE Trans. Ind. Informat.* **2020**, *16*, 4447–4455. [CrossRef]
20. Wang, G.; Xu, Q. Design and Precision Position/Force Control of a Piezo-Driven Microinjection System. *IEEE/ASME Trans. Mechatron.* **2017**, *22*, 1744–1754. [CrossRef]
21. Permana, S.; Grant, E.; Walker, G.M.; Yoder, J.A. A Review of Automated Microinjection Systems for Single Cells in the Embryogenesis Stage. *IEEE/ASME Trans. Mechatron.* **2016**, *21*, 2391–2404. [CrossRef]
22. Xu, Q. Design and Development of Flexure-Based Dual-Stage Nanopositioning System with Minimum Interference Behavior. *IEEE Trans. Autom. Sci. Eng.* **2012**, *9*, 554–563. [CrossRef]
23. Michellod, Y.; Mullhaupt, P.; Gillet, D. Strategy for the Control of a Dual-Stage Nano-Positioning System with a Single Metrology. In Proceedings of the 2006 IEEE Conference on Robotics, Automation and Mechatronics, Bangkok, Thailand, 1–3 June 2006; pp. 1–8.
24. Dong, W.; Tang, J.; ElDeeb, Y. Design of a Linear-Motion Dual-Stage Actuation System for Precision Control. *Smart Mater. Struct.* **2009**, *18*, 095035. [CrossRef]
25. Zhu, H.; Pang, C.K.; Teo, T.J. Integrated Servo-Mechanical Design of a Fine Stage for a Coarse/Fine Dual-Stage Positioning System. *IEEE/ASME Trans. Mechatron.* **2016**, *21*, 329–338. [CrossRef]
26. Long, Y.Q.; Bao, S.H.; Yuan, S. *Structural Mechanics*, 3rd ed.; The Higher Education Press: Beijing, China, 2012.
27. Zhu, W.; Zhu, Z.; Shi, Y.; Wang, X.; Guan, K.; Ju, B. Design, Modeling, Analysis and Testing of a Novel Piezo-Actuated XY Compliant Mechanism for Large Workspace Nano-Positioning. *Smart Mater. Struct.* **2016**, *25*, 115033. [CrossRef]
28. Koseki, Y.; Tanikawa, T.; Koyachi, N.; Arai, T. Kinematic Analysis of Translational 3-DOF Micro Parallel Mechanism Using Matrix Method. In Proceedings of the 2000 IEEE/RSJ International Conference on Intelligent Robots and Systems, Takamatsu, Japan, 31 October–5 November 2000; pp. 786–792.
29. Ling, M.; Cao, J.; Li, Q.; Zhuang, J. Design, Pseudo-Static Model and PVDF-Based Motion Sensing of a Piezo-Actuated XYZ Flexure Manipulator. *IEEE/ASME Trans. Mechatron.* **2018**, *23*, 2837–2848. [CrossRef]
30. Zhu, W.; Zhu, Z.; To, S.; Liu, Q.; Ju, B.; Zhou, X. Redundantly Piezo-Actuated XYθz Compliant Mechanism for Nano-Positioning Featuring Simple Kinematics, Bi-Directional Motion and Enlarged Workspace. *Smart Mater. Struct.* **2016**, *25*, 125002. [CrossRef]
31. Guo, Z.; Tian, Y.; Liu, C.; Wang, F.; Liu, X.; Shirinzadeh, B.; Zhang, D. Design and Control Methodology of a 3-DOF Flexure-Basedmechanism for Micro/Nano Positioning. *Robot. Comput-Integr. Manuf.* **2015**, *32*, 93–105. [CrossRef]
32. Chen, M. Research on Mechanical Properties Test and Dynamic Material Model of TC4 Titanium Alloy. Master's Thesis, Nanjing University of Aeronautics and Astronautics, Nanjing, China, 2012.
33. Wang, D.; Yang, Q.; Dong, H. A Monolithic Compliant Piezoelectric-Driven Microgripper: Design, Modeling, and Testing. *IEEE/ASME Trans. Mechatron.* **2013**, *18*, 138–147. [CrossRef]

Article

Development of a Novel Dual Servo Magnetic Levitation Stage

Dahoon Ahn [1], Ji-Won Jin [2], Hyeeun Yun [1] and Jaeheon Jeong [3,*]

[1] Department of Mechanical System Design Engineering, Seoul National University of Science and Technology, Seoul 01811, Korea; dhahn@seoultech.ac.kr (D.A.); hye1871@seoultech.ac.kr (H.Y.)
[2] Green Mobility R&D Center, Jeonbuk Institute of Automotive Convergence Technology, Gunsan 54158, Korea; jinjiwon@jiat.re.kr
[3] Agency for Defense Development, Daejeon 34186, Korea
* Correspondence: pkwogjs007@gmail.com

Abstract: The main objective of this paper is to propose, design, and control a novel dual servo magnetic levitation stage which is precise and vacuum compatible. The dual servo mechanism, comprising a coarse stage and a fine stage, was applied to a magnetic levitation stage system for the first time. The dual servo stage achieves high precision and a long stroke at the same time. The fine stage, which comprises voice coil motors, achieves high-precision motion by overcoming the limit of the coarse stage, the form of which is a planar motor. The planar motor was mathematically modeled and analyzed with respect to the main design parameters, after which the fine stage was optimally designed to be driven by high force. Both stages including a common heat exchanger were manufactured, and the heat exchanger cools down the heat given off from the planar motor and voice coil motors. The position measuring system consisted of laser interferometers and capacitive sensors, and the integrated dual servo stage was controlled with a master–slave control scheme. The experimental results showed a precision of 10 nm, thus confirming the suitability of the developed magnetic levitation stage for a high-precision fabrication process such as wafer lithography.

Keywords: magnetic levitation; dual servo; planar motor; voice coil motor; high precision

Citation: Ahn, D.; Jin, J.-W.; Yun, H.; Jeong, J. Development of a Novel Dual Servo Magnetic Levitation Stage. *Actuators* **2022**, *11*, 147. https://doi.org/10.3390/act11060147

Academic Editors: Limin Zhu, Yuen Kuan Yong, Yanling Tian and Yingxiang Liu

Received: 28 April 2022
Accepted: 25 May 2022
Published: 30 May 2022

1. Introduction

A stage is a system that makes a specimen track desired positions in real time. Numerous types of stage systems and their applications have been researched by many research groups [1–7], as described below. As stage systems—especially for semiconductor lithography—must be highly precise and have a long range of motion, there have been many studies on high-precision motion.

Although the rotary servo motor with a lead screw appeared to be an easy way to implement a stage system, it had a serious problem with backlash, resulting in poor precision [1]. A linear motor capable of transmitting actuation force directly to a moving object was used, but the friction from guides such as ball bearings, cross rollers, and dovetails was the main reason for the low precision [2]. A combination of linear motors with air bearings was used to eliminate the mechanical connection between the actuators and moving objects. By eliminating friction with non-contact configuration, high-precision motion was achieved [3,4].

In order to achieve a high degree of precision coupled with long-range motion, a dual servo mechanism was introduced. The dual servo stage is the combination of a coarse stage with a low degree precision but a long range of motion and a fine stage with a short range of motion range but a high degree of precision. Thanks to the air bearing and dual servo mechanism, the stage showed both high precision and a long range of motion [8]. A stage fitted with air bearings, however, cannot be used in next-generation semiconductor lithography processes. Some processes require a vacuum environment because irradiated light emanating from certain types of sources, such as the extreme ultraviolet (EUV) laser, is scattered through the air. As such, the stages used in the lithography process must be vacuum compatible. Magnetic levitation (Maglev) mechanisms that have vacuum compatibility, non-contact actuators, and non-contact guides have been introduced to stage systems [9–11]. Moreover, maglev stages equipped with a planar motor can achieve long-range, high-speed motion due to their relatively light moving body and rapid response [12–19]. Maglev planar motors generate actuation force between a two-dimensional magnet and coil arrays. They are categorized into two broad groups, i.e., the moving magnet type [19] and the moving coil type [20]. The former type does not have any wires attached to the mover, so there is no wire disturbance, but complex coil switching dependent on the mover position is necessary for control. Conversely, the moving coil type is disturbed by wire tension, but complex coil switching is not required.

For maglev planar motors, many academic trials have been carried out with the aim of improving its performance. One such trial involved the modification of the coil. More than two layers of coil were used [21–23] to increase the actuation force, and a new winding coil shape was tried [24]. Another approach consisted of enhancing the magnetic flux. Usually, the magnetic flux is increased with the use of the Halbach array and the variation in the magnet shape [14]. However, the maglev planar motor uses a highly complex scheme to control the current, which somewhat degrades the precision, despite the fact that many studies have attempted to develop an accurate current model [25–27].

This paper, however, proposes a novel maglev dual servo stage whose coarse stage is driven by a maglev planar motor and whose fine stage is driven by voice coil motors (VCMs). The fine stage realizes a very high degree of precision, while the coarse stage enables a long range of motion. In addition, the features of non-contact and vacuum compatibility contribute to the verification of the stage's suitability for the wafer lithography process. The rest of the paper is organized as follows. Section 2 introduces the structure and the design results of the dual servo maglev stage; Section 3 presents the fabrication of the dual servo maglev stage and an explanation of the experimental setup; Section 4 presents the evaluation of the performance of the dual servo maglev stage by experiments; and, lastly, Section 5 presents the conclusion.

2. Structure of the Dual Servo Maglev Stage

The dual servo maglev stage is composed of a coarse stage and a fine stage, as shown in the schematic diagram of Figure 1. They are driven by two different types of electromagnetic actuators, namely, a planar motor and VCMs. An electromagnetic actuator employing Lorentz force has two parts, a stator and a mover. The dual servo maglev stage developed in this paper consists of three layers: the bottom is the coarse stage stator of the planar motor magnets; the middle is the coarse stage mover of the planar motor coils assembled with the fine stage stator of the VCM coils; and the top is the fine stage mover of the VCM magnets. The three layers are mechanically separated by a maglev mechanism which ensures that the stage has fast dynamics due to the absence of friction. Compared to the dual servo stage with an air bearing, in which the stages are separated without any mechanical contact, the dual servo stage equipped with a maglev mechanism allows the stage to be used in a vacuum environment.

For the VCMs of the fine stage, the magnets belong to the mover and the coils belong to the stator in order to be resistant to thermal deformation and easy to control without force coupling or wire disturbance. If the coils are placed on the moving body of the fine stage, the heat generated at the coils is directly transferred to the moving body and then to the specimen and the measurement system, making it a potential source of poor precision. In addition, parasitic force is induced when the moving body of the fine stage has yaw motion [7]. Finally, the wires connected to the external power supply are a source of disturbance for the moving body. Therefore, the moving magnet type was adopted for the structure of the fine stage.

For the planar motor of the coarse stage, the coils belong to the mover and the magnets belong to the stator, as this makes it easier to control and requires fewer current drivers than the moving magnet-stationary coil structure. Unlike the VCM, the planar motor generates a six-degrees-of-freedom (DOF) coupled force between the magnets and the coils, so the relationship between the current of each coil and the generated force is required to control the planar motor. We call that relationship "actuator kinematics". The actuator kinematics relationship varies with respect to the position of the mover, since it depends on the magnetic field made by the permanent magnet array, and the magnetic field has spatial variation. When the coil array is the mover, the actuator kinematics relationship varies periodically in accordance with the repeated pattern of the stator and the magnet array. Thus, the actuator kinematics can be obtained by analytic or experimental calculation for one period of the magnet array. If the magnet array is the mover, the actuator kinematics relationship varies periodically in accordance with the repeated pattern of the coil array, which has a longer period than the magnet array. Furthermore, the finite size of the magnet array has the end-effect of showing a distorted magnetic field near the edge of the magnet array. Thus, the size of the real-time computation of the actuator kinematics is very large. In addition, a large number of current drivers are used, since the area of the stator is wider than that of the mover. Furthermore, because only the coils under the magnets can generate force, the coils should be turned on and off according to the position of the moving magnets. All of these factors increase the complexity of the control and power consumption. Considering the characteristics of the planar motor, the coarse stage was designed as a moving coil type.

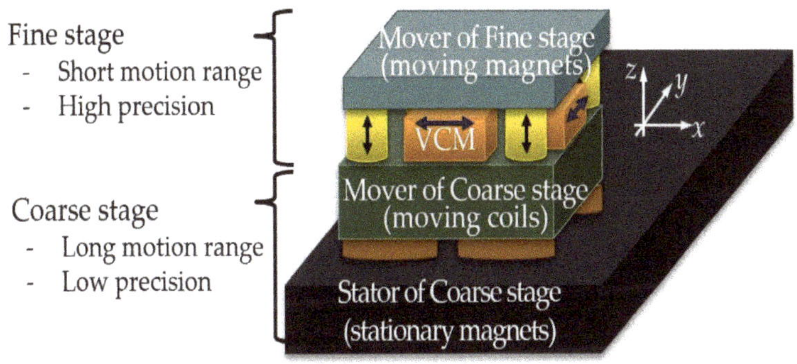

Figure 1. Schematic diagram showing the structure of the dual servo maglev stage.

According to the decision on the structure, all coils that are a source of the heat are placed on the middle layer. This arrangement separates the heat source from the moving part that carries the specimens. Additionally, it ensures that the top layer is not disturbed by wire tension related to the wires and coolant-carrying tubes.

2.1. Fine Stage

The fine stage requires at least six VCMs since it realizes six-DOF motion. In this paper, however, eight VCMs were used for symmetry. The fine stage mover is controlled so as to move on the XY plane by constraining the motion of the other three DOF. Therefore, it is useful to separate the role of the VCMs according to their use, which means decoupling each VCM force in accordance with the desired directions of motion. Two VCMs for motion along the *x*-axis and another two VCMs for motion along the *y*-axis were placed, as shown in Figure 2a. They are denoted by HVCM, which means the VCM for horizontal motion, i.e., the in-plane motion. Since the arrangement of the VCMs is symmetric, the forms of kinetics are the same for both directions of motion. In addition, the effort to regulate the yaw motion is even for all HVCMs. It is very beneficial when a practical control algorithm is applied, and it also simplifies the design process because one design result can be applied to all four HVCMs. VVCM means the VCM for vertical motion, i.e., the out-of-plane motion. The VVCMs could also be placed in the same manner as the HVCMs.

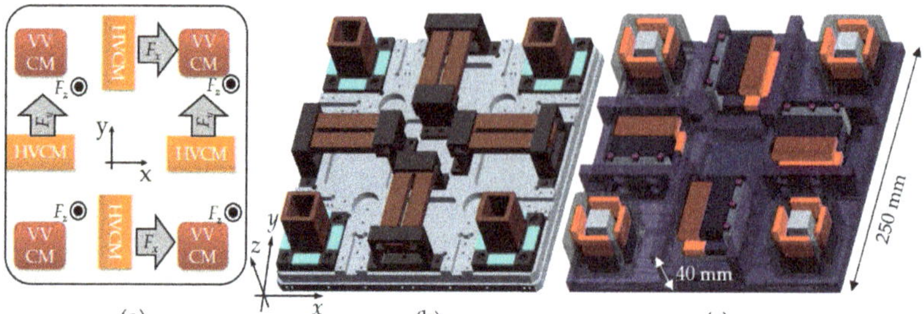

(a) (b) (c)

Figure 2. Design of the fine stage: (**a**) The topology of the VCMs; (**b**) the fine stage stator with coils; and (**c**) the fine stage mover with magnets (inverted) [27]. Reprinted from International Journal of Applied Electromagnetics and Mechanics, 62, Dahoon Ahn, et al., Design process of square column-shaped voice coil motor design for magnetic levitation stage, 517–540, Copyright (2020), with permission from IOS Press.

Each HVCM was designed to generate a force of at least 50 N, while each VVCM was designed to generate a force of at least 25 N to support and drive a fine stage mover with an expected weight of 10 kgf. The HVCMs and VVCMs have different sizes and structures. Each VCM was mathematically modeled, analyzed, and optimized to have high force. The design results were verified by FE (finite element) simulations and experiments measuring the force constant. The simple and compact structure of the VCMs exerted high force and uniform force that was constant throughout the entire range of motion. The detailed design process is presented in the previous work [27]. The final design of the fine stage is shown in Figure 2b,c and Table 1. The remainder of the space not occupied by the VCMs is used by sensors and mechanical motion stoppers. The final size of the designed fine stage is 250 mm in width and length and 52 mm in height. In the fine stage, the width and the length of the mover and the stator are the same, the height of the mover is 52 mm, and the height of the stator is 40 mm. If the mover is put on the stator like a lid, the overall height of the fine stage is 52 mm, which is the same as the height of the mover. After the fabrication of the dual servo stage, the mass of the mover of the fine stage was measured to be 8.89 kg, and the mass of the stator was measured to be 4.74 kg.

Table 1. Design specifications of the VCMs of the fine stage [27]. Reprinted from International Journal of Applied Electromagnetics and Mechanics, 62, Dahoon Ahn, et al., Design process of square column-shaped voice coil motor design for magnetic levitation stage, 517–540, Copyright (2020), with permission from IOS Press.

Design Results	HVCM	VVCM
Size	$95 \times 50 \times 40$ mm^3	$55 \times 55 \times 40$ mm^3
Wire diameter	0.6 mm	0.4 mm
Number of turns	297 turns	738 turns
Electric resistance	3.29 Ω	11.29 Ω
Max. voltage	26.5 V	36.9 V
Max. current	3.43 A	1.24 A
Max. power	17.5 W	17.5 W
Max. Force	52.4 N	28.2 N
Force constant	15.6 N/A	22.3 N/A

2.2. Coarse Stage

The coarse stage is levitated and driven by a maglev planar motor. The planar motor generates force between the two-dimensional magnet array and the three-phase coil array. The magnet array creates the spatially periodic magnetic field, and the coil array within the magnetic field generates force via the provided electric current. The main components of the force generated by one set of three-phase coils are parallel and perpendicular to the top surface of the magnet array. In order to regulate the two components of force, the magnitude and the phase of the electric current are controlled. Force generation using a multi-phase coil and a magnet array has been dealt with in detail in many previous studies [10,12,16,19,22,26]. In this section, the topology of the coil and the magnet array to realize the six-DOF motion of the coarse stage is presented. In addition, based on the electromagnetics and kinetics model, the design parameters are determined for the coarse stage so as to show optimal performance.

2.2.1. Coil and Magnet Array Topology

The coarse stage requires at least three sets of 3-phase coils to implement the six-DOF motions. In this research, however, four sets of 3-phase coils were used for structural symmetry, as in the case of the VCMs of the fine stage. As shown in Figure 3, two coil sets generate driving force along the x-axis and levitation force along the z-axis, whereas the other sets generate driving force along the y-axis and levitation force along the z-axis. For the benefit of the coil and magnet arrangement, the design result of one coil set can also be applied to the other coil set.

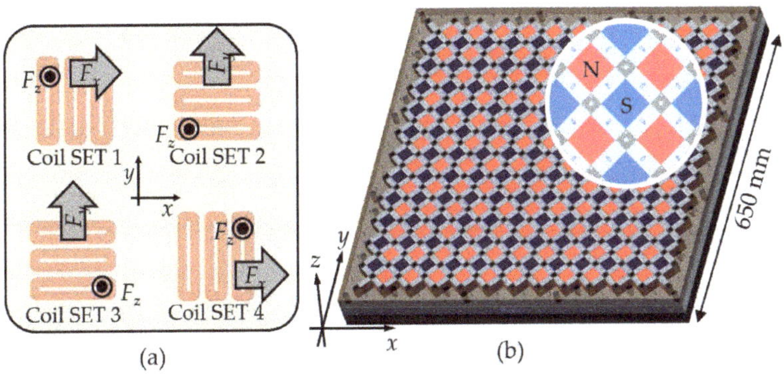

Figure 3. Design of the coarse stage: (**a**) The topology of the 3-phase coils; (**b**) the coarse stage stator with a two-dimensional magnet array.

The magnet array generates a magnetic field in the space where the coils are placed. The spatial period of the magnetic field (i.e., the spatial distance between magnets) is dependent on the size of the coils. In order to obtain a higher Lorentz force, higher magnetic flux density is desirable. Since the magnet array is the stator of the coarse stage, there is no limitation on the weight. Thus, a steel back-yoke and a Halbach array of tall magnets were used.

The stator of the coarse stage is 650 mm in width and length and 65 mm in height. This enables a fine stage of 250 mm in width and length to have a stroke of about 200 mm. The mover of the coarse stage is the same as the stator of the fine stage, so it is 250 mm in width and length. The height is 52 mm, including the coolant heat exchanger. After the fabrication of the dual servo stage, the mover mass of the coarse stage was measured to be 5.3 kg. The total mass of the magnetically levitated mover, including about 0.1 kg of the coolant mass, is 19.03 kg.

2.2.2. Design Parameter Analysis

Figure 4 shows the design parameters of the magnets and coils. There are six independent parameters, five dependent parameters, and three pre-defined parameters, which are the dimensions of the magnets and coils. Descriptions of each parameter and the relationships between them are shown in Table 2, along with Equations (1)–(5).

$$g_3 = \frac{3\sqrt{2}}{4}\tau \quad (1)$$

$$\tau = \frac{l_w^c - (g_3 + 2w_c^c + g_4)}{4\sqrt{2}} \quad (2)$$

$$\tau_m = \alpha\tau \quad (3)$$

$$l_c^c = 2\sqrt{2}\tau \quad (4)$$

$$g_1 = \frac{2\sqrt{2}}{3}\tau - 2w_c^c - g_2 \quad (5)$$

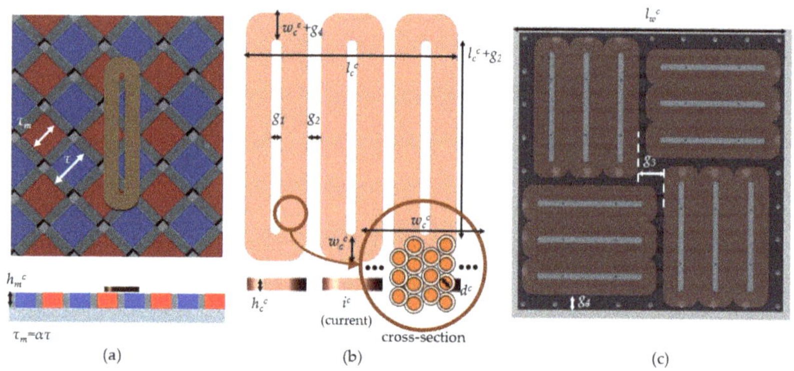

Figure 4. Design parameters of the planar motor, (**a**) magnet array, (**b**) coil array, and (**c**) four sets of 3-phase coils.

Table 2. Design parameters of the planar motor.

Design Parameters		Description
Independent parameters	$h_m{}^c$	Height of the magnet array
	α	Ratio of the length of the main magnet to the half-pitch of the magnet array
	$w_c{}^c$	Width of a bundle of coil threads with the same current direction
	$h_c{}^c$	Height of the coil array
	i^c	Current through the coil
	d^c	Diameter of the core of the coil (without sheath)
Dependent parameters	τ	Half-pitch of the magnet array
	τ_m	Length of the main magnet
	$l_c{}^c$	Effective length of the coil
	g_1	Center gap of the coil winding
	g_3	Distance between the coil sets which generate forces in the same direction
Pre-defined parameters	g_2	Pre-defined to match the phase of the current and the magnetic flux density: 0 mm
	g_4	Pre-defined to consider manufacturing the tolerance of coil winding: 8 mm
	$l_w{}^c$	Pre-defined by the size of the mover of the coarse stage: 250 mm
Materials	Magnet	NdFeB, N-38H, magnetization of 8.99×10^5 A/m
	Coil	Copper wire, resistivity of 1.793×10^{-8} Ωm
	Yoke	AISI 1020

Using design parameter analysis, the effect of the dependent design parameters on the important indices was inspected. The generated force, the mass of the mover, and the ohmic loss were observed when the height of the magnet, the ratio of the magnet length, the width of a coil bundle, the height of the coil, the current, and the diameter of the coil all varied from their nominal values. To calculate the indices, the number of coil turns and the electric resistance were obtained using Equations (6) and (7), respectively. ρ_e is the electric resistivity of the coil wire.

$$n = \left(\frac{h_c{}^c}{1.1d_c}\right)\left(\frac{2}{\sqrt{3}}\left(\frac{w_c{}^c}{1.1d_c} - 1\right) + 1\right) \tag{6}$$

$$R = \rho_e \frac{n(2l_c^c + w_c^c + g_1)}{\pi d_c{}^2/4} \tag{7}$$

For the maglev planar motor, the thrust and levitation forces are the most important factors. Since the magnetic flux density has a harmonic form, the thrust and levitation forces of the equations can be approximated to first-order harmonic terms [9,15]. The amplitude of the force is represented by the product of the current and the force constant, k_{fx} and k_{fz}, as shown in Equations (8) and (9).

$$F_{coil,x} = i^c k_{fx} cos(\frac{2\pi}{\sqrt{2}\tau} x_p) \tag{8}$$

$$F_{coil,z} = i^c k_{fz} sin(\frac{2\pi}{\sqrt{2}\tau} x_p) \tag{9}$$

If we use three coil windings to create the constant force of a 3-phase coil set, the current can be provided as a harmonic function, while the resultant thrust and levitation forces are as shown in Equations (10) and (11). The magnitude of the generated force of a 3-phase coil can be defined as shown in Equation (12), where i^c is the amplitude and ϕ is the phase of the provided current required to control the magnitude and the ratio of the thrust and levitation forces.

$$F_x = \sum_{k=1}^{3}\left[I^c cos(\frac{2\pi}{\sqrt{2}\tau} x_p + \frac{4(k-1)\pi}{3} + \varphi) k_{fx} cos(\frac{2\pi}{\sqrt{2}\tau} x_p + \frac{4(k-1)\pi}{3})\right] = \frac{3}{2} I^c k_{fx}(cos\varphi) \tag{10}$$

$$F_z = \sum_{k=1}^{3}\left[I^c cos\left(\frac{2\pi}{\sqrt{2}\tau} x_p + \frac{4(k-1)\pi}{3} + \varphi\right) k_{fz} sin(\frac{2\pi}{\sqrt{2}\tau} x_p + \frac{4(k-1)\pi}{3})\right] = -\frac{3}{2} I^c k_{fz}(sin\varphi) \tag{11}$$

$$F^c = \sqrt{(F_x^2 + F_z^2)} \tag{12}$$

Then, the ohmic loss of (13) can be obtained from (7), (10), and (11). The mass is given by (14). M_0 is the mass of the coarse stage mover, excluding the coils.

$$P^c = \frac{2}{3} I^{c2} R = \frac{2R}{3}\left(\left(\frac{F_x}{k_{fx}}\right)^2 + \left(\frac{F_z}{k_{fz}}\right)^2\right) \tag{13}$$

$$M^c = M_0 + 12\rho_c(2w_c^c l_c^c h_c^c + 2w_c^c(w_c^c + g_1)h_c^c) \tag{14}$$

To compare the indices, the values were normalized by the nominal performance values obtained from the nominal values of the design parameters. Table 3 shows the nominal values and the variation in the design parameters. Figure 5 shows the tendency and sensitivity of the indices to the design parameters. Figure 5a shows that the force is saturated to a certain level when the height of the magnet array increases. The magnitude of the force with an h_m of 20 mm is about 99.2% of the magnitude of the force with an h_m of 45 mm. The magnetic flux density also shows similar behavior when the height of a magnet increases [15]. From Figure 5a, the design parameter h_m was fixed to 20 mm for the later optimization process, because magnets that are larger than 20 mm barely increase the force and also make manufacturing and assembly difficult. Figure 5b shows that the Halbach magnet array is more advantageous than the normal magnet array corresponding to a ratio of 1. Figure 5b also shows that there is an optimal ratio which can be used to determine the size of the Halbach magnet array. Thus, the design variable α, which determines the sizes of the magnets, should be determined in the design process and handled well in the manufacturing process. From Figure 5c–f, it was found that there is a trade-off between heat generation and force generation. If the width and height of the coil winding are increased, the Lorentz force and electric resistance are also increased. An increase in force is desirable, but the accompanying increase in heat generation is unfavorable. The generated force and heat also increase simultaneously when the electric current increases. This phenomenon clearly shows that force is proportional to the current and that heat is proportional to the square of the current. Figure 5f shows that heat generation is very sensitive to variations in the coil diameter below the value of about 0.8 mm. Figure 5c,d show that mass is relatively insensitive to the design parameters, since the portion of the coil is small compared to the main body of the coarse stage mover. The mass is changed by less than 5% or so when the design parameters related to the coil winding vary. However, the other performance indices show a trade-off, and the optimal performance and the optimal design parameters should be determined through an optimization process. The graphs in Figure 5c,d,f are not smooth because the number of coil turns does not vary continuously with respect to the coil's width, height, and diameter. Thus, the design parameters w_c, h_c, and d are handled in a discretized manner through the optimization process.

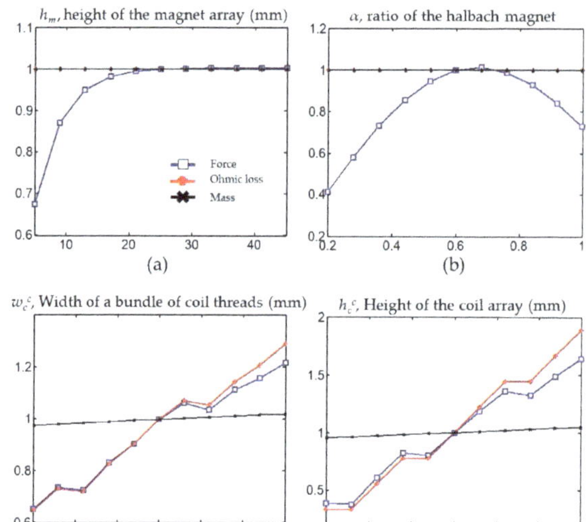

Figure 5. Results of the sensitivity analysis. (**a**) effect of height of the magnet array; (**b**) effect of the Halbach magnet ratio; (**c**) effect of bundle width of the coil; (**d**) effect of height of the coil array; (**e**) effect of electric current of the coil; (**f**) effect of height of the coil array.

Table 3. Variations in the design parameters of the planar motor for sensitivity analysis.

Independent Design Parameters	Nominal Value	Variation
h_m	25 mm	5 mm–45 mm
α	0.6	0.2–1
$w_c{}^c$	8 mm	5 mm–11 mm
$h_c{}^c$	5 mm	2 mm–8 mm
i^c	2 A	1 A–3 A
d^c	0.8 mm	0.3 mm–1.3 mm

2.2.3. Optimization

The objective of the optimization is set to maximize the force and minimize the mass of the coarse stage mover. The production of high force is advantageous not only for acceleration and deceleration within a limited range of motion but also for a fast dynamic response. The mover of the coarse stage is always levitated during the operation, which causes constant power consumption. However, the amount of power consumption can be decreased if the mover is light. Furthermore, the lighter the mover, the easier it is for the stages to achieve the specified acceleration with a smaller inertial force. The objective and the constraints are presented in Table 4.

Table 4. Objective and constraints of optimization for the 3-phase coil set of the planar motor.

Objective	$\min\left[(M^c/F^c)^2\right]$	
Constraints	Vertical force	>50 N
	Horizontal force	>100 N
	Ohmic loss	<130 W

Since the planar motor must support the weight of the mover of the coarse stage and the whole fine stage, each 3-phase coil set should be able to generate at least 50 N in the vertical direction for the expected weight of 20 kgf. At the same time, the planar motor must accelerate the mover in the horizontal direction. Therefore, one set of coils should be able to generate at least 100 N, because two sets of coils should exert at least 200 N in the horizontal direction. The details of the values are described in a previous work [27]. In addition, a larger ohmic loss means that more electric energy is converted to heat, which is then transferred to the surrounding components, causing a negative effect. The value of the ohmic loss is constrained to 130 W by the simulation that is explained in the next section.

The optimization was conducted using the SQP (Sequential Quadratic Programming) algorithm of MATLAB. Since the coil diameter is available only in a discrete size in practice, optimized solutions were obtained for each coil diameter. As shown in Table 5, the results show similar levels of force, ohmic loss, and mass for the different coil diameters. There is no feasible solution for coil diameters that are more than 0.9 mm or less than 0.5 mm in diameter. Among the optimization results in Table 5, a coil with a diameter of 0.6 mm was chosen for the optimized value due to the limited features of the current amplifier. The maximum current output and the terminal voltage of the current amplifier were 8 A and 48 V, respectively. In cases in which a coil diameter is more than 0.6 mm, a current exceeding 8 A should be used, while a terminal voltage of 56.5 V is needed when a coil diameter is 0.5 mm; for a coil diameter of 0.6 mm, however, the terminal voltage should be 39.5 V.

Table 5. Results of optimization with respect to the various coil diameters.

	Design Parameters				Constraints		
α	$w_c{}^c$ (mm)	$h_c{}^c$ (mm)	i^c (A)	d^c (mm)	F^c (N)	P^c (W)	m^c (kg)
0.668	13.9	5.98	5.17	0.5	154.9	130	19.17
0.671	13.9	6.06	7.44	0.6	155.8	130	19.20
0.672	13.9	6.63	9.68	0.7	154.2	130	19.23
0.646	13.9	5.83	13.62	0.8	151.3	130	19.11
0.671	13.9	8.08	14.54	0.9	160.0	130	19.97

3. Realization of the Dual Servo Maglev Stage

3.1. Heat Exchanger

The generation of heat by the coils of the fine stage and the coarse stage can be a very serious problem. An increase in temperature due to heat can cause the self-bonded coil wires to become unglued, the magnets to lose their magnetism above the Curie temperature, and the thermal deformation of the stage structure, leading to the propagation of motion errors to a specimen or sensors. Therefore, most of the heat generated by the coils should be transferred to a heat sink, for which purpose a heat exchanger was used.

A heat exchanger with a coolant situated right behind the coils was designed. Water was used as the coolant, and two water channels were engraved on the bottom surface of the fine stage stator. The material of the stator is aluminum with very high thermal conductivity. The water channels are closed by a ceramic block on which the coil array for the planar motor is placed. A ceramic coil bearer was chosen due to its high thermal conductivity for heat transfer and to the fact that it does not cause eddy currents induced by the magnet array. The coolant circulates through each channel, and the heat from the coils is transferred to the coolant. The pressure and temperature of the water at the channel inlet are maintained by a chilling circulator (RW-2040G, JEIO Tech., Daejeon, Korea).

The heat transfer from the coils of both the fine and coarse stages to the water coolant was verified by FE simulation (Ansys Icepak, Canonsburg, PA, US). In the analysis model, the coil windings were set to the source of constant heat generation. Heat generation of 150 W was applied to each coil set of the planar motor, while 17.5 W was applied to each coil of the VCM. A combined total of 740 W of heat was assumed to be generated by all the coils. The cross section of the channel was a square, with an area of 5×5 mm^2, and the water flow rate from the chilling circulator was 0.5 L/min. The results of the simulation show that the total heat generation of 740 W is well dissipated to the water flowing through the channel. The simulation results in Figure 6 show that the coils, ceramic block, and stator of the fine stage and the coolant have a temperature of less than 50 °C. The difference in temperature within the structure is less than 10 °C. In consideration of the fact that the temperature endurance limit is 100 °C for the coil windings and 90 °C for the magnets, it is concluded that there is no thermal problem, based on the results of the simulation. Based on these results, the constraint of optimization was set to 130 W instead of 150 W, taking into account a margin of roughly 10%.

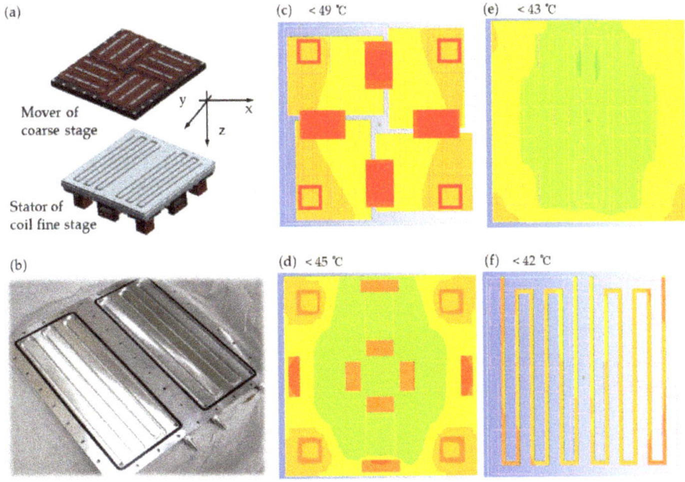

Figure 6. FE analysis results of the water-cooling heat exchanger: (**a**) configuration of the heat exchanger; (**b**) coolant channel engraved on the stator of the fine stage (**c**) temperature at the coils; (**d**) temperature at the aluminum block (the stator of the fine stage); (**e**) temperature at the ceramic block (the stator of the coarse stage) (**f**) temperature at the coolant.

3.2. Experiment Setup

3.2.1. Sensors

Two sets of six sensors were employed to realize the six-DOF motions of the dual servo maglev stage. The motion of the fine stage was measured globally using three laser interferometers and three capacitive sensors installed outside the motion system. The laser interferometers (RLE10, Renishaw, Gloucestershire, UK), which are denoted by SG1~SG3, were used to measure the in-plane motion of the fine stage by targeting the bar mirrors placed on the sides of the fine stage. The capacitive sensors (C5S, Lion precision, Oakdale, MN, USA), which are denoted by SG4~SG6, were used to measure the out-of-plane motion of the fine stage by targeting the top surface of the fine stage. The motion of the coarse stage relative to that of the fine stage was measured locally using the six capacitive sensors installed between the stages. The capacitive sensors (C9.5R and C8S, Lion Precision, Oakdale, MN, USA), denoted by SL1~SL6, measured the motion of the coarse stage relative to that of the fine stage. The arrangement of the sensors is shown in Figure 7, and the specifications of the sensors are shown in Table 6.

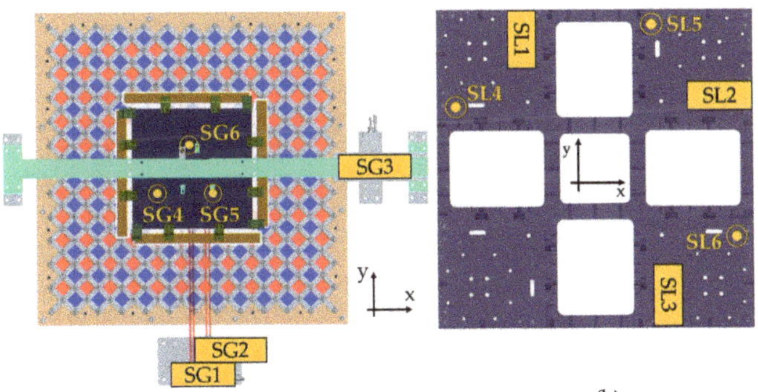

Figure 7. Arrangement of the motion feedback sensors: (**a**) global sensors for the fine stage, (**b**) local sensors for the coarse stage.

Table 6. Specifications of the sensors.

Sensors		Measurement Range	Theoretical Resolution (16-bit A/D)
Laser interferometers	RLE10	1 m	38.63 pm
Capacitive sensors	C5S	10 μm	150 pm
	C9.5R	1250 μm	19.07 nm
	C8S	2000 μm	30.52 nm

3.2.2. Electric Devices and Data Acquisition

All the coils of the VCMs and the planar motor are driven by linear current amplifiers (TA115, Trust automation, San Luis Obispo, CA, USA), and all the amplifiers are powered by a customized power supply (OPM505, ODA Technologies, Incheon, Korea). A controller (DS1005, dSPACE, Paderborn, Germany) for the high-level control algorithm was used for feedback control. The dual servo maglev system is shown in Figure 8.

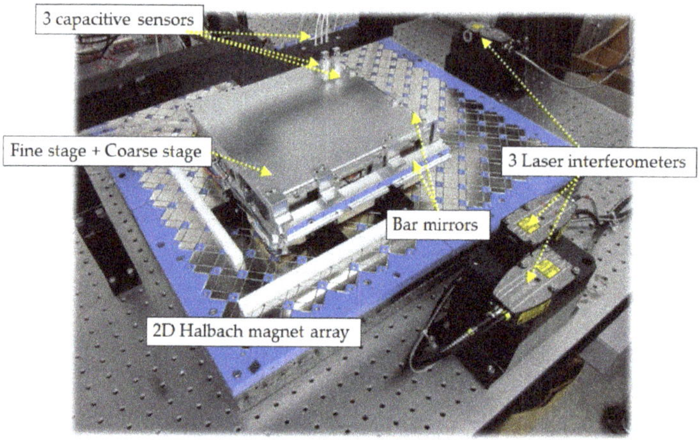

Figure 8. The realized dual servo maglev stage system.

3.2.3. Control Strategy

The control strategy for the dual servo stage is the master–slave control algorithm. The fine stage, whose position is measured by the global sensors, is the master, while the coarse stage, whose position relative to the fine stage is measured by the local sensors, is the slave. The control block diagram is shown in Figure 9. The master is controlled to follow the main command r_f, and the slave is controlled to follow the master. The command of the slave, r_c, is always the zero vector, since the feedback is the relative motion between the fine stage and the coarse stage.

By utilizing the pseudo inverse of the actuator kinematics of the fine stage, K^{f^+}, the coupled plant model can be converted to a simple decoupled plant model of six-DOF in accordance with the Cartesian coordinates. Thus, six simple PID controllers were used based on the six-SISO (Single Input Single Output) model. The position vectors $\vec{x_a^f}$ and $\vec{x_m^f}$ denoted in Figure 9 represent the actual and measured positions of the mass center of the fine stage. The measured position, $\vec{x_m^f}$, is obtained from the sensor signals, $\vec{S^f}$, by the sensor transformation matrix, H^f, which is determined by the position of the sensors. The coarse stage is controlled in the same manner as the fine stage. However, in this case, the actuator kinematics relationship is dependent on the position of the coarse stage. Thus, the pseudo inverse of the actuator kinematics $K^{c+}(\vec{x^c})$ is determined at every position in real time. By utilizing the pseudo inverse, the PID controller has six output signals of control force as well as six feedback signals of displacement in the Cartesian coordinates. The position vector $\overrightarrow{x_a^c}$ represents the actual position of the mass center of the coarse stage in the global sense, and $\overrightarrow{x_m^c}$ represents the measured position of the mass center of the coarse stage relative to that of the fine stage. The sensor transformation matrix H^c is obtained from the sensor signals and the measured positions, $\overrightarrow{S^c}$ and $\overrightarrow{x_m^c}$, respectively.

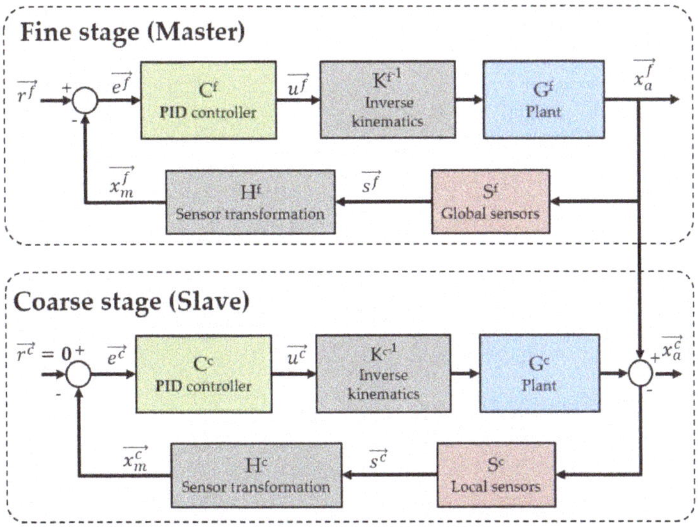

Figure 9. Control block diagram of the master–slave control scheme of the dual servo maglev stage.

4. Results

The dual servo maglev stage was controlled using the master–slave control scheme, and the basic positioning performances were evaluated accordingly.

The motion stroke of the maglev dual servo stage was evaluated. As a result of the position-controllable range, the x-axis is 100 mm and the y-axis is 140 mm, as shown in Figure 10. Since the size of the coarse stage mover is 250 mm and the size of the magnet array of the coarse stage stator is 600 mm, the motion range based on simple geometry was expected to be 350 mm at the design stage. However, the length of the mirror located above the fine stage mover targeted by the laser interferometers used as a feedback sensor is 260 mm. Considering the distance between the double path beams, the position-controllable range is about 200 mm. Additionally, as the mover approaches the outer edge of the permanent magnet array, the periodicity of the magnetic flux density decreases due to the edge effect of the permanent magnets, and the accuracy of the kinematics model obtained using the harmonic model for control decreases. For this reason, the x-axis and y-axis motion strokes were lower than the design value, and two laser interferometers were installed along the x-axis to measure the yaw motion, which further reduced the motion stroke in the x-axis.

The in-position stability of the dual servo stage in each axis was evaluated, and the corresponding root-mean-square (RMS) values are shown in Figure 11 and Table 7. The coarse stage was levitated 1 mm from the surface of the magnet array, and the fine stage was levitated 0.5 mm from the coarse stage. The stages were controlled to be in position, and the feedback sensor data were measured for 20 s. As shown in Figure 11, the positioning stability of the fine stage is greatly superior to that of the coarse stage. The in-plane position of the fine stage was kept to within 10 nm in the translational directions and to around 0.01 arcsec in the rotational direction, while the in-plane position stability of the coarse stage was larger than 100 nm in the translational directions and larger than 0.2 arcsec in the rotational directions. The positioning stability in the y direction is superior to that of the x direction because the sensor signals from the two laser interferometers are averaged. During the experiment, the temperature was measured with a non-contact infrared thermometer while the stage levitated in place was being cooled by 10 °C water. The temperature did not exceed 30 °C. In the FE analysis, the maximum power consumption was assumed in all coils, but in practice, the temperature was observed to be much lower by using less power.

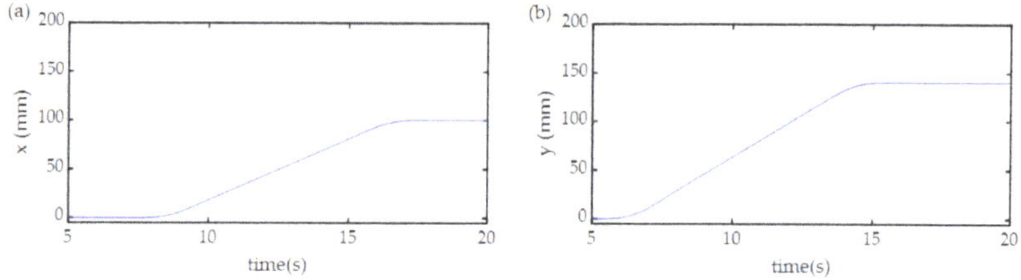

Figure 10. The evaluated motion stroke data of the dual servo stage: (**a**) x direction, (**b**) y direction.

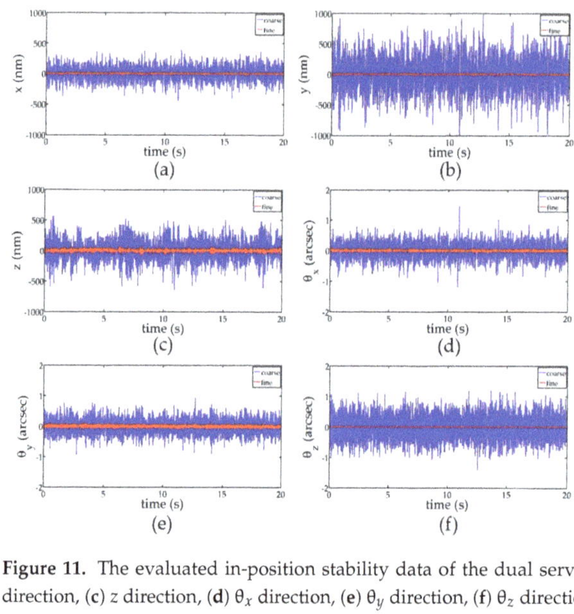

Figure 11. The evaluated in-position stability data of the dual servo stage: (**a**) x direction, (**b**) y direction, (**c**) z direction, (**d**) θ_x direction, (**e**) θ_y direction, (**f**) θ_z direction.

Table 7. The evaluated in-position stability of the dual servo stage with RMS values.

(RMS Value)	x (nm)	y (nm)	z (nm)	θ_x (arcsec)	θ_y (arcsec)	θ_z (arcsec)
Fine stage	9.29	7.85	23.2	0.0323	0.0400	0.0106
Coarse stage	115.94	275.87	170.73	0.3640	0.2662	0.2435
Fine-to-coarse ratio (%)	8.01	2.85	13.59	8.87	15.03	4.35

In order to evaluate the minimum positioning resolution, stepped trajectories were commanded. Figure 12 shows the evaluated result. In the translational direction, 40 nm, 30 nm, and 100 nm steps were clearly resolved in the x, y, and z directions, respectively. In the rotational direction, 0.15 arcsec, 0.15 arcsec, and 0.04 arcsec steps were clearly resolved in the θ_x, θ_y, and θ_z directions, respectively.

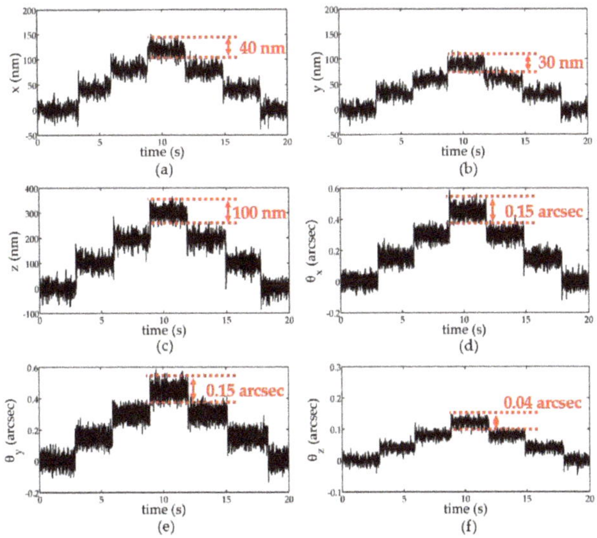

Figure 12. The evaluated minimum positioning resolution data of the dual servo stage: (**a**) x direction, (**b**) y direction, (**c**) z direction, (**d**) θ_x direction, (**e**) θ_y direction, (**f**) θ_z direction.

Figure 13 shows the settling performance of the dual servo maglev stage. The magnitude of the step command was 1 mm, and the time taken for the stage to settle to within 1% of the command was measured. The fine stage settled within 200 ms for both the x and y directions, while the coarse stage showed a settling time of 470 ms.

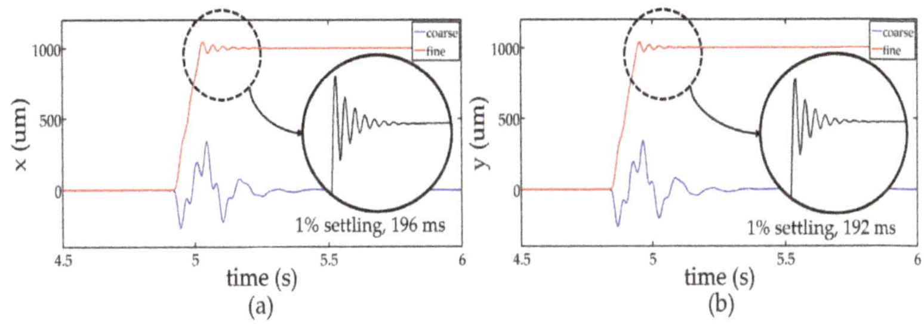

Figure 13. The evaluated settling time data of the dual servo stage: (**a**) x direction, (**b**) y direction.

The current provided to the coarse and fine stages was measured while moving the stage by 50 mm in the x and y directions. The distance of 50 mm corresponds to the length of one spatial period of both the magnet array and the magnetic field. As shown in Figure 14, while the dual servo maglev stage moves in the x direction, the current provided to the six coils of the planar motor placed along the y-axis varies considerably, generating a levitation force according to the spatially varying magnetic field. The variation in the current provided to the other six coils placed along the x-axis is small because the variation in the magnetic field affecting the coils is small when the stage is moving in the x direction. The same phenomenon is observed when the stage moves in the y direction. The current provided to the VVCMs is steady because the vertical level of the fine stage is maintained while the stage moves in the x and y directions; however, the current provided to the HVCMs varies, generating force in the x or y directions and moving the fine stage.

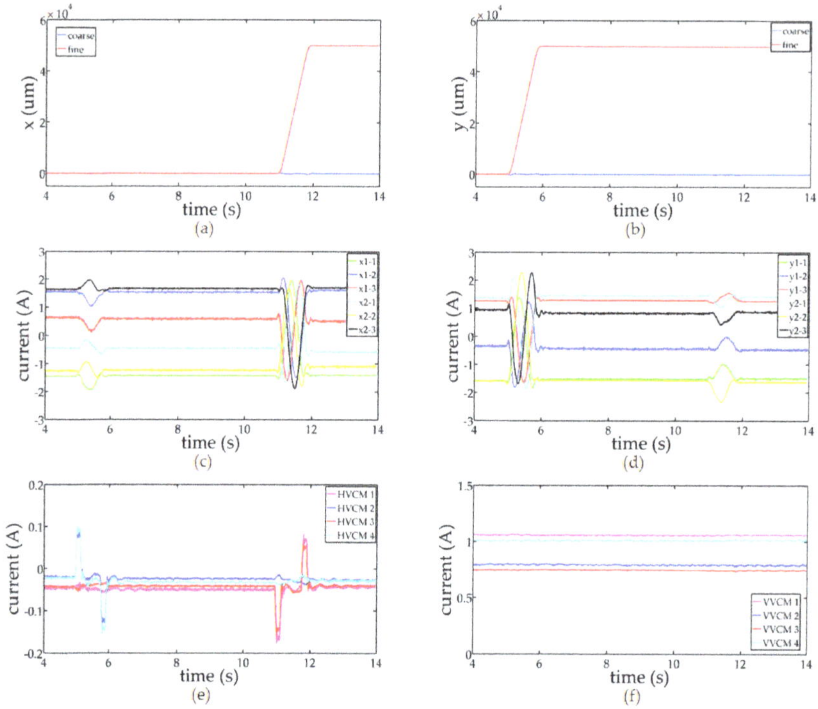

Figure 14. The current provided to the coils when the dual servo maglev stage moves 50 mm along the x-axis and y-axis: (**a**) measured position along x direction, (**b**) measured position along y direction, (**c**) electric current applied to the coils placed along the y-axis, (**d**) electric current applied to the coils placed along the x-axis, (**e**) electric current applied to the HVCMs, (**f**) electric current applied to the VVCMs.

5. Conclusions

This paper proposes a novel dual servo magnetic levitation stage designed to achieve a high degree of precision in the next-generation semiconductor lithography process. It is also vacuum compatible and allows the range of motion to be expanded easily due to the planar motor scheme. The structure of the novel dual servo magnetic levitation stage is composed of three layers: the bottom is the coarse stage stator of the planar motor magnets; the middle is the coarse stage mover of the planar motor coils assembled with the fine stage stator of the VCM coils; and the top is the fine stage mover of the VCM magnets. The layers are separated by the maglev mechanism, which removes mechanical connection and friction between the layers. Therefore, the stage offers a fast dynamics response and a high degree of precision.

As the middle layer has all the coils of the coarse and fine stages, the coarse stage becomes a moving coil type and the fine stage becomes a moving magnet type. The coarse stage belongs to the moving coil type in the form of a planar motor, which makes it easy to control due to the simple actuator kinematics, and it is free of the end-effect of the magnet array. With this type, the coarse stage requires fewer current drivers and has an easier control scheme without complex coil switching. The fine stage of the moving magnet type is free from direct heat transfer and is easy to control without force coupling and wire disturbance. Moreover, the placement of the coils allows only the middle layer to cool, and the coolant-carrying tubes do not disturb the top layer.

Both stages were analytically modeled with respect to the design parameters. Electromagnetics, kinetics, and electrics were used to create a mathematical model. The effect of the design parameters on the performance of the stage has been inspected. Based on the model and the study, the optimization process determines the design parameters and the final performances of the stage. The fine stage and the coarse stage are integrated into the dual servo maglev stage, and they are equipped with sensors for position feedback. The master–slave concept was used for the control strategy regarding the coordinate of the fine stage as the master and the relative coordinate between the fine stage and the coarse stage as the slave. Linear current amplifiers are utilized to input current to the coils of the planar motor and the voice coil motors. The real-time controller, with several data acquisition boards and a processor, is used to realize the 12-SISO PID control algorithms.

The results of the performance evaluation show that the dual servo stage has an in-position stability of 10 nm along the motion directions of the x-axis and the y-axis. Additionally, the experiment on a 1% settling time for a 1 mm step command shows that the fine stage settles within 200 ms for both the x and y directions, while the coarse stage exhibits a settling time of 470 ms. Therefore, it is concluded that the novel dual servo magnetic levitation stage designed and manufactured for this paper delivers a high degree of precision and fast dynamics.

Author Contributions: Conceptualization and methodology, D.A.; software, J.J.; simulation and optimization, J.-W.J.; formal analysis and experiments, J.-W.J., H.Y. and J.J.; data curation and writing—original draft preparation, D.A.; writing—review and editing, H.Y. and J.J. All authors have read and agreed to the published version of the manuscript.

Funding: This work was supported by a Korea Institute for Advancement of Technology (KIAT) grant funded by the Korean Government (MOTIE) (P0012744, The Competency Development Program for Industry Specialist).

Institutional Review Board Statement: Not applicable.

Informed Consent Statement: Not applicable.

Data Availability Statement: Not applicable.

Conflicts of Interest: The authors declare no conflict of interest.

References

1. Otsuka, J. Nanometer level positioning using three kinds of lead screws. *Nanotechnology* **1992**, *3*, 29–36. [CrossRef]
2. Futami, S.; Furutani, A.; Yoshida, S. Nanometer positioning and its micro-dynamics. *Nanotechnology* **1990**, *1*, 31–37. [CrossRef]
3. Ro, S.-K.; Park, J.-K. Development of a Miniature Air-bearing Stage with a Moving-magnet Linear Motor. *Int. J. Precis. Eng. Manuf.* **2008**, *9*, 19–24.
4. Ro, S.-K.; Kim, S.; Kwak, Y.; Park, C.H. A linear air bearing stage with active magnetic preloads for ultraprecise straight motion. *Precis. Eng.* **2010**, *34*, 186–194. [CrossRef]
5. Wang, H.; Zhang, X. Input coupling analysis and optimal design of a 3-DOF compliant micro-positioning stage. *Mech. Mach. Theory* **2008**, *32*, 400–410. [CrossRef]
6. Kim, H.Y.; Ahn, D.H.; Gweon, D.G. Development of a novel 3-degrees of freedom flexure based positioning system. *Rev. Sci. Instrum.* **2012**, *83*, 055114. [CrossRef]
7. Kim, K.H.; Gweon, D.G.; Jung, H.S.; Lee, S.H.; Hong, M.S.; Lee, M.-G. A Stage Based on Voice Coil Motor with High Speed and Long Range for Laser Micro/Nano Fabrication. *Key Eng. Mat.* **2007**, *345*, 757–760. [CrossRef]
8. Kim, K.H.; Choi, Y.M.; Nam, B.-U.; Lee, M.-G. Dual servo stage without mechanical coupling for process of manufacture and inspection of flat panel displays via modular design approach. *Int. J. Precis. Eng. Manuf.* **2012**, *13*, 407–412. [CrossRef]
9. Compter, J.C. Electro-dynamic planar motor. *Precis. Eng.* **2004**, *28*, 171–180. [CrossRef]
10. Peijnenburg, A.T.A.; Vermeulen, J.P.M.; van Eijk, J. Magnetic levitation systems compared to conventional bearing systems. *Microelectron. Eng.* **2006**, *83*, 1372–1375. [CrossRef]
11. Jansen, J.W.; van Lierop, C.M.M.; Lomonova, E.A.; Vandenput, A.J.A. Ironless magnetically levitated planar actuator. *J. Appl. Phys.* **2008**, *103*, 07E905. [CrossRef]
12. Boeij, J.D.; Lomonova, E. Experimental verification of look-up table based real-time commutation of 6-DOF planar actuators. *J. Syst. Des. Dyn.* **2009**, *3*, 563–571. [CrossRef]
13. Ueda, Y.; Ohsaki, H. Six-Degree-of-Freedom Motion Analysis of a Planar Actuator with a Magnetically Levitated Mover by Six-Phase Current Controls. *IEEE Trans. Magn.* **2008**, *44*, 4301–4304. [CrossRef]
14. Kim, J.J.; Lee, M.G.; Jeong, J.; Gweon, D.G. Design of a magnetically levitated six degrees-of-freedom planar motor using T-shape halbach magnet array. In Proceedings of the Euspen International Conference, Delft, The Netherland, 31 May 2010.

15. Jansen, J.W. Magnetically Levitated Planar Actuator with Moving Magnets: Electromechanical Analysis and Design. Ph.D. Thesis, Technical University Eindhoven, Eindhoven, The Netherlands, 2007.
16. Zhu, H.; Pang, C.K.; Law, T.L.; Teo, T.J. Design and Control of a Six Degrees-of-Freedom Magnetically Levitated Positioning System. *IFAC Pap.* **2016**, *49*, 127. [CrossRef]
17. Zhu, H.; Teo, T.J.; Pang, C. K, Conceptual design and modeling of a six degrees-of-freedom unlimited stroke magnetically levitated positioner. In Proceedings of the IEEE/ASME International Conference on Advanced Intelligent Mechatronics, Besacon, France, 8–11 July 2014. [CrossRef]
18. Zhang, L.; Kou, B.; Zhang, H.; Guo, S. Characteristic Analysis of a Long-Stroke Synchronous Permanent Magnet Planar Motor. *IEEE Trans. Magn.* **2012**, *48*, 4658–4661. [CrossRef]
19. Lu, X.; Usman, I.U.R. 6D direct-drive technology for planar motion stages. *CIRP Ann.* **2012**, *61*, 359–362. [CrossRef]
20. Rovers, J.M.M.; Janse, J.W.; Lomonova, E.A. Design and measurements of the Double Layer Planar Motor. In Proceedings of the International Electric Machines & Drives Conference, Chicago, IL, USA, 12–15 May 2013. [CrossRef]
21. Guo, L.; Zhang, H.; Galea, M.; Li, J.; Lu, W.; Gerada, C. Analysis and Design of a Magnetically Levitated Planar Motor with Novel Multilayer Windings. *IEEE Trans. Magn.* **2015**, *51*, 8106909. [CrossRef]
22. Kou, B.; Xing, F.; Zhang, C.; Zhang, L.; Zhou, Y.; Wang, T. Improved ADRC for a Maglev Planar Motor with a Concentric Winding Structure. *Appl. Sci.* **2016**, *6*, 419. [CrossRef]
23. Zhang, L.; Kou, B.; Xing, F.; Zhang, H. Analysis and comparison of two two-dimensional Halbach permanent magnet arrays for magnetically levitated planar motor. *J. Appl. Phys.* **2014**, *115*, 17E704. [CrossRef]
24. Zhang, L.; Kou, B.; Li, L.; Zhao, B. Modeling and design of an integrated winding synchronous permanent magnet planar motor. In Proceedings of the 16th International Symposium on Electromagnetic Launch Technology, Beijing, China, 15–19 May 2012. [CrossRef]
25. Usman, I.U.R.; Lu, X. Force Ripple Attenuation of 6-DOF Direct Drive Permanent Magnet Planar Levitating Synchronous Motors. *IEEE Trans. Magn.* **2015**, *51*, 8208708. [CrossRef]
26. Xing, F.; Kou, B.; Zhang, L.; Yin, X.; Zhou, Y. Design of a Control System for a Maglev Planar Motor Based on Two-Dimension Linear Interpolation. *Energies* **2017**, *10*, 1132. [CrossRef]
27. Ahn, D.; Kim, H.; Choi, K.; Choi, Y.-M.; Lim, J.-Y. Design process of square column-shaped voice coil motor design for magnetic levitation stage. *Int. J. Appl. Electrom.* **2020**, *62*, 517–540. [CrossRef]

actuators

Article

Position Tracking for Multi-Channel Double-Crystal Monochromator Scanning Based on Iterative Learning Control

Siyu He, Haolin Lu, Zhao Feng and Xiaohui Xiao *

School of Power and Mechanical Engineering, Wuhan University, Wuhan 430072, China; syhewhu@whu.edu.cn (S.H.); luhaolin@whu.edu.cn (H.L.); fengzhaozhao7@whu.edu.cn (Z.F.)
* Correspondence: xhxiao@whu.edu.cn

Abstract: As a core component of the X-ray absorption fine structure spectroscopy (XAFS) system, the multi-channel double-crystal monochromator (DCM) can improve the time resolution of the system significantly. In contrast to the conventional single-channel DCM, the multi-channel DCM includes more pairs of crystals that are located separately in the master and slave motor axis with the same driving direction. However, a mismatched parallelism in the pitch direction, which can result from the manual mounting operation between the two separated crystals, directly affects the performance of the flux and the angular stability of the monochromatic beam. This poses a significant challenge to the precision position tracking of this system. In this paper, the mounting errors were translated into repetitive errors in the slave motor when the master motor was rotated at a constant velocity. Therefore, the iterative learning control (ILC) was considered in order to improve the tracking accuracy of the slave motor motion. The zero-magnitude error controller (ZMETC) was used to calculate the learning function to accelerate the convergence of the control inputs, and the convergence conditions of the control signal and error were also given. To validate the effectiveness of the proposed method, comparative experiments were performed on the motor motion platform. Experimental results indicated that the ILC effectively decreased the parallelism errors of the multi-channel DCM under various trajectories by comparing them with feedback controllers and the ZMETC, respectively.

Keywords: multi-channel crystal monochromator; feedforward control; iterative learning control; fourth-order trajectory tracking

Citation: He, S.; Lu, H.; Feng, Z; Xiao, X. Position Tracking for Multi-Channel Double-Crystal Monochromator Scanning Based on Iterative Learning Control. *Actuators* **2022**, *11*, 177. https://doi.org/10.3390/act11070177

Academic Editor: Nariman Sepehri

Received: 5 May 2022
Accepted: 20 June 2022
Published: 23 June 2022

1. Introduction

With the development of synchrotron light source technology, the study of the real-time structure of materials under in situ conditions has been widely applied in the fields of life and material science [1–3]. The experimental method of time-resolved X-ray absorption fine structure (XAFS) spectroscopy is an important technical method for realizing such research [4,5]. As a core component of the XAFS system, the double-crystal monochromator (DCM) plays an essential role in the stable output of the outgoing beam's position, energy, and flux. In order to reflect the dynamic process of the sample's structure in real time with more detailed test results [6], the improvement of time resolution for the monochromator spectrum is particularly important. It should be pointed out that time resolution is determined by the rotation frequency of the DCM; thus, it is necessary to improve its rotation precision and speed.

According to the Bragg diffraction formula, with a varied angle between the incident beam and the diffractive crystal plane, single-wavelength X-rays are separated from the continuous spectrum of synchrotron radiation, and the outgoing beam is diffracted into a fixed position on the sample. The parallelism in the pitch direction between the two separated crystals directly affects the performance of the flux and the angular stability of the monochromatic beam, as demonstrated in Figure 1. For a conventional DCM, the parallelism of the pair of crystals is manually adjusted by the stepping motor before actual experiments [7]. However, due to the backlash in the structure of a DCM, the angular stability is destroyed significantly, and, consequently, the rotation velocity is limited during such a motion process. In addition, due to the incident beam being concentrated on the single crystal in both traditional and channel-cut DCMs, the crystals need to be cooled by costly liquid nitrogen [8,9], which increases the complexity of the system. In order to address the mentioned problems, a novel monochromator with multi-channel crystals was designed with two independent polyhedral spindles that were separately driven by master and slave motors. As the crystals were mounted on the surface of the polyhedral spindles, the parallelism of the crystals was controlled by the relative precision rotation of the motors. However, the manual mounting errors of the crystals could degrade the positioning accuracy significantly, resulting in the failure of the output of the monochromatic beam. In particular, it was worth considering that the mounting errors could be regarded as stable errors that repetitively appeared during the rotation process when the master motor was rotated at a constant velocity. Therefore, the mounting errors could be reduced by improving the reference tracking accuracy, just for the slave motor on the fly.

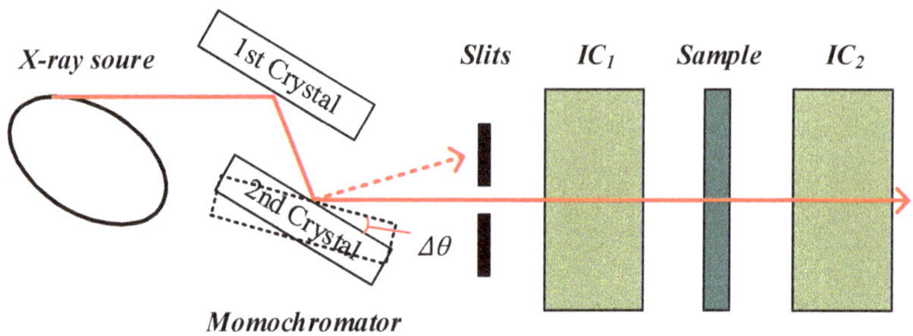

Figure 1. The structure of the XAFS.

There are some motion control methods to improve the position accuracy for single motor motion [10–12]. Feedback controllers are usually used to implement motion control in order to improve the tracking accuracy, including proportion–integration–differentiation (PID) control, robust control, and sliding mode control (SMC), etc. [13–15]. For example, the servo control system, combined with position–speed–current loops, is usually used to track the defined trajectory via a PID controller [16]. The SMC controller is also used to decrease the tracking errors caused by external disturbances or parameter uncertainty in a dynamic motion system [17,18]. Although these methods can improve precision to some extent, their performance is limited by the phase delay of the feedback controller and the achieved bandwidth [19–21]. Some model inversion-based feedforward control methods have been proposed to improve the tracking precision, including the zero-phase error tracking controller (ZPETC) and the zero-magnitude error controller (ZMETC) [22–24], where the control inputs are pre-calculated with the known trajectory via the inverse of the identified system model. Nevertheless, the effectiveness of the feedforward controllers relies heavily on the accuracy of the identified model, and the variation of the plant directly reduces the tracking precision significantly when the models are not identified accurately enough [22,25,26]. It is worthy of being noted that tracking errors, as well as the above-mentioned translated mounting errors, appeared repetitively. Thus, it was natural to use iterative learning control (ILC), a popular learning-based feedforward controller, to improve the repetitive reference tracking performance [27]. ILC can minimize repetitive errors through updating the control input through learning the historical tracking errors and input signals [28]. The combination of feedback and ILC is a promising control scheme, where unknown disturbances are decreased by the feedback controller, and the feedforward controller can compensate for the repetitive errors and known disturbances. Therefore, the design of the ILC was a crucial component for significant performance enhancement due to the existing repetitive errors in the multi-channel

DCM. The ILC consists of a learning function to determine the learning method and a low-pass Q-filter to balance the robustness and precision [29–31]. Although the proportion (P-), differentiation (D-), and proportion-differentiation (PD-) type ILC can be implemented without an accurate model, the convergence speed is lower than the model-inversion ILC [32,33]. To this end, the model-inversion ILC was considered to tackle the repetitive errors along with the slave motor rotation.

Motivated by the aforementioned essential issues, the objective of this paper was to compensate for the repetitive errors, followed by the slave motor on the fly, for a multi-channel DCM, which has rarely been researched in the field of control for DCM equipment. According to the repetitive property, ILC was adopted to guarantee the parallelism between a pair of crystals. In order to accelerate the convergence speed, the method named ZMETC was used as the learning function, and a Q-filter was delicately designed to improve the robustness of the whole system. Furthermore, a single-axis motor motion platform was built to verify the tracking performance of the references for tracking triangular waves and fourth-order trajectories, respectively, and comparative results using PID control and ZPETC were demonstrated to evaluate the effectiveness of the used method in the improvement of parallelism accuracy in the application of a multi-channel DCM.

The rest of the paper is organized as follows. In Section 2, the principle of a multi-channel monochromator is described, and the tracking problem is analyzed. The design of an ILC and learning function are described in Section 3. The experiment using the single-axis motor motion platform and a comparison of the results are discussed in Section 4. The thesis summary and subsequent research plans are included in Section 5.

2. System Description

The comparisons of the principles between the conventional single-channel and multi-channel DCMs are illustrated in Figure 2, where it is obvious that the scanning frequency of the single-channel DCM was determined by the velocity of the reciprocal rotation. However, the backlash problem seriously limited the accuracy of the scanning angle with high velocity due to the driven principle, i.e., the direction needed to be switched frequently at the corners. Moreover, the same working area of the crystal plane was always exposed on the incident beam during the reciprocal rotation so that the temperature was varied, which might lead to an uneven crystal surface along with undesired outcomes of the DCM.

In contrast to the conventional single-channel DCM, the multi-channel DCM included more pairs of crystals located at a separate motor axis driven in the same direction to ensure the parallelism of the crystals on the fly. As illustrated in Figure 2b, the scanning frequency could be improved through more crystal pairs and a high rotation speed, and the working crystals varied during the rotation process, which was particularly important for the improvement of the thermal stability of the multi-channel DCM.

As multiple pairs of crystals were manually mounted on the surface of the polyhedral spindles, the parallelism errors were the main concern that hindered the desired performance. Thus, for the multi-channel DCM, the master motor needed to be controlled at a constant velocity, while the slave motor was conducted with a varied velocity to follow the master motor, as displayed by the different trajectories in Figure 2c. Therefore, a single-axis motor motion platform was designed to verify the ILC performance for the slave motor, which could be further applied to the multi-channel DCM scanning process.

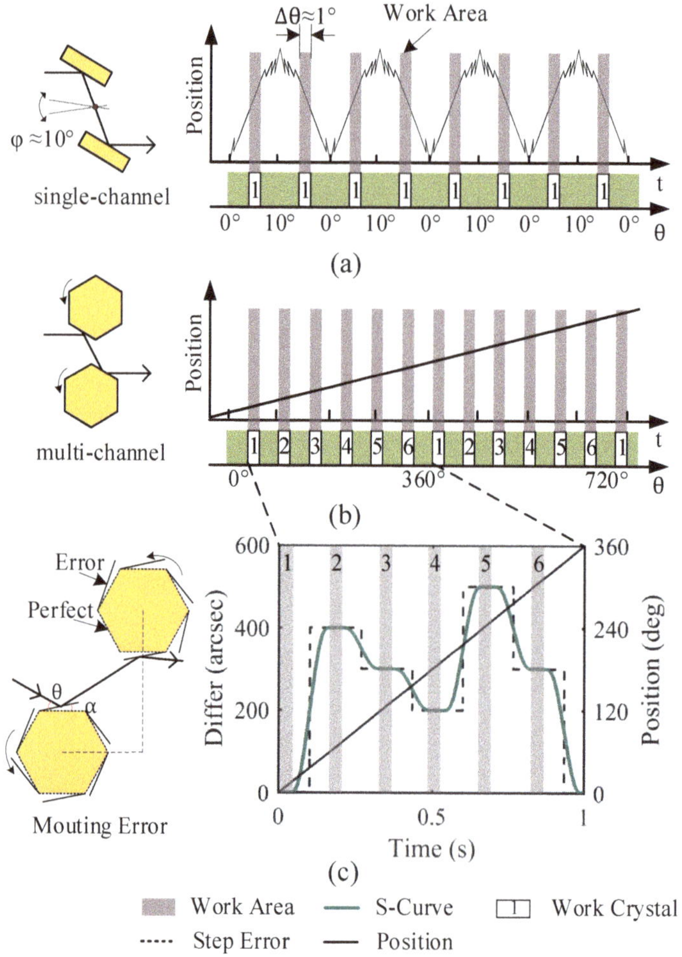

Figure 2. The comparison between the single-channel and multi-channel DCMs. The working principles of single-channel and multi-channel DCMs are separately represented by (**a**,**b**), the different trajectories between the master and slave motor are displayed in (**c**).

3. Controller Design

3.1. Iterative Learning Control

ILC is an efficient algorithm to improve the performance of a system by learning from previous information, i.e., errors and control inputs. In this paper, the ILC method was utilized to improve the performance of a multi-channel DCM system.

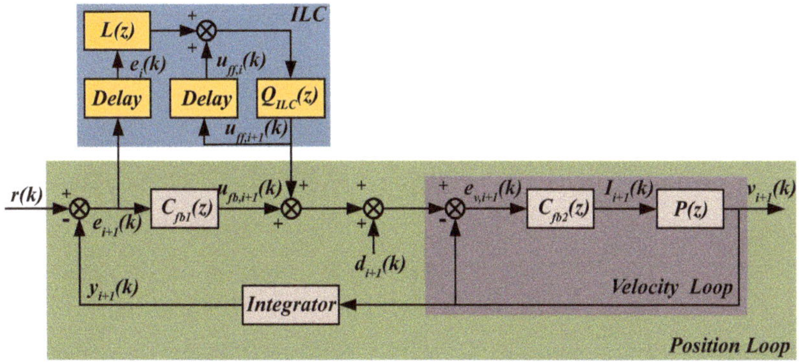

Figure 3. The controller framework of ILC.

The block diagram of the controller is demonstrated in Figure 3. $r(k)$ represents the repeat reference, $C_{fb1}(z)$ and $C_{fb2}(z)$ are respectively represented as the position feedback controller and velocity feedback controller, $P(z)$ refers to the combination of the motor driver and the DCM stage, $u_{ff,i+1}(k)$ is the feedforward control input based on the previous (i)-th iteration with i indicating the number of iteration, and the external disturbance is represented by $d_i(k)$. It should be noted that $d_i(k)$ includes repetitive and random interference. According to the configuration of the multi-channel DCM system, the repetitive disturbance comes from the mounting errors of the crystals and torque ripple caused by brushless motor pole stitching, and the environmental vibrations, such as mounting base or coolant flow, results in random interference.

Based on Figure 3, the output $y_{i+1}(k)$ is given by

$$y_{i+1}(k) = T(z)r(k) + J(z)(u_{ff,i+1}(k) + d_{i+1}(k)) \tag{1}$$

where $T(z)$, $J(z)$ are separately represented as the transfer functions of complementary sensitivity and process sensitivity, which are given as

$$T(z) = \frac{G_p(z)C_{fb1}(z)}{1 + G_p(z)C_{fb1}(z)} \tag{2}$$

$$J(z) = \frac{G_p(z)}{1 + G_p(z)C_{fb1}(z)} \tag{3}$$

where $G_p(z)$ is the closed-loop transfer function of the velocity loop with an integrator, which is calculated as

$$G_p(z) = \frac{1}{s}\frac{C_{fb2}P(z)}{1 + C_{fb2}P(z)} \tag{4}$$

The output of error is defined as follows:

$$e_{i+1}(k) = r(k) - y_{i+1}(k) \tag{5}$$

The loop of the ILC algorithm is made up of a learning function $L(z)$ and a low-pass filter $Q_{ILC}(z)$. According to the control law of ILC, the $(i+1)$-th control input is calculated by

$$u_{ff,i+1}(k) = Q_{ILC}(z)(u_{ff,i}(k) + L(z)e_i(k)) \tag{6}$$

where Q_{ILC} is the low-pass filter that is used to reduce the influence of high-frequency noise, $L(z)$ is the learning function.

With the combination of Equations (1), (5), and (6), the output of error at the $(i+1)$-th iteration is calculated by

$$u_{ff,i+1}(k) = Q_{ILC}(z)(1 - L(z)J(z))u_{ff,i}(k) + Q_{ILC}(z)[1 - (T(z)r(k) - J(z)d_i(k))] \quad (7)$$

To ensure the convergence of the control inputs in ILC when the number of iterations tends to infinity, the convergence condition is given by Equation (8):

$$||Q_{ILC}(1 - L(z)J(z))||_\infty < 1 \quad (8)$$

where $||\cdot||_\infty$ represents the infinite norm of the above transfer function.

The control input $u_{ff}(k)_\infty$ is calculated based on Equation (7):

$$u_{ff,\infty}(k) = \frac{Q_{ILC}L(z)[(1 - T(z))r(k) - J(z)d_i(k)]}{1 - Q_{ILC}(1 - L(z)J(z))} \quad (9)$$

After the $(i+1)$-th iteration, the output of error $e_{i+1}(k)$ is given by

$$\begin{aligned} e_{i+1}(k) &= r(k) - y_{i+1}(k) \\ &= (1 - T(z))r(k) - J(z)(u_{ff,\infty}(k) + d_{i+1}(k)) \\ &= \frac{(1 - T(z))(1 - Q_{ILC})}{1 - Q_{ILC}(1 - L(z)J(z))}r(k) + \\ J(z)&[\frac{Q_{ILC}L(z)J(z)}{1 - Q_{ILC}(1 - L(z)J(z))}d_i(k) - d_{i+1}(k)] \end{aligned} \quad (10)$$

According to the Equations (8) and (10), the error converges to zero with $L(z) = J^{-1}(z)$ and $d_i(k) = d_{i+1}(k)$, which indicates that ILC could compensate for the repetitive errors. On the other hand, the set value of Q_{ILC} significantly influences the convergence of the control input. If the cut-off frequency was set too high, the high-frequency noise would lead to divergence during the iteration process; otherwise, if the value was set too low, it would spend more time trying to achieve the convergence goal, and the precision of the tracking trajectory was also limited. Hence, it was essential to delicately design the cut-off frequency during the implementation of the iterative controller.

3.2. Learning Function

According to inequality (8), restrictions were easy to meet when the learning function was designed by the inverse of the process sensitivity transfer function $J(z)$. However, because the platform was a non-minimum-phase system, the model inversion was difficult to solve due to the presence of the unstable zeros. In this paper, the zero-magnitude error tracking controller (ZMETC) method was used to obtain the approximate inverse model of $J(z)$. The equivalent representation of $J(z)$ is given by

$$J(z) = \frac{B_s(z)B_u(z)}{A_s(z)} \quad (11)$$

where $B_s(z)$ and $B_u(z)$ respectively include the stable and unstable zeros, and $A_s(z)$ includes all the stable poles. $B_u(z)$ can also be translated to the expression of n-th order polynomial with n non-minimal phase zeros which is given by

$$B_u(z) = b_{un}z^n + b_{u(n-1)}z^{n-1} + \ldots + b_{u0} \quad (12)$$

According to the law of ZMETC, the approximate inverse model of $J(z)$ is given by

$$J^{-1}(z) = \frac{A_s(z)B_u^f(z)z^{-(n+d)}}{B_s(z)(B_u(z)|_{z=1})^2} \quad (13)$$

where d changes the symbol d such that n represents the relative orders of the model. The expression of $B_u^f(z)$ is solved by reversing the order of the coefficients in Equation (12) as

$$B_u^f(z) = b_{u0}z^n + b_{u1}z^{n-1} + \ldots + b_{un} \tag{14}$$

4. Experiment Results and Analysis

4.1. Experimental Setup

A single-axis motor motion system was used for the experiments, and the experimental setup is shown in Figure 4, which included a Maxon EC-Flat brushless motor, Elmo's motor driver, a 29-bit Heidenhain absolute value encoder, and an aluminum metal shaft. The motor driver was driven by the input current signal generated by the 16-bit digital to analog converters (DAC) on the dSPACE-MicroLabBox platform. The real-time positions were acquired by the absolute value encoder and transferred via the Endat 2.2 protocol port. The control system was established in the Simulink software environment and executed in a real-time application on the dSPACE-MicroLabBox platform. Meanwhile, the sampling frequency of the controller system was set to 5 kHz.

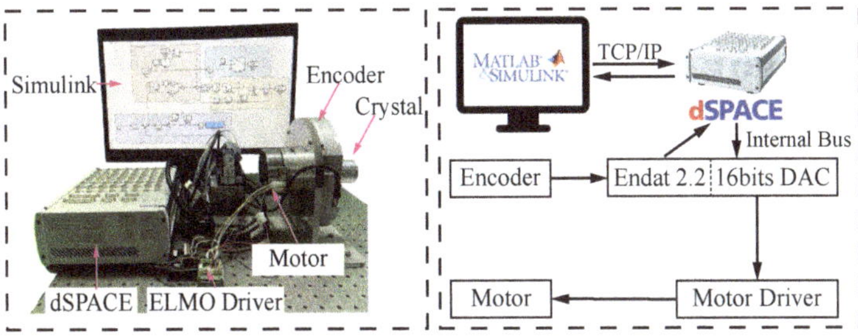

Figure 4. The experimental setup of the single axis motion platform.

In order to identify the linear dynamic model of the experiment platform, a set of the sin-sweep current signal was applied to the motor driver. To avoid the backlash of the motor, the direction of the motor motion was set to be same, and the current driven signal was expressed as

$$u(t) = 1.5 + 0.5 \times sin(2\pi f t) \tag{15}$$

where $u(t)$ represents the input current signal, and the frequency of the current signal was expressed as f, which varied between 0.01 and 500 Hz for the system identification.

A continuous-time transfer function model was identified by the system identification toolbox in the MATLAB software and was subsequently discretized via the zero-order holder (ZOH) method to facilitate the controller design, given as

$$P(z) = \frac{0.4491z^2 - 1.207z + 0.972}{z^3 - 2.809z^2 + 2.785z - 0.9757} \tag{16}$$

Figure 5 shows the match between the measured and identified open-loop frequency responses. It was clear that the first-order resonance point existed at 300 Hz for the experimental platform.

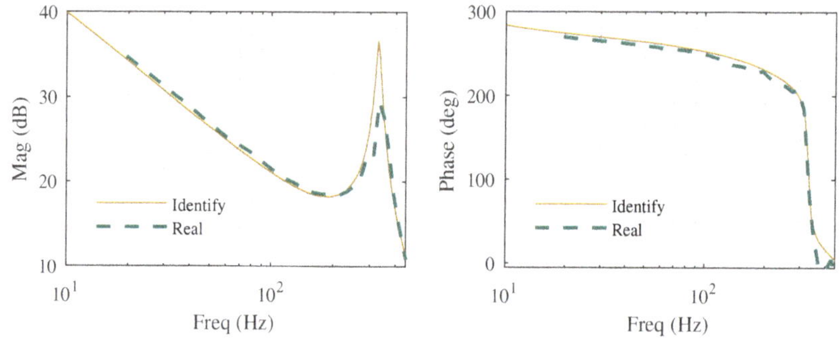

(a) (b)

Figure 5. Frequency responses of the single axis motion platform, amplitude and phase frequency responses are separately represented by (**a**,**b**).

4.2. Tracking Results

A set of triangular waves and fourth-order reference trajectories were applied to verify the performance of ILC in its application to multi-channel DCM scanning [34]. The example of the defined references for experiments is illustrated in Figure 6.

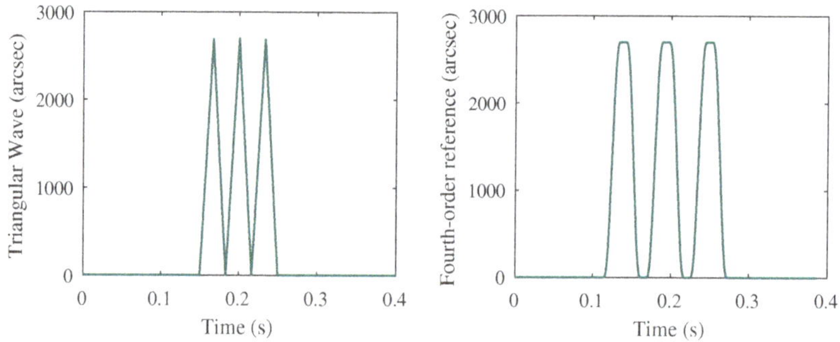

(a) 30 Hz triangular waves. (b) 36 Hz fourth-order reference trajectories.

Figure 6. The example of reference trajectory.

In order to evaluate the performance of the designed controller, the following controllers were also used for comparison:

1. C_1: feedback controller PID;
2. C_2: model-based feedforward controller ZMETC;
3. C_3: ILC which is illustrated in Figure 3.

For the PID controller, the parameters were set as follow: $K_p = 53.0469$, $K_i = 3633.5934$, $K_d = 1.9348$ for the position loop, and $K_p = 0.0034$, $K_i = 0.1703$ for the velocity loop.

As shown in Figure 7, the inversion of model $G_p^{-1}(z)$ used in the ZMETC controller is expressed as

$$G_p^{-1}(z) = \frac{155.16 \times (z - 0.9991)^4 (z^2 - 1.987z + 0.9875)(z^2 - 1.823z + 0.9849)}{z(z - 0.9901)(z - 0.9973)^2 (z - 1)^2 (z^2 - 1.242z + 0.4624)} \quad (17)$$

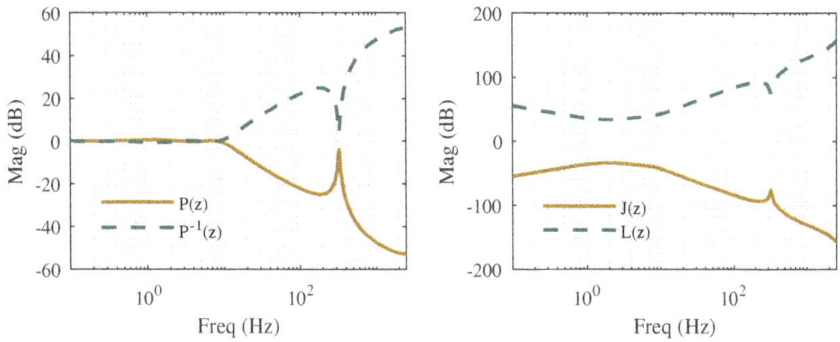

(a) The inverse model of the plant (b) The learning function

Figure 7. Amplitude frequency responses for the inverse model of the plant and the learning function.

Figure 8 shows the frequency spectrum of tracking errors that were measured before the ILC algorithm was loaded, and it was obvious that the errors were mainly located in the region of 30–50 Hz. It should be noted that high-frequency interference might result in the divergence of the control inputs, and thus the bandwidth of the low-pass filter could be set at three to five times the central frequency regions of the errors [32]. Therefore, the cut-off frequency of the Q-filter was set to 100 Hz and realized by the butter function in MATLAB software in this paper.

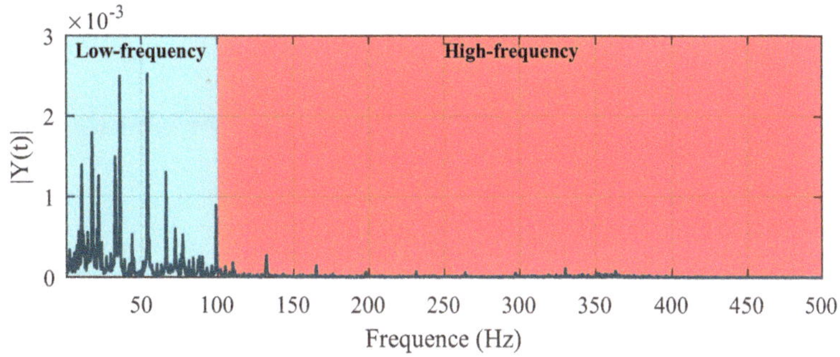

Figure 8. The FFT spectrum of tracking errors.

In order to further improve the convergence speed, the learning function $L(z)$ was designed by utilizing the inversion of $J(z)$, and here ZMETC was adopted to obtain the approximated inversion as

$$L(z) = J^{-1}(z) = \frac{2.57 \times 10^6 (z - 0.9991)^4 (z^2 - 1.987z + 0.9875)(z^2 - 1.823z + 0.9849)}{z(z + 0.7705)(z - 0.9901)(z - 1)^3 (z^2 - 1.253z + 0.4663)} \quad (18)$$

4.2.1. Tracking Triangular Wave

The results of fifty iterations are illustrated in Figure 9 for the reference tracking with 30 Hz triangular waves, and the statistical results for different tracking frequencies are listed in Table 1. The results showed that the tracking errors decreased with the increment of iterations, indicating the ILC's ability to compensate for the repetitive errors during the iteration process, as demonstrated in Figure 9a,b. It was clear that the errors tended to converge after 20 iterations, and the RMS error was much smaller than other controllers. Figure 9d shows the comparison between the reference and actual motion tracking at the 20th iteration. It was evident that the ILC achieved the best tracking performance among the three controllers, with an RMS error of 80.02″. The performance was worst for the feedback PID controller with RMS errors of 545.27″ for the phase delay and no ability to compensate for repetitive errors along with the reference tracking. Although the performance of feedforward ZMETC was improved significantly in comparison with PID control, the tracking precision depended on accurate model identification, and no learning mechanism was integrated to handle the repetitive errors.

Table 1. Error results of triangular reference tracking.

Frequency (Hz)	RMS Error (arcsec)			MAX Error (arcsec)		
	PID	ZMETC	ILC	PID	ZMETC	ILC
10	593.28	57.60	52.56	779.4	188.64	116.28
20	545.27	91.33	80.02	1366.6	293.57	185.17
30	474.42	151.93	103.27	1736.4	560.97	306.14

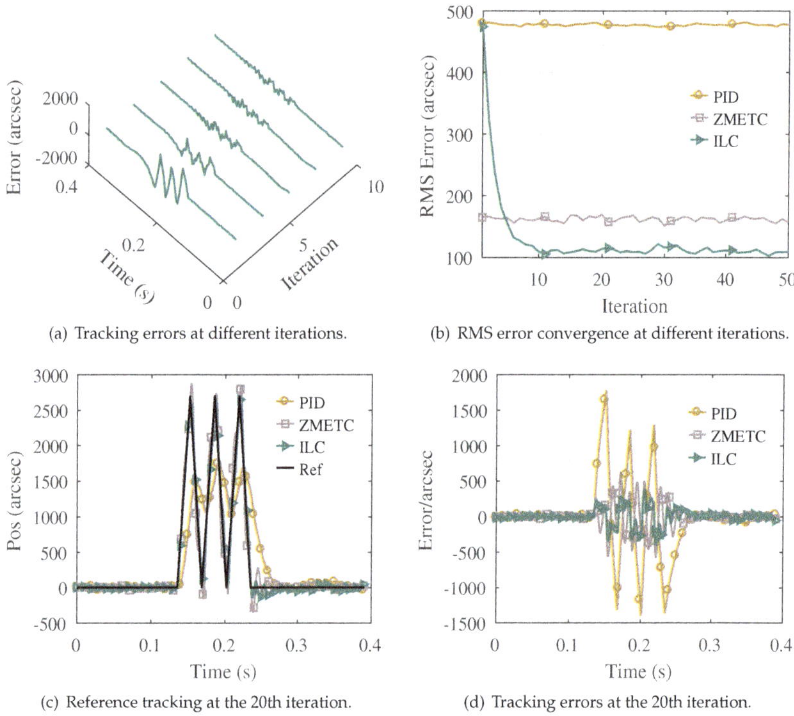

(a) Tracking errors at different iterations.

(b) RMS error convergence at different iterations.

(c) Reference tracking at the 20th iteration.

(d) Tracking errors at the 20th iteration.

Figure 9. Triangular wave reference tracking performance compared with existing methods.

4.2.2. Tracking Fourth-Order Motion Reference Trajectory

In this paper, the fourth-order reference was used for trajectory planning for the difference in trajectories between the master and slave motors, which could significantly decrease the impact of the variable acceleration speed. The results of the 50th iteration are illustrated in Figure 10 for the reference tracking with a 36 Hz fourth-order motion reference, and the statistical results for variable frequency tracking are listed in Table 2. The tracking errors decreased with an increase in the iterations, as shown in Figure 10a, which proved the ability of the ILC to track the fourth-order reference. It was clear that the ILC achieved the best performance with RMS errors of 127.89″, compared with 758.09″ for the PID controller, and 220.98″ for the ZMETC controller. The PID controller achieved the worst performance due to phase delay, which was difficult to eliminate.

In addition, the operating interval during the crystal rotation was particularly displayed as the red shaded area in Figure 10c, and the errors were controlled within the limits of 36″ for the ILC, which indicated that the ILC could contribute to decreasing the parallelism errors in the area for the working crystals.

Table 2. Error results of fourth-order motion reference tracking.

Frequency (Hz)	RMS Error (arcsec)			MAX Error (arcsec)		
	PID	ZMETC	ILC	PID	ZMETC	ILC
18	819.25	68.26	76.83	1581.1	137.38	142.56
24	842.11	102.29	89.48	1761.4	237.37	203.77
36	758.09	220.98	127.89	1925.0	716.16	406.96

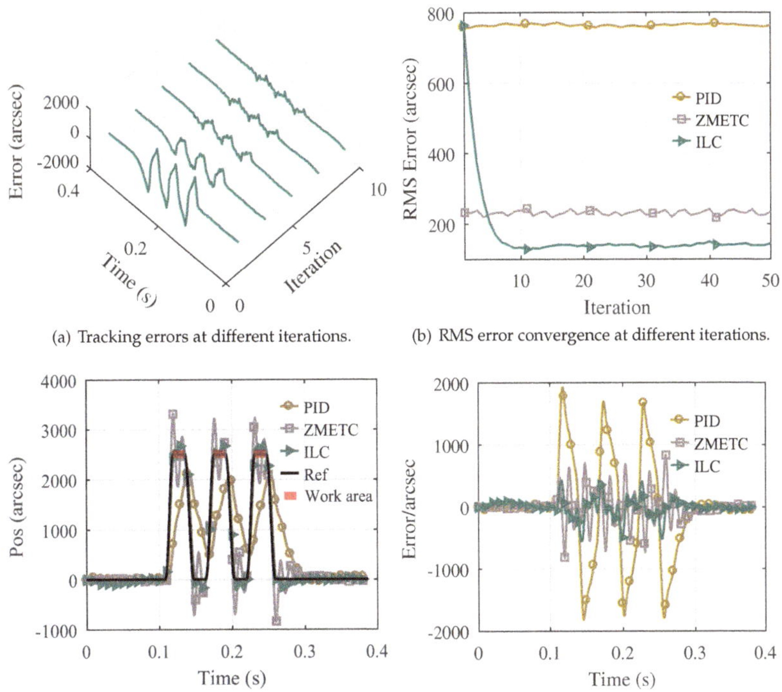

(a) Tracking errors at different iterations. (b) RMS error convergence at different iterations.

(c) Reference tracking at the 20th iteration. (d) Tracking errors at the 20th iteration.

Figure 10. Fourth-order motion reference tracking performance compared with existing methods.

5. Conclusions

In this paper, the tracking controller design for a multi-channel DCM was researched to improve the parallelism between the pairs of crystals. As the mounting errors could be translated into the repetitive errors in the slave motor when the master motor was rotated at a constant velocity, ILC was designed to improve the tracking precision by dealing with the repetitive trajectories and disturbances. The control law, as well as the convergence condition, was also given and discussed. To accelerate the convergence of control inputs, the learning function chosen as the inverse of the model was calculated by the method of ZMETC, and the bandwidth of the Q-filter was delicately designed through the frequency spectrum analysis of the measured errors. A set of controllers (PID, ZMETC, and ILC) were tested on the developed single-axis motion platform to evaluate the performance.

Experimental results demonstrated that the best performance was achieved by the ILC, both for triangular wave and fourth-order motion trajectory, which indicated its ability to improve the parallelism accuracy for multi-channel DCM on the fly. Future work will concentrate on improving parallelism accuracy for the coupled crystals in the multi-channel DCM system.

Author Contributions: Conceptualization, S.H.; methodology, S.H. and Z.F.; software, S.H.; validation, S.H. and H.L.; formal analysis, S.H. and Z.F.; investigation, S.H.; resources, X.X.; data curation, Z.F.; writing—original draft preparation, S.H.; writing—review and editing, S.H. and Z.F.; visualization, H.L.; supervision, X.X.; project administration, X.X. All authors have read and agreed to the published version of the manuscript.

Funding: This research received no external funding.

Institutional Review Board Statement: Not applicable.

Informed Consent Statement: Not applicable.

Data Availability Statement: Data sharing is applicable.

Conflicts of Interest: The authors declare no conflict of interest.

References

1. Kunz, C. Synchrotron radiation: Third generation sources. *J. Phys. Condens. Matter* **2001**, *13*, 7499. [CrossRef]
2. Yabashi, M.; Tanaka, H. The next ten years of X-ray science. *Nat. Photon.* **2017**, *11*, 12–14. [CrossRef]
3. Ezquerra, T.A.; García-Gutiérrez, M.C.; Nogales, A.; Müller, A.J. Introduction to the special issue on "Applications of synchrotron radiation in polymers science". *Eur. Polym. J.* **2016**, *81*, 413–414. [CrossRef]
4. Frahm, R.; Richwin, M.; Lützenkirchen-Hecht, D. Recent advances and new applications of time-resolved X-ray absorption spectroscopy. *Phys. Scr.* **2005**, *2005*, 974. [CrossRef]
5. Müller, O. Hard X-ray Synchrotron Beamline Instrumentation for Millisecond Quick Extended X-ray Absorption Spectroscopy. Ph.D. Thesis, Universität Wuppertal, Fakultät für Mathematik und Naturwissenschaften, Wuppertal, Germany, 2018.
6. Yamazaki, H.; Matsuzaki, Y.; Shimizu, Y.; Tsuboki, I.; Ikeya, Y.; Takeuchi, T.; Tanaka, M.; Miura, T.; Kishimoto, H.; Senba, Y.; et al. Challenges toward 50 nrad-stability of X-rays for a next generation light source by refinements of SPring-8 standard monochromator with cryo-cooled Si crystals. *AIP Conf. Proc.* **2019**, *2054*, 60018.
7. Richwin, M.; Zaeper, R.; Lützenkirchen-Hecht, D.; Frahm, R. Piezo-XAFS-time-resolved x-ray absorption spectroscopy. *Rev. Sci. Instruments* **2002**, *73*, 1668–1670. [CrossRef]
8. Sergueev, I.; Döhrmann, R.; Horbach, J.; Heuer, J. Angular vibrations of cryogenically cooled double-crystal monochromators. *J. Synchrotron Radiat.* **2016**, *23*, 1097–1103. [CrossRef] [PubMed]
9. Chumakov, A.I.; Sergeev, I.; Celse, J.P.; Rüffer, R.; Lesourd, M.; Zhang, L.; Sánchez del Río, M. Performance of a silicon monochromator under high heat load. *J. Synchrotron Radiat.* **2014**, *21*, 315–324. [CrossRef] [PubMed]
10. Boeren, F.; Bruijnen, D.; van Dijk, N.; Oomen, T. Joint input shaping and feedforward for point-to-point motion: Automated tuning for an industrial nanopositioning system. *Mechatronics* **2014**, *24*, 572–581. [CrossRef]
11. Ling, J.; Feng, Z.; Zheng, D.; Yang, J.; Yu, H.; Xiao, X. Robust adaptive motion tracking of piezoelectric actuated stages using online neural-network-based sliding mode control. *Mech. Syst. Signal Process.* **2021**, *150*, 107235. [CrossRef]
12. Ming, M.; Liang, W.; Feng, Z.; Ling, J.; Al Mamun, A.; Xiao, X. PID-type sliding mode-based adaptive motion control of a 2-DOF piezoelectric ultrasonic motor driven stage. *Mechatronics* **2021**, *76*, 102543. [CrossRef]
13. Bai, Y.; Hu, J.; Yao, J. Adaptive neural network output feedback robust control of electromechanical servo system with backlash compensation and disturbance rejection. *Mechatronics* **2022**, *84*, 102794. [CrossRef]
14. Loof, J.; Besselink, I.; Nijmeijer, H. Automated lane changing with a controlled steering-wheel feedback torque for low lateral acceleration purposes. *IEEE Trans. Intell. Veh.* **2019**, *4*, 578–587. [CrossRef]
15. Feng, Z.; Liang, W.; Ling, J.; Xiao, X.; Tan, K.K.; Lee, T.H. Integral terminal sliding-mode-based adaptive integral backstepping control for precision motion of a piezoelectric ultrasonic motor. *Mech. Syst. Signal Process.* **2020**, *144*, 106856. [CrossRef]
16. Makarem, S.; Delibas, B.; Koc, B. Data-driven tuning of PID controlled piezoelectric ultrasonic motor. *Actuators* **2021**, *10*, 148. [CrossRef]
17. Wang, W.; Ma, J.; Cheng, Z.; Li, X.; De Silva, C.; Lee, T.H. Global iterative sliding mode control of an industrial biaxial gantry system for contouring motion tasks. *IEEE/ASME Trans. Mechatronics* **2021**, *27*, 1617–1628. [CrossRef]
18. Li, L.; Huang, W.W.; Wang, X.; Zhu, L.M. Dual-Notch Based Repetitive Control for Tracking Lissajous Scan Trajectories with Piezo-Actuated Nano-Scanners. *IEEE Trans. Instrum. Meas.* **2022**, *71*, 1–12.
19. Mohammadi, A.; Fowler, A.G.; Yong, Y.K.; Moheimani, S.R. A feedback controlled MEMS nanopositioner for on-chip high-speed AFM. *J. Microelectromech. Syst.* **2013**, *23*, 610–619. [CrossRef]

20. Liu, Y.; Yan, J.; Wang, L.; Chen, W. A two-DOF ultrasonic motor using a longitudinal–bending hybrid sandwich transducer. *IEEE Trans. Ind. Electron.* **2018**, *66*, 3041–3050. [CrossRef]
21. Liu, Y.; Chen, W.; Liu, J.; Shi, S. A cylindrical traveling wave ultrasonic motor using longitudinal and bending composite transducer. *Sens. Actuators A Phys.* **2010**, *161*, 158–163. [CrossRef]
22. Butterworth, J.A.; Pao, L.Y.; Abramovitch, D.Y. Analysis and comparison of three discrete-time feedforward model-inverse control techniques for nonminimum-phase systems. *Mechatronics* **2012**, *22*, 577–587. [CrossRef]
23. Tomizuka, M. Zero phase error tracking algorithm for digital control. *J. Dyn. Sys. Meas. Control* **1987**, *109*, 65–68. [CrossRef]
24. Qin, Y.; Tian, Y.; Zhang, D.; Shirinzadeh, B.; Fatikow, S. A novel direct inverse modeling approach for hysteresis compensation of piezoelectric actuator in feedforward applications. *IEEE/ASME Trans. Mechatron.* **2012**, *18*, 981–989. [CrossRef]
25. Li, L.; Fleming, A.J.; Yong, Y.K.; Aphale, S.S.; Zhu, L. High performance raster scanning of atomic force microscopy using Model-free Repetitive Control. *Mech. Syst. Signal Process.* **2022**, *173*, 109027. [CrossRef]
26. Liu, Y.; Li, J.; Jin, Z. Trajectory Tracking Control for Reaction–Diffusion System with Time Delay Using P-Type Iterative Learning Method. *Actuators* **2021**, *10*, 186. [CrossRef]
27. Wu, M.; Yu, P.; Chen, X.; She, J. Design of repetitive-control system with input dead zone based on generalized extended-state observer. *J. Dyn. Syst. Meas. Control* **2017**, *139*, 071008. [CrossRef]
28. Bazaei, A.; Yong, Y.K.; Moheimani, S.R.; Sebastian, A. Tracking of triangular references using signal transformation for control of a novel AFM scanner stage. *IEEE Trans. Control Syst. Technol.* **2011**, *20*, 453–464. [CrossRef]
29. Song, F.; Liu, Y.; Shen, D.; Li, L.; Tan, J. Learning Control for Motion Coordination in Wafer Scanners: Towards Gain Adaptation. *IEEE Trans. Ind. Electron.* **2022**. [CrossRef]
30. Kim, K.S.; Zou, Q. A modeling-free inversion-based iterative feedforward control for precision output tracking of linear time-invariant systems. *IEEE/ASME Trans. Mechatron.* **2012**, *18*, 1767–1777. [CrossRef]
31. Bolder, J.; Kleinendorst, S.; Oomen, T. Data-driven multivariable ILC: Enhanced performance by eliminating L and Q filters. *Int. J. Robust Nonlinear Control* **2018**, *28*, 3728–3751. [CrossRef]
32. Bristow, D.A.; Tharayil, M.; Alleyne, A.G. A survey of iterative learning control. *IEEE Control Syst. Mag.* **2006**, *26*, 96–114.
33. van Zundert, J.; Oomen, T. On inversion-based approaches for feedforward and ILC. *Mechatronics* **2018**, *50*, 282–291. [CrossRef]
34. Fang, Y.; Hu, J.; Liu, W.; Shao, Q.; Qi, J.; Peng, Y. Smooth and time-optimal S-curve trajectory planning for automated robots and machines. *Mech. Mach. Theory* **2019**, *137*, 127–153. [CrossRef]

actuators

MDPI

Article

Integrated Development of a Topology-Optimized Compliant Mechanism for Precise Positioning

Yaoyuan Hu, Bingfeng Ju and Wule Zhu *

The State Key Laboratory of Fluid Power Transmission and Control, School of Mechanical Engineering, Zhejiang University, Hangzhou 310027, China; yyhu@zju.edu.cn (Y.H.); mbfju@zju.edu.cn (B.J.)
* Correspondence: wulezhu@zju.edu.cn

Abstract: A scheme for modelling and controlling a two-dimensional positioning system with a topology-optimized compliant mechanism is presented. The system is designed to ensure a relatively large workspace and exhibit robustness against system nonlinearities. A detailed design procedure based on topology optimization is presented, and a nonlinear description of the designed mechanism is developed as a starting point for further precise position control. The theoretical model is shown to be suitable for a considerably larger working range without losing consistency. A backstepping controller is employed to manipulate the nonlinearities in the model resulting from the geometrical and material nonlinearity of the mechanical structure. The hysteresis of the piezoelectric actuator is also taken into consideration. An experimental verification of the controller demonstrates that the proposed design approach improves the performance of compliant mechanism and satisfies the needs for precision positioning.

Keywords: topology optimization; backstepping robust control; system identification; compliant mechanism

Citation: Hu, Y.; Ju, B.; Zhu, W. Integrated Development of a Topology-Optimized Compliant Mechanism for Precise Positioning. *Actuators* **2022**, *11*, 179. https://doi.org/10.3390/act11070179

Academic Editor: Eniko T. Enikov

Received: 11 April 2022
Accepted: 16 June 2022
Published: 27 June 2022

1. Introduction

Compliant mechanisms [1,2] have been widely used over the last few decades as a promising routine for transforming motions [3], forces [4], or energy [5,6] from input to output. Unlike rigid-body mechanisms, compliant mechanisms are monolithic and gain their mobility from the deflection of flexible members rather than movable joints. This offers increased precision and reliability combined with reduced wear, eliminating the need for lubrication. Such systems are often designed with certain aims, not only at the macro-scale, such as large-input displacement amplification [7] or sufficiently high output stiffness [8,9], but also at the micro-scale, such as micro-electromechanical systems (MEMS) [10] and surgical applications [11]. As design problems become more complicated, the continuum topology optimization (TO) freeform design methodology has also become a popular routine for such mechanisms [4,12–16]. In TO, the traditional trial-and-error design approach is replaced by an automated iterative design approach, which determines the optimal material distribution in a finite-element model while minimizing a given cost function [17,18]. This allows for the design of components or systems based on prescribed loads and boundary conditions, which harness rigorous optimization and simulation schemes to achieve superior performance.

Piezoelectric-actuated compliant mechanisms composed of piezoelectric actuators (PEA) and TO compliant mechanisms (CM) are one of the promising applications in precision positioning, owing to the fast response and extreme positioning resolution of PEAs. These systems also gain flexibility in applications from the TO method because, theoretically, they can take arbitrary design domains and boundary conditions.

However, unlike traditional kinematics-based approaches [1], which allow the designer to apply knowledge and approximations developed for rigid-body mechanisms kinetically [19,20] or dynamically [21,22], the dynamics of topology-optimized CMs are usually nonlinear and complicated. This drawback of the TO method, together with the inherent hysteresis nonlinearity of PEAs and system uncertainties, leads to challenges in controlling such topology-optimized CMs, consequently limiting their practical applications. Further, although researchers have made significant advancements in topology-optimized CM design with multiple conditions, constraints, or output ports to composite motion [23–25] or to generate paths [26], the design of an optimization objective for maximum output workspace, which is a crucial attribute for multi-axis positioning stages, remains undone.

In this study, we aim to develop an integrated effective design and control scheme for multi-DOF positioning stage based on topology-optimized compliance mechanisms with a novel TO objective function designed for large output workspace. Firstly, the TO process is performed based on a nonlinear finite-element analysis. Secondly, a reduced spring-mass model with nonlinear disturbances is developed to cover the dynamics of the designed CM. Then, a MIMO backstepping robust controller is employed to achieve a high trajectory tracking performance. The output motions are not deliberately decoupled in the TO process; rather, the CM gains the ability to composite motions by tangent space modeling of the controller. The controller design process starts from the bounded-input-bounded-output PEA systems, which then serve as an ideal input for a new controller that stabilizes the subsequent compliant mechanism, making up a robust backstepping controller [27]. An experiment verification process is also given afterwards. The resulting system is intended to greatly extend the capabilities of such TO compliant mechanisms.

2. Topology Optimization

The topology optimization process in this study is based on the solid isotropic material with penalization (SIMP) [28,29] approach. The basic idea is that each finite element is associated with a fictitious pseudo-density variable ρ, such that $0 \leq \rho \leq 1$, which parameterizes the topology optimization procedure. The general algorithm is shown in Algorithm 1, and in this section, we will follow the design steps.

Algorithm 1 SIMP algorithm for topology optimization. SIMP Algorithm

Initialization:
Generate mesh and boundary conditions, define material properties, initialize pseudo-density
Main iteration loop:
While not convergent do
Finite element analysis
Objective and constraints evaluation
Update density
End while
Evaluation of results

2.1. Boundary Conditions and Model Specifications

The first thing to do is to specify the boundary conditions of the design domain. Consider a standard compact setup from previous research [30,31] for a piezoelectric-actuated compliance mechanism as the input unit of the design domain, as shown in Figure 1a, and assume the specifications of the PEAs, as shown in Table 1. The total dynamics of a thus-designed PEA set can be described as a single-DOF spring–mass mechanical system, given by:

$$m_{in}\ddot{x}_{in} = F_{in} - f - F_{res}, \tag{1}$$

where m_{in} is the mass of the output blocking plate; F_{in} is the input force produced by the PEA unit; $F_{res}(x_{in}, \dot{x}_{in}) \approx K_{in}x_{in} + C_{in}\dot{x}_{in}$ is the lumped restraint force with K_{in} and C_{in} the spring stiffness and the damping coefficient of the flexure blocking plate, respectively; and f is the output force of the total set, which is the opposite direction of the force to be applied on the compliant mechanism. The force F_{in} generated by the PEA can be modeled by a compressive spring according to the following:

$$F_{in} = K_{pea} \cdot \delta x, \tag{2}$$

where K_{pea} is the stiffness of the PEA and δx is the difference between the nominal load-free output x_{nom} and the actual output x_{in}. Substituting Equation (2) into Equation (1) yields the total dynamic model of a single PEA set, as follows:

$$m_{in}\ddot{x}_{in} = K_{pea} \cdot (x_{nom} - x_{in}) - f - F_{res}(x_{in}, \dot{x}_{in}). \tag{3}$$

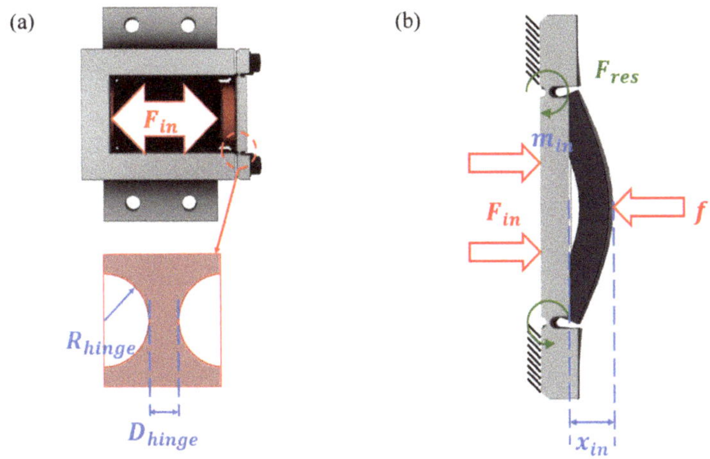

Figure 1. (**a**) Schematic illustration of a single PEA set; (**b**) Design parameters and simplified model of the output port.

Table 1. Specifications of the PEAs.

Symbol	Quantity	Value
K_{pea}	Stiffness	250 [N/μm]
max x_{nom}	Max stroke [a]	60 [μm]
f_{block}	Blocking force	1.5×10^4 [N]
R_{hinge}	Hinge radius	0.75 [mm]
D_{hinge}	Hinge thickness	0.5 [mm]

[a] The max stroke is the maximum of the nominal output x_{nom} of a given PEA.

In this work, we focus on the study of a simplified 2-DOF positioning application. This process can be easily generalized to higher DOF and higher dimensional designs [32], and the basic ideas are similar. The whole setup is shown in Figure 2a.

The input forces applied to the design space are equal in magnitude but opposite in direction to the reaction force of the compliant mechanism. Because the two PEAs are fixed to the platform bed and the output of the PEAs are most likely to be horizontal, the inputs to the design domain can be assumed as forces parallel to the x-axis on roll supports, as shown in Figure 2b. In addition, we make the following assumptions concerning the input forces and the reaction forces (f_is) of the compliant mechanism to simplify further discussions:

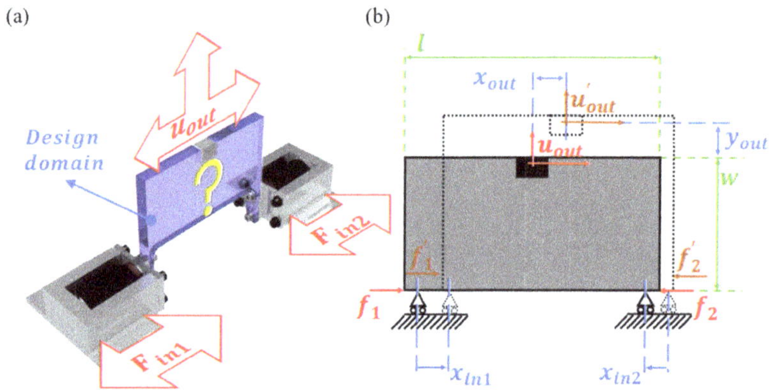

Figure 2. (**a**) Schematic illustration of a 2-DOF compliance set; (**b**) Simplification of the design domain and boundary conditions.

Assumption 1. *The input forces to the design domain are balanced by the internal residual forces at the corresponding input nodes.*

In other words, f_is can be represented by the deformations and accelerations of the compliance mechanism. We will return to this assumption in the next subsection.

Assumption 2. *The resonance frequency of the design domain is much higher than the input frequency; therefore, the dynamics of the PEAs and the designed compliant mechanism can be separated without loss of generality.*

Suppose that the output vector $\boldsymbol{u}_{out}$ is decomposed into orthogonal vectors $\boldsymbol{x}_{out}$ and $\boldsymbol{y}_{out}$ along the x-axis and y-axis, respectively, and that the PEAs are almost identical with only minor differences. Typical working conditions, as shown in Table 2, summarize the relationship between the PEA stroke direction ($\rightarrow$ or $\leftarrow$, i.e., along or against the direction of the positive x-axis) and the output $\boldsymbol{y}_{out}$.

Table 2. Typical working conditions.

	Working Conditions	PEA1 Stroke x_{in1}	PEA2 Stroke x_{in2}
Regular configuration	Maximum $\boldsymbol{y}_{out}$	$\rightarrow$	$\leftarrow$
	Minimum $\boldsymbol{y}_{out}$	$\leftarrow$	$\rightarrow$
Reversed configuration	Maximum $\boldsymbol{y}_{out}$	$\leftarrow$	$\rightarrow$
	Minimum $\boldsymbol{y}_{out}$	$\rightarrow$	$\leftarrow$

The reversed configuration assumes that $\boldsymbol{y}_{out}$ is along the negative y-axis under compressive inputs, useful when design domain width w is limited.

2.2. Topology Optimization Process

Consider the 2D design problem of compliant mechanisms to maximize the positioning workspace, which occurs when the mechanism is used for precision positioning [33,34], and the whole process is performed in a static sense. The selection of optimization and material parameters is given in Table 3 [35].

Table 3. Algorithm and material specifications for the TO method.

Optimization specifications	**Penalty p**	**3**
	Max iteration	1000
	Pseudo density lower bound ρ_{min}	1×10^{-6}
	Discretized steps p_m	40
Mesh specifications	Length l	12 [cm]
	Width w	6 [cm]
	Thickness t	1 [cm]
	Discretization	$\text{n}_\text{l} \times \text{n}_\text{w} = 120 \times 60$
Material properties	Young's modulus E	73 [GPa]
	Min modulus E_{min} [a]	1 [kPa]
	Poisson ratio ν	0.33
	Material density ρ_0	2.71×10^3 [kg/m^3]
	Initial hyper − elastic coefficient c_1	1×10^{-5}E [GPa]
	Hyper − elastic strain threshold $\varepsilon^\star$	0.5

[a] A minimum modulus for the design domain material is assigned to void regions to prevent the stiffness matrix from becoming singular.

Suppose that the inputs given as in the previous section can be discretized into p_m steps and can simultaneously satisfy the extreme working conditions listed in Table 2. Then, each pair of inputs at step p yields a specific output vector, as shown in the schematic illustration in Figure 3. The aim of maximizing a reachable set of the mechanism is equivalently turned into maximizing the interior area bounded by the static trajectory, which is, in general, an irregular polygon. The area of the polygon is given by the g-determinant of the individual output vectors $\boldsymbol{u}_{oj}$, $j = 1, 2, \ldots, p_m$ at each sampled step along the perimeter of the workspace, as the following [36]:

$$J = \text{gdet}(\boldsymbol{u}_{o1}, \ldots, \boldsymbol{u}_{opm}) = \frac{1}{2}\sum_{j}^{p_m} \boldsymbol{u}_{oj} \times \boldsymbol{u}_{oj+1}, \tag{4}$$

where the summation is cyclic, such that $\boldsymbol{u}_{op_m+1} := \boldsymbol{u}_{o1}$.

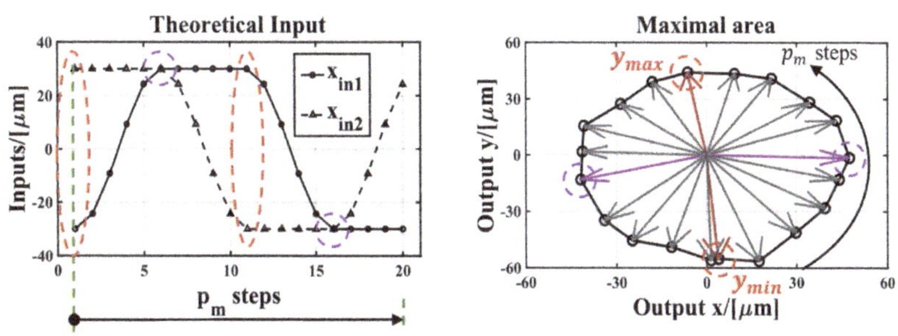

Figure 3. Proposed discretized input in this study and the corresponding L_o output vectors on the perimeter of the workspace.

The constraints for the optimization process are imposed on both the input displacements and the total strain energy: the maximum stroke of the PEAs is limited to $\max x_{in} = 60$ μm, and the stroke is smaller in practice because of the assembly error; the strain energy, $\boldsymbol{\Sigma}$, is also constrained, with a coarse upper bound, $\boldsymbol{\Sigma}_0$, in a manner similar to that described in previous research [37,38] to avoid fatigue failures. As a result, the TO problem of maximizing J can be formulated as follows:

$$\begin{aligned} \max_{\rho} \quad & J \\ \text{s.t.} \quad & \sum_{j}^{p_m} \left| \boldsymbol{P}_j - \boldsymbol{R}_j(\boldsymbol{u}_j, \boldsymbol{\rho}_j) \right| = 0 \\ & \boldsymbol{\Sigma} \leq \boldsymbol{\Sigma}_0 \\ & u_{ini} < \max x_{nomi}, \quad i = 1,2 \\ & \boldsymbol{0} < \rho_{min} \leq \boldsymbol{\rho} \leq \boldsymbol{1} \end{aligned} \quad . \tag{5}$$

In that the summation element $\left| \boldsymbol{P}_j - \boldsymbol{R}_j(\boldsymbol{u}_j, \boldsymbol{\rho}_j) \right|$ in the first constraint of the above Equation (5) is always positive, the following holds for all independent variables $\boldsymbol{u}_j$ and $\boldsymbol{\rho}_j$:

$$\boldsymbol{P}_j - \boldsymbol{R}_j(\boldsymbol{u}_j, \boldsymbol{\rho}_j) = 0. \tag{6}$$

The above Equation (6) is a nonlinear ordinary differential equation system for the unknown displacement field $\boldsymbol{u}$ and design parameter $\boldsymbol{\rho}$ within the calculation domain. The solution of this differential equation system is usually not analytical, and a conventional way is to approximate the displacement field by performing iterative Newton–Raphson method on the spatial discretization of the design domain [39,40]. A brief introduction of the nonlinear finite element synthesis including the additive hyper-elastic element modification to suppress instability during the optimization process [41] can be found in Appendix A.

Now consider the discretized finite element expression of the design domain. A vector $\boldsymbol{L}_o^T$ consisting of all zeros but 1 at the index of the output nodes can be used to select the desired output nodes from the discretized displacement field $\hat{\boldsymbol{u}}$, represented by:

$$\hat{\boldsymbol{u}}_o = \boldsymbol{L}_o^T \hat{\boldsymbol{u}}. \tag{7}$$

Since the following derivations will be mainly on this discretized representation, we will drop the hat representing the discretization process for simplicity. Thus, the sensitivity of the objective function can be represented by the summation of the sensitivities of the individual cross products, as follows:

$$\frac{\partial J}{\partial \rho_e} = \frac{1}{2} \sum_{j}^{p_m} \frac{\partial (\boldsymbol{u}_{oj} \times \boldsymbol{u}_{oj+1})}{\partial \rho_e}. \tag{8}$$

By introducing a set of vectors of the Lagrangian multipliers λ_j, $j = 1, 2, \ldots, p_m$, for each output $\boldsymbol{u}_{oj}$, and assuming that the solution of Equation (6) has already been found via the Newton–Raphson method, the term $\lambda_j^T (\boldsymbol{R}_j(\boldsymbol{u}_j) - \boldsymbol{P}_j)$ is equal to zero and can be added to the displacement vector without changing the result, yielding:

$$\frac{\partial J_j}{\partial \rho_e} = \frac{1}{2} \frac{\partial \left[\left(\boldsymbol{u}_{oj} + \lambda_j^T (\boldsymbol{R}_j - \boldsymbol{P}_j) \right) \times \left(\boldsymbol{u}_{oj+1} + \lambda_{j+1}^T (\boldsymbol{R}_{j+1} - \boldsymbol{P}_{j+1}) \right) \right]}{\partial \rho_e}. \tag{9}$$

We define the total residual for the jth load step as follows:

$$\boldsymbol{\Re}_j(\boldsymbol{u}_j, \boldsymbol{\rho}) = \boldsymbol{P}_j - \boldsymbol{R}_j(\boldsymbol{u}_j, \boldsymbol{\rho}). \tag{10}$$

Then, the sensitivity of this modified function can be given by:

$$\begin{aligned}\frac{\partial J_j}{\partial \rho_e} = \frac{1}{2}\left(\boldsymbol{L}_{oj}\frac{\partial \boldsymbol{u}_j}{\partial \rho_j} + \lambda_j^T\left(\frac{\partial \boldsymbol{\Re}_j}{\partial \boldsymbol{u}_j}\frac{\mathrm{d}\boldsymbol{u}_j}{\mathrm{d}\rho_e} + \frac{\partial \boldsymbol{\Re}_j}{\partial \rho_e}\right)\right) \times \left(\boldsymbol{u}_{oj+1} + \lambda_{j+1}^T \boldsymbol{\Re}_{j+1}\right) \\ + \frac{1}{2}(\boldsymbol{u}_{oj} + \lambda_j^T \boldsymbol{\Re}_j) \times \left(\boldsymbol{L}_{oj+1}\frac{\partial \boldsymbol{u}_{j+1}}{\partial \rho_e} + \lambda_{j+1}^T\left(\frac{\partial \boldsymbol{\Re}_{j+1}}{\partial \boldsymbol{u}_{j+1}}\frac{\mathrm{d}\boldsymbol{u}_{j+1}}{\mathrm{d}\rho_e} + \frac{\partial \boldsymbol{\Re}_{j+1}}{\partial \rho_e}\right)\right).\end{aligned} \tag{11}$$

in accordance with Ref. [42]. By doing some arithmetic and transformations, and introducing the tangent stiffness matrix $\boldsymbol{K}_T$ as the linearized approximation of $\frac{\partial \boldsymbol{\Re}_j}{\partial \boldsymbol{u}_j}$, we have:

$$\begin{aligned}\frac{\partial J_j}{\partial \rho_e} &= \frac{1}{2}\left(\left(\boldsymbol{L}_{oj} + \lambda_j^T\frac{\partial \boldsymbol{\Re}_j}{\partial \boldsymbol{u}_j}\right)\frac{\mathrm{d}\boldsymbol{u}_j}{\mathrm{d}\rho_e} + \lambda_j^T\frac{\partial \boldsymbol{\Re}_j}{\partial \rho_e}\right) \times (\boldsymbol{u}_{oj+1} + \lambda_{j+1}^T\boldsymbol{\Re}_{j+1}) + \\ &\quad \frac{1}{2}(\boldsymbol{u}_{oj} + \lambda_j^T\boldsymbol{\Re}_j) \times \left(\left(\boldsymbol{L}_{oj+1} + \lambda_{j+1}^T\frac{\partial \boldsymbol{\Re}_{j+1}}{\partial \boldsymbol{u}_{j+1}}\right)\frac{\mathrm{d}\boldsymbol{u}_{j+1}}{\mathrm{d}\rho_e} + \lambda_{j+1}^T\frac{\partial \boldsymbol{\Re}_{j+1}}{\partial \rho_e}\right) \\ &= \frac{1}{2}\left((\boldsymbol{L}_{oj} + \lambda_j^T\boldsymbol{K}_{Tj})\frac{\mathrm{d}\boldsymbol{u}_j}{\mathrm{d}\rho_e} + \lambda_j^T\frac{\partial \boldsymbol{\Re}_j}{\partial \rho_e}\right) \times \left(\boldsymbol{u}_{oj+1} + \lambda_{j+1}^T\boldsymbol{\Re}_{j+1}\right) + \\ &\quad \frac{1}{2}\left(\boldsymbol{u}_{oj} + \lambda_j^T\boldsymbol{\Re}_j\right) \times \left((\boldsymbol{L}_{oj+1} + \lambda_{j+1}^T\boldsymbol{K}_{Ti})\frac{\mathrm{d}\boldsymbol{u}_{j+1}}{\mathrm{d}\rho_e} + \lambda_{j+1}^T\frac{\partial \boldsymbol{\Re}_{j+1}}{\partial \rho_e}\right)\end{aligned} \tag{12}$$

Notice that the Lagrange multiplier λ_is can be chosen freely, such that $\boldsymbol{L}_{oi} + \lambda_i^T\boldsymbol{K}_{Ti} = \boldsymbol{0}$ is easily satisfied, then the coefficient term before $\frac{\mathrm{d}\boldsymbol{u}_\bullet}{\mathrm{d}\rho_e}$ can be eliminated immediately. As a result, we can write the sensitivity of the optimization objective regarding working area J in the following form:

$$\frac{\partial J}{\partial \rho_e} = \frac{1}{2}\sum_{j}^{p_m}\left(\lambda_j^T\frac{\partial \boldsymbol{\Re}_j}{\partial \rho_e} \times \left(\boldsymbol{u}_{oj+1} + \lambda_{j+1}^T\boldsymbol{\Re}_{j+1}\right) + \left(\boldsymbol{u}_{oj} + \lambda_j^T\boldsymbol{\Re}_j\right) \times \lambda_{j+1}^T\frac{\partial \boldsymbol{\Re}_{j+1}}{\partial \rho_e}\right). \tag{13}$$

Since the evaluation of the gradients only requires the accuracy of a tangent space approximation [37], the approximated $\frac{\partial \boldsymbol{\Re}_j}{\partial \rho_e}$ can be given by the following:

$$\frac{\partial \boldsymbol{\Re}_j}{\partial \rho_e} = -p\rho_e^{p-1}(E_0 - E_{min})\boldsymbol{K}_{Tj}\boldsymbol{u}_i. \tag{14}$$

The design parameter is updated using the multi-criterion method of asymptotes (MMA) [35,43], in which case the sensitivities of the constraint functions are also needed. In this study, the input displacement sensitivity is almost identical to the expression given by Equation (13) [17], except that the selecting vector is changed to $\boldsymbol{L}_i$, which is defined for the input nodes in a similar sense to Equation (7). The total strain energy is given by the following:

$$\boldsymbol{\Sigma} = \boldsymbol{u}^T\boldsymbol{R}(\boldsymbol{u}) \approx \rho^p\hat{\boldsymbol{u}}^T\boldsymbol{K}_T\hat{\boldsymbol{u}}. \tag{15}$$

The sensitivity of Equation (15) is approximated by its linear part in the tangent space, determined elementwise [44] as follows:

$$\frac{\partial \boldsymbol{\Sigma}_e}{\partial \rho_e} = -p\rho_e^{p-1}\hat{\boldsymbol{u}}_e{}^T\boldsymbol{K}_{eT}\hat{\boldsymbol{u}}_e. \tag{16}$$

By performing the iterative optimization process as given in Algorithm 1, the results of the TO are given in Figure 4. The initial calculated boundary of the output workspace is indicated by the dark blue parallelograms, and the final optimization results are indicated by the red parallelograms. The area of the viable output region is enlarged by a factor of over two with respect to the initial configuration. This intermediate result gives us an initial perspective on how the designed compliant reacts to certain input loads and will serve as the start point of the order reduction and controller design process in the following sections.

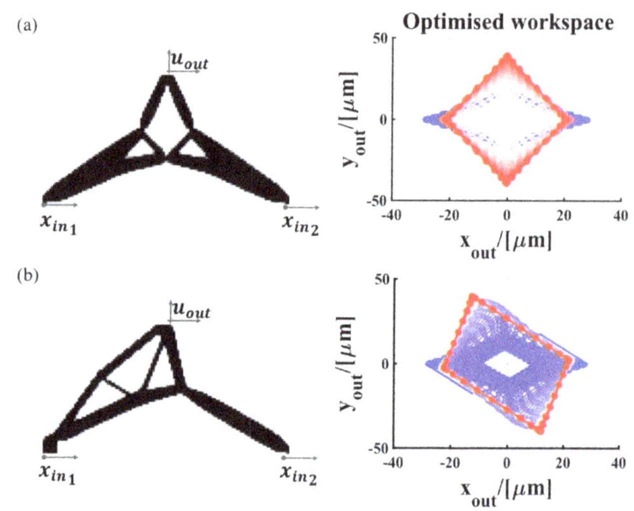

Figure 4. (**a**) TO results with extreme points set in accordance with the normal configuration in Table 2, and (**b**) TO results with a tilted output region w.r.t (**a**), and the calculated output workspace distribution evolution during the iterative optimization process.

3. System Modeling and Controller Design

The basic idea of the controller design for the whole positioning system can be outlined by simultaneously suppressing the influence of the PEA units and the nonlinearities of the designed CM. Considering the cascade nature of the system, i.e., the output displacement of a single PEA is first described as a reacting force on the output plate of the unit, which then serve as the input load to the CM, a robust backstepping controller is a suitable candidate for the controller structure design; by applying such controllers, the dynamics of the PEA and the passive CM part can be treated separately.

3.1. Inverse Multiplicative Compensation Scheme for PEA

By a careful selection of the compensation controller, the outputs of the PEA units can be seen as ideal displacement sources. Feedforward control is a common approach to compensate for the hysteresis effect [45,46]. The Prandtl–Ishlinskii, Duhem, and Bouc–Wen models were developed to describe the hysteresis effect, which is then used to construct the inverse controller for compensation of effects of hysteresis [47–50]. These nonlinear models motivated researchers to develop robust feedback controllers such as the sliding-mode controller [51,52], damping controller [53], disturbance observer [31], and adaptive robust controller [54,55]. In these control approaches, the hysteresis nonlinearity is treated as a disturbance to the tangent space systems. Specifically, in this section, an asymmetric Bouc–Wen model [49,56] is chosen to simulate the nonlinear hysteresis of the PEAs.

The Bouc–Wen model is based on an artificial state variable h. The model represents the hysteresis relationship between an excitation F and the state h, according to the following differential equation:

$$\begin{aligned}\frac{\mathrm{d}h}{\mathrm{d}t} &= A_{bw}\frac{\mathrm{d}F}{\mathrm{d}t} - B_{bw}\left|\frac{\mathrm{d}F}{\mathrm{d}t}\right|h|h|^{n-1} - \Gamma_{bw}\frac{\mathrm{d}F}{\mathrm{d}t}|h|^{n} + \delta_{bw}F\,\mathrm{sgn}\left(\frac{\mathrm{d}F}{\mathrm{d}t}\right),\\ \mathrm{sgn}(\cdot)\& &= \begin{cases} 1, & x > 0 \\ 0, & x = 0 \\ -1, & x < 0 \end{cases},\end{aligned} \tag{17}$$

where A_{bw} is the amplitude of the restoring force, B_{bw} and Γ_{bw} control the shape of the hysteresis loop, n controls the smoothness of the transition from elastic-to-plastic response, and δ is the non-symmetrical factor.

For PEAs, the excitation input F in Equation (17) is replaced by the applied voltage U, and n is set to 1 in accordance with the conventional practice. As a result, the Bouc–Wen model for PEAs can be expressed as follows:

$$\begin{cases} x_{nom}(t) = d_p U(t) - h(t) \\ \frac{dh}{dt} = A_{bw}\frac{dU}{dt} - B_{bw}\left|\frac{dU}{dt}\right|h - \Gamma_{bw}\frac{dU}{dt}|h| + \delta_{bw} U \,\mathrm{sgn}\left(\frac{dU}{dt}\right) \end{cases}, \quad (18)$$

where x_{nom} is the displacement output. The parameter d_p represents the piezoelectric coefficient and is strictly positive. A detailed parameter identification process of the involved parameters for the PEAs used in this study can be found in Appendix B; a list of the estimation results is also given.

The output of the PEA unit can be compensated separately following the inverse multiplicative scheme given by previous research [31,57,58]. Note that the state variable h can be rewritten as:

$$h = H(U), \quad (19)$$

where $H(U)$ is a nonlinear operator characterized by the second equation in Equation (18). Then, the Bouc–Wen model can be reduced to the following:

$$x_{nom} = d_p U - H(U). \quad (20)$$

Suppose that we have a desired reference x_{nom_d}. Extracting the value of U that meets the reference yields the following:

$$U = \frac{1}{d_p}\left(x_{nom_d} + H(U)\right). \quad (21)$$

An outline of the total inverse multiplicative compensation diagram is given in Figure 5. Thus, in the following designing process, the output of the PEA units will be simplified to ideal displacement sources with bounded disturbances and uncertainties, which can be addressed suitably by a robust controller.

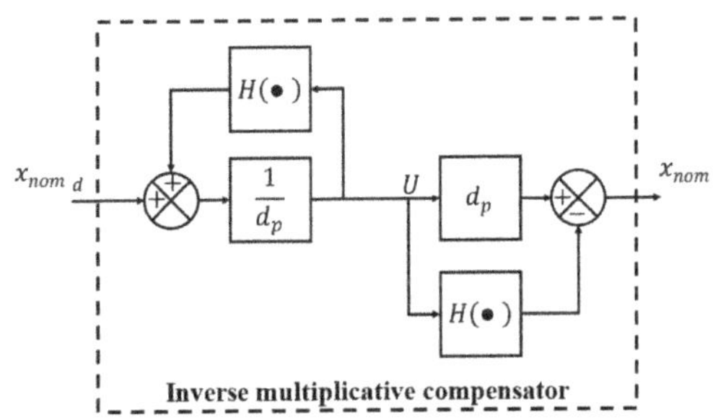

Figure 5. Inverse multiplicative hysteresis compensation diagram for the PEAs.

3.2. Controller Design

The controller design process is based on the total system dynamics considering both the PEA units and the TO designed compliant mechanism. Since the PEA units can be treated as ideal displacement sources with bounded disturbances, the discussion of this subsection will be focused on the dynamics of the compliant mechanism and the deduction of the actual form of the robust controller.

3.2.1. Reduced Order System Dynamics

Although there are handy controllers designed for finite element models considering fuzzy logic [57] or for set-invariance under uncertain constraints [58], it is more straightforward to consider the tangent space of the compliant mechanism and regard the bounded nonlinearities as disturbances and deviations.

Take the second result in Figure 4 as an example, we denote by $\boldsymbol{e} = \boldsymbol{T}(\theta)\begin{bmatrix}\boldsymbol{E}_1\\ \boldsymbol{E}_2\end{bmatrix}$ the local basis of a tilted output configuration with $\boldsymbol{T}(\theta) = \begin{bmatrix}\cos\theta & \sin\theta\\ -\sin\theta & \cos\theta\end{bmatrix}$ the rotation matrix depicted in Figure 6a. We intuitively guess the geometric relation between the input displacement $\boldsymbol{x}_{in} = \sum_i x_{ini}\boldsymbol{E}_i$, the output compliant mechanism deformation $\boldsymbol{u}_o = \sum_i u_{oi}\boldsymbol{E}_i$, and the actual output $\boldsymbol{q} = \sum_i q_i\boldsymbol{e}_i$ in a static sense. The actual output is then represented as a vector sum of the moving spatial frame $\boldsymbol{x} = \begin{bmatrix}\frac{1}{2}(x_{in1} + x_{in2})\\ 0\end{bmatrix}$ and the system deformation $\boldsymbol{u}_o$ as follows:

$$\boldsymbol{q} = \boldsymbol{x} + \boldsymbol{u}_o = \begin{bmatrix}\frac{1}{2}(x_{in1} + x_{in2}) + u_{o1}\\ u_{o2}\end{bmatrix}. \tag{22}$$

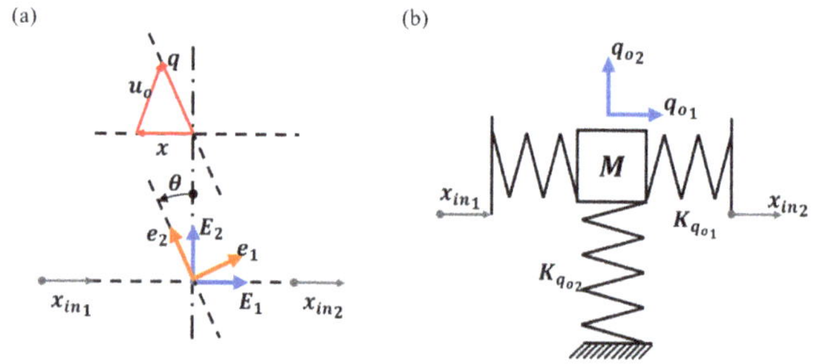

Figure 6. Schematic illustrations of (**a**) The 2nd-order spring–mass model of the compliant mechanism and (**b**) The tilted output configuration with rotation angle of θ.

Consider the geometric properties of the compliant mechanism setup, the output $\boldsymbol{q}$ reaches its extreme value when the input boundary displacements satisfy $x_{in1} \approx -x_{in2}$. In this case, the theoretical output along the x-axis of the spatial frame also vanishes, and the output $\boldsymbol{u}_o = [u_{o1}, u_{o2}]^T$ can be approximated by its discretized tangent space approximation in accordance with the solution of Equation (6). Consequently, the original load $\boldsymbol{P}$ can also be replaced by a complementary displacement constraint [59,60], satisfying:

$$\boldsymbol{P} = \begin{bmatrix}\boldsymbol{P}_f\\ \boldsymbol{P}_s\end{bmatrix} \approx \begin{bmatrix}\boldsymbol{K}_{Tff} & \boldsymbol{K}_{Tfs}\\ \boldsymbol{K}_{Tsf} & \boldsymbol{K}_{Tss}\end{bmatrix}\begin{bmatrix}\boldsymbol{u}_f\\ \boldsymbol{u}_s\end{bmatrix}, \tag{23}$$

where the matrices $\boldsymbol{K}_{Tff}$, $\boldsymbol{K}_{Tfs}$, $\boldsymbol{K}_{Tsf}$, and $\boldsymbol{K}_{Tss}$ are partitions of the original tangent stiffness $\boldsymbol{K}_T$ introduced in Equation (11). Specifically, $\boldsymbol{K}_{Tff}$ is the restrained structural stiffness matrix, which is square and symmetric, $\boldsymbol{K}_{Tfs}$ and $\boldsymbol{K}_{Tsf}$ are the off-diagonal submatrices relating to the known nodal displacement $\boldsymbol{u}_s$ to the unknown reaction forces $\boldsymbol{P}_f$ on the free nodes. Then, the output deformation of the compliant mechanism can be written as follows:

$$\begin{aligned}[\boldsymbol{u}_o]^{2\times 1} &= \boldsymbol{u}_{fo} = \boldsymbol{L}_o|_f\boldsymbol{u}_f\\ &= \left[\boldsymbol{L}_o|_f\right]^{2\times(N-2)}\left[(\boldsymbol{K}_T)_{ff}^{-1}\right]^{(N-2)\times(N-2)}\left[\boldsymbol{K}_{fs}\right]^{(N-2)\times 2}[\boldsymbol{x}_{in}]^{2\times 1}\\ &= [\boldsymbol{\alpha}]^{2\times 2}\boldsymbol{x}_{in},\end{aligned} \tag{24}$$

where $L_o|_f$ is the selection vector given in Equation (7) restricted to the subset $\boldsymbol{u}_f \subset \boldsymbol{u}$, $\boldsymbol{\alpha} = \begin{bmatrix} \alpha_{11} & \alpha_{12} \\ \alpha_{21} & \alpha_{22} \end{bmatrix}$ is the transfer matrix from the input to the output, N is the total number of finite elements of the design domain (including the elements with $\rho_e \approx 0$), and $s = 2$ corresponding to the two input displacements in our case. Thus, the elementary form of Equation (24) is given as follows:

$$\begin{array}{l} u_{o1} = \alpha_{11}x_{in1} + \alpha_{12}x_{in2} \\ u_{o2} = \alpha_{21}x_{in1} + \alpha_{22}x_{in2} \end{array}. \tag{25}$$

We consider the following conditions:

i. $x_{in1} = x_{in2}$, no compressive force exists, and $\boldsymbol{u}_o = \mathbf{0}$;
ii. $x_{in1} = -x_{in2}$, the spatial frame $\boldsymbol{x} = \mathbf{0}$, and $\boldsymbol{u}_o = \boldsymbol{q}_o$.

When the output displacement reaches a maximum, the angle between $\boldsymbol{u}_o$ and the y-axis is presumed to be θ, in which case $\frac{u_{o1}}{u_{o2}} = \tan\theta$. We then have the following two sets of linear equations:

$$\begin{array}{l} \left\{ \begin{array}{l} \alpha_{11}x_{in1} + \alpha_{12}x_{in1} = 0 \\ \alpha_{21}x_{in1} + \alpha_{22}x_{in1} = 0 \end{array} \right. \Rightarrow \left\{ \begin{array}{l} \alpha_{11} = -\alpha_{12} \\ \alpha_{21} = -\alpha_{22} \end{array} \right. \\ \left\{ \begin{array}{l} \alpha_{11}\max x_{in1} - \alpha_{12}\max x_{in2} = q_{o1} \\ \alpha_{21}\max x_{in1} - \alpha_{22}\max x_{in2} = q_{o2} \end{array} \right. \Rightarrow \alpha_{11} = -\tan\theta\alpha_{22}. \end{array} \tag{26}$$

This result indicates that the actual value of the matrix $[\boldsymbol{\alpha}]^{2\times 2}$ can be represented by a single value of α_{22}. Thus, we can conclude the static relation between the input and the output by an approximation of α_{22}, which is given by the ratio of the maximum output and the maximum compressive input as follows:

$$\alpha := \alpha_{22} \approx \frac{\max q_{o2}}{\max(x_{in1} - x_{in2})} = \frac{\max q_{o2}}{2\max x_{in1}}, \tag{27}$$

yielding

$$\boldsymbol{\alpha}(\theta) \approx \begin{bmatrix} \tan\theta\alpha & -\tan\theta\alpha \\ \alpha & -\alpha \end{bmatrix}. \tag{28}$$

This result is especially useful when the full systematic model is unknown, so that one can quickly provide an initial guess to the system's kinematic properties. We provide an approximation of the actual output of our application in the following form:

$$\boldsymbol{q} := \begin{bmatrix} q_1 \\ q_2 \end{bmatrix} = \boldsymbol{A}(\theta)\boldsymbol{x}_{in}, \tag{29}$$

where $A(\theta) = \begin{bmatrix} \frac{1}{2} + \tan\theta\alpha & \frac{1}{2} - \tan\theta\alpha \\ \alpha & -\alpha \end{bmatrix}$

When the system is working dynamically, i.e., tracking a certain output trajectory, the whole compliant mechanism can be reduced to a dynamic spring–mass system in correspondence with Equation (29) and Assumption 1, given by:

$$\begin{array}{l} M\ddot{q}_{o1} = 2K_{q_{o1}}(q_1 - q_{o1}) + 2C_{q_{o1}}(\dot{q}_1 - \dot{q}_{o1}) + \text{cross terms} \\ M\ddot{q}_{o2} = K_{q_{o2}}(q_2 - q_{o2}) + C_{q_{o2}}(\dot{q}_2 - \dot{q}_{o2}) + \text{cross terms} \end{array}, \tag{30}$$

where $K_{q_{oi}}$s and $C_{q_{oi}}$s are the lumped stiffness and damping coefficients of the spring–mass system along the direction of q_{oi}, $i = 1, 2$, respectively, M is the mass of the system, and the cross terms denote the influence of the spring between the two output directions. An illustration of this reduced system is given in Figure 6b. We can treat the cross terms in Equation (30) as a disturbance term in the form of $\boldsymbol{\Delta}_q = \begin{bmatrix} \delta_1(q_{o2}) \\ \delta_2(q_{o1}) \end{bmatrix}$, and we leave the structural analysis of the cross terms for future research. For simplicity, we now consider the normal configuration, where $\tan\theta = 0$. The expansion of Equation (29) yields a direct tangent space approximation of the relation between the input x_{in} and the system output q_o as follows:

$$M\ddot{q}_o \& = \Delta_q + K(A_0 x_{in} - q_o) + C(A_0 \dot{x}_{in} - \dot{q}_o), \tag{31}$$

where

$$\begin{gathered} M = \begin{bmatrix} M & 0 \\ 0 & M \end{bmatrix}, K = \begin{bmatrix} 2K_{q_{o1}} & 0 \\ 0 & K_{q_{o2}} \end{bmatrix} \\ C = \begin{bmatrix} 2C_{q_{o1}} & 0 \\ 0 & C_{q_{o2}} \end{bmatrix}, A_0 = A(0) = \begin{bmatrix} \frac{1}{2} & \frac{1}{2} \\ \alpha & -\alpha \end{bmatrix}. \end{gathered} \tag{32}$$

By including the developed model for the PEA units, we denote by f the interactive force between the inputs $x_{ini}, i = 1, 2$, and the output q_o as follows:

$$f = \begin{bmatrix} f_1 \\ f_2 \end{bmatrix} = \begin{bmatrix} K_{q_{o1}}(x_{in1} - q_{o1}) + C_{q_{o1}}(\dot{x}_{in1} - \dot{q}_{o1}) \\ K_{q_{o1}}(x_{in2} - q_{o1}) + C_{q_{o2}}(\dot{x}_{in2} - \dot{q}_{o1}) \end{bmatrix}. \tag{33}$$

For the two PEA sets, we already have the matrix form of Equation (3):

$$\begin{gathered} \ddot{x}_{in} = M_{in}^{-1} K_{pea} x_{nom} - M_{in}^{-1} K_{in} x_{in} - M_{in}^{-1} C_{in} \dot{x}_{in} \\ + M_{in}^{-1} K_{in,q} q_o + M_{in}^{-1} C_{in,q} \dot{q}_o, \end{gathered} \tag{34}$$

where

$$\begin{gathered} x_{in} = \begin{bmatrix} x_{in1} \\ x_{in2} \end{bmatrix}, x_{nom} = \begin{bmatrix} x_{nom1} \\ x_{nom2} \end{bmatrix} \\ M_{in} = \begin{bmatrix} m_{in1} & 0 \\ 0 & m_{in2} \end{bmatrix}, K_{pea} = \begin{bmatrix} K_{pea1} & 0 \\ 0 & K_{pea2} \end{bmatrix} \\ K_{in} = \begin{bmatrix} K_{q_{o1}} + K_{in1} + K_{pea1} & 0 \\ 0 & K_{q_{o1}} + K_{in2} + K_{pea2} \end{bmatrix} \\ C_{in} = \begin{bmatrix} C_{q_{o1}} + C_{in1} & 0 \\ 0 & C_{q_{o1}} + C_{in2} \end{bmatrix} \\ K_{in,q} = \begin{bmatrix} K_{q_{o1}} & 0 \\ K_{q_{o1}} & 0 \end{bmatrix}, C_{in,q} = \begin{bmatrix} C_{q_{o1}} & 0 \\ C_{q_{o1}} & 0 \end{bmatrix} \end{gathered} \tag{35}$$

are the actual output vectors, mass matrix, stiffness matrices, and damping coefficient matrices, respectively.

Thus, the total system dynamics can be written as a combination of Equations (31) and (34), based on Assumption 2. Denoting the state variable vector by $z = \begin{bmatrix} q_o & \dot{q}_o & x_{in} & \dot{x}_{in} \end{bmatrix}^T$ and the input vector $u = \begin{bmatrix} x_{nom1} & x_{nom2} \end{bmatrix}^T$, we obtain the following linearized state-space form of the total system dynamics:

$$\dot{z} = Bz + \begin{bmatrix} 0 \\ D \end{bmatrix} u + \Delta, \tag{36}$$

where B and D are the corresponding system matrices, whose details can be found in Appendix C, and $\Delta = \begin{bmatrix} \Delta_\eta & \Delta_\xi \end{bmatrix}^T$ is the model error, attributable mainly to the following:

i. finite-element modelling error ϵ_{FE}
ii. piezoelectric modelling error ϵ_{PE}
iii. asymmetric modelling error ϵ_{ASYM}
iv. systematic noise n_{SYS}
v. measurement error δ_M.

Additional uncertainties, such as assembly error and manufacturing defects, are also possible in practice. In this study, we only treat these errors as bounded (i.e., $\|\Delta\| \leq \delta_I$), matched, and Gaussian for simplicity.

3.2.2. Backstepping Robust Controller

In a practical setup, not all state variables in Equation (36) are measurable. In our case, specifically, only the output q is measured. To build an explicit control scheme, estimates of the state variables are essential. Therefore, we utilize K-filters [27] to provide exponentially convergent estimates of the unmeasured states. We design the observer to obtain the estimate $\hat{z}$ as follows:

$$\begin{aligned} \dot{\hat{z}} &= \mathbf{B}_0\hat{z} + k_{e1}z_1 + k_{e2}z_2 + bu + \Delta \\ y &= \begin{bmatrix} z_1 & z_2 \end{bmatrix}, \end{aligned} \tag{37}$$

where

$$\begin{aligned} \mathbf{B}_0 = \mathbf{B} - k_{e1}\underbrace{\begin{bmatrix}1 & 0 & \cdots\end{bmatrix}}_{8} - k_{e2}\underbrace{\begin{bmatrix}0 & 1 & 0 & \cdots\end{bmatrix}}_{8}, \\ b = \begin{bmatrix} 0 \\ D \end{bmatrix}. \end{aligned} \tag{38}$$

By choosing a suitable k_{e_d}, we maintain the stability of observer matrix $\mathbf{B}_0$. Thus, there exists a symmetric and positive definite matrix P such that:

$$P\mathbf{B}_0 + \mathbf{B}_0^T P = -I, \quad P = P^T > 0. \tag{39}$$

Following the design procedure in [27,61], the K-filters are given by:

$$\hat{z} = \zeta + \Omega^{\mathrm{T}} b, \tag{40}$$

where

$$\begin{aligned} \dot{\zeta} &= \mathbf{B}_0\zeta + k_{e1}z_1 + k_{e2}z_2 \\ \dot{\Omega}^T &= \mathbf{B}_0\Omega^{\mathrm{T}} + \begin{bmatrix} d_{11} & 0 \\ 0 & d_{22} \end{bmatrix}\begin{bmatrix} u_1 e_7 \\ u_2 e_8 \end{bmatrix}, \end{aligned} \tag{41}$$

and where e_i denotes the ith standard basis vector. The state-estimation error $\varepsilon = z - \hat{z}$ is readily shown to satisfy the following:

$$\dot{\varepsilon} = \mathbf{B}_0\varepsilon, \tag{42}$$

and will decay exponentially to zero.

Equation (33) can be modified into a state-space form representing a cascaded connection of two subsystems by selecting the state variables as $\eta = \begin{bmatrix} z_1 & z_2 & z_3 & z_4 & z_5 & z_6 \end{bmatrix}^T$ and $\xi = \begin{bmatrix} z_7 & z_8 \end{bmatrix}^T$.

$$\begin{aligned} \dot{\eta} &= \mathbf{B}_{\eta\eta} + \mathbf{B}_{\eta\xi}\xi + \Delta_\eta, \\ \ddot{\xi} &= \mathbf{D}u + \mathbf{B}_{\xi\eta}\eta + \mathbf{B}_{\xi\xi}\dot{\xi} + \Delta_\xi. \end{aligned} \tag{43}$$

We use the following nonlinear input transformation:

$$u = D^{-1}\left[-\left(\hat{\mathbf{B}}_{\xi\eta}\hat{\eta} + \hat{\mathbf{B}}_{\xi\xi}\dot{\hat{\xi}} \right) + \dot{v} \right], \tag{44}$$

where $\hat{\mathbf{B}}_{\xi\eta}$ and $\hat{\mathbf{B}}_{\xi\xi}$ are the estimation of the coefficient matrices of $\mathbf{B}_{\xi\eta}$ and $\mathbf{B}_{\xi\xi}$ evaluated at states $\hat{\eta} = \begin{bmatrix} z_1 & z_2 & \hat{z}_3 & \hat{z}_4 & \hat{z}_5 & \hat{z}_6 \end{bmatrix}^T$ and $\dot{\hat{\xi}} = \begin{bmatrix} \hat{z}_7 & \hat{z}_8 \end{bmatrix}^T$. All the unmeasured state variables, i.e., $\hat{z}_i$s, are replaced by their estimations for the controller input calculations. Since the estimators are designed to converge asymptotically to the real value of the state variables, we will drop the hat above the state variables η and ξ unless specifically mentioned. To reduce Equation (43) to the pure integrator from the new input v to ξ, such that:

$$\dot{\xi} = v, \tag{45}$$

we must introduce an stabilizing control law $\boldsymbol{a}(\boldsymbol{\eta})$ such that the subsystems in Equation (43) can be stabilized within finite time, as suggested in ref. [27]. The idea behind the selection of this intermediate control law is straightforward:

i. $\boldsymbol{\xi}$ is the actual input of the first subsystem in Equation (43) and is controlled by $\boldsymbol{v}$; $\boldsymbol{a}$ is the desired input control law of the first subsystem in Equation (43). Therefore, if we can find some $\boldsymbol{v}$ such that $\boldsymbol{\xi}$ is able to track $\boldsymbol{a}$ very closely, then the first subsystem in Equation (43) is automatically stabilized.
ii. The same for the second subsystem in Equation (43), where the problem turns into finding $\boldsymbol{u}$ from the same $\boldsymbol{v}$ based on the relation given in Equation (44).

Therefore, the ultimate goal is to find this specific $\boldsymbol{v}$ satisfying both conditions. However, before that, we need to solve for the exact form of the ideal control law $\boldsymbol{a}(\boldsymbol{\eta})$. We denote $\epsilon_i = \eta_i - \eta_{di} = z_i - z_{di}$, $i = 1, 2$, the tracking error between the output node $\eta_i = z_i$, and the desired trajectory $\eta_{di} = z_{di}$. The relative degree of the total system is 2 [62]. A selection of the feedback error term $\boldsymbol{s} = \dot{\boldsymbol{\epsilon}} + \boldsymbol{k}_\epsilon \boldsymbol{\epsilon}$ yields the total error dynamic in the form of:

$$\dot{\boldsymbol{s}} = \begin{bmatrix} s_1\left(\boldsymbol{\eta}, \boldsymbol{a}(\boldsymbol{\eta}), \dot{\hat{\boldsymbol{\xi}}}, \boldsymbol{\eta}_d\right) + \Delta_1 \\ s_2\left(\boldsymbol{\eta}, \boldsymbol{a}(\boldsymbol{\eta}), \dot{\hat{\boldsymbol{\xi}}}, \boldsymbol{\eta}_d\right) + \Delta_2 \end{bmatrix}, \tag{46}$$

where $\boldsymbol{a}(\boldsymbol{\eta})$ is the desired input for the subsystem in Equation (43), as discussed; the $\dot{\hat{\xi}}$ is used specifically to avoid ambiguity that considers the control Lyapunov function (CLF):

$$\boldsymbol{V}(\boldsymbol{\eta}) = \frac{1}{2}\begin{bmatrix} s_1^2 \\ s_2^2 \end{bmatrix}. \tag{47}$$

The path derivative of the CLF with respect to the solution $\boldsymbol{\eta}(t)$ is as follows:

$$\dot{\boldsymbol{V}}(\boldsymbol{\eta}) = \begin{bmatrix} s_1\dot{s}_1 \\ s_2\dot{s}_2 \end{bmatrix}. \tag{48}$$

The desired controller output $\boldsymbol{a}$ can be determined further as the combination of a linear feedback term $\boldsymbol{a}_m$ that expresses exponential suppression of the tracking errors and a robust term $\boldsymbol{a}_s$ to compensate for the systematic errors, in the form of:

$$\boldsymbol{a} = \boldsymbol{a}_m + \boldsymbol{a}_s. \tag{49}$$

This kind of separation gives us additional flexibility to adjust the control law for better performance. Thus, the expression of $\boldsymbol{a}$ can be deduced by solving the inequality assuming that $\dot{\boldsymbol{V}}(\boldsymbol{\eta}) < 0$ always holds in Equation (48), satisfying the Lyapunov stability criterion.

Note that there is also a deviation between the actual input $\boldsymbol{\xi}$ and its desired value $\boldsymbol{a}$; let $\boldsymbol{\epsilon}_a$ be this deviation:

$$\boldsymbol{\epsilon}_a = \boldsymbol{\xi} - \boldsymbol{a}. \tag{50}$$

Augmenting Equation (47) with a quadratic term of the error variable $\boldsymbol{\epsilon}_a$, we obtain a CLF for the whole system:

$$\boldsymbol{V}_\alpha(\boldsymbol{\eta}, \boldsymbol{\xi}) = \boldsymbol{V}(\boldsymbol{\eta}) + \frac{1}{2}\begin{bmatrix} \epsilon_{\alpha 1}^2 \\ \epsilon_{\alpha 2}^2 \end{bmatrix}. \tag{51}$$

The path derivative of $\boldsymbol{V}_\alpha$ is computed as follows:

$$\dot{\boldsymbol{V}}_\alpha = \dot{\boldsymbol{V}}(\boldsymbol{\eta}) + \begin{bmatrix} \epsilon_{\alpha 1}\dot{\epsilon}_{\alpha 1} \\ \epsilon_{\alpha 2}\dot{\epsilon}_{\alpha 2} \end{bmatrix}. \tag{52}$$

Again, we choose to separate the controller input v into two parts, such that:

$$v = v_m + v_s. \tag{53}$$

where, same as Equation (49), v_m is the proportional feedback stabilizing controller input and v_s is a robust controller that eliminates the effects of the modelling errors. The expression can also be deduced by assuming that Equation (52) always holds. We denote the approximation of the calculation of the system input as $\hat{v} = v_m + v_s$. The system input follows from Equation (44) and this completes the design of a backstepping robust controller for the simplified spring–mass model that was developed for the compliant mechanism. A schematic illustration of the whole closed-loop system with the backstepping robust control law is shown in Figure 7. A detailed deduction process for the controller design can be found in Appendix D.

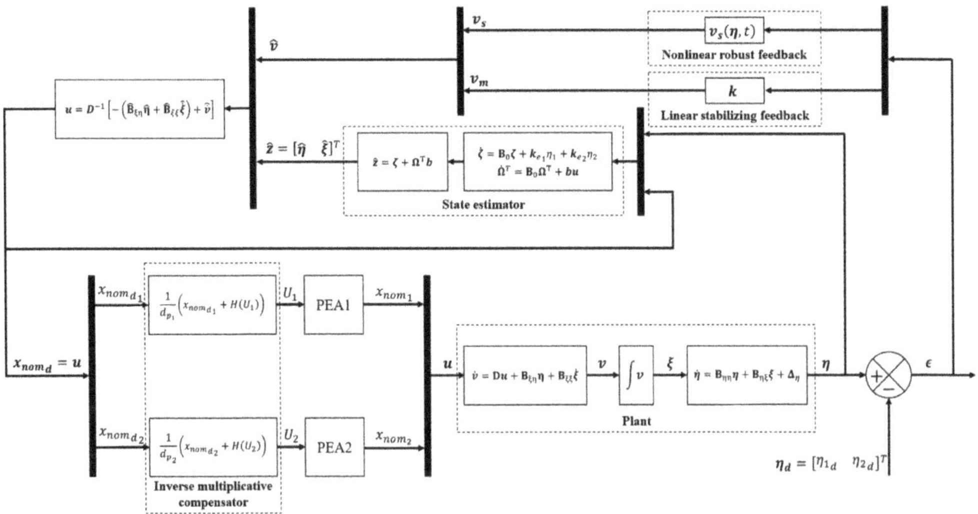

Figure 7. Schematic illustration of the total closed-loop system. The controller consists of a general state feedback controller to suppress the tracking errors and a robust controller to compensate for the nonlinearities occurring in the system.

4. Experimental Results

4.1. Experimental Setup

Experiments were carried out on a 2-DOF compliant setup, as shown in Figure 8. The compliant mechanism was manufactured using a wire-cutting method and made of aluminum alloy 6061. The PEA set output plates were made of spring steel. Initial guesses of the material property values are the same as those listed in Table 3. The adopted PEAs have a stroke of more than 60 [μm]. Other specifications are listed in Table 1. To accomplish the full cycle shown in Figure 4, the PEA sets were preloaded with a biased input voltage yielding a nominal output displacement of ±30 [μm]. The drivers of the PEAs were linear amplifiers (Type E-472.20, PI Inc.). The position sensors of the output were two capacitive distance sensors, one with a resolution of 7.5 [nm] and one with a resolution of 10 [nm]. The velocity signal was obtained from the difference between two consecutive position measurements, and the horizontal output displacement was derived by vector decomposition using the Pythagorean theorem. Note that in this setup the deformation of the joint will affect the precision of the measurement results, and so the control system is actually following a "nominal" trajectory with some minor differences to the actual desired displacement. This has little effect on a proof-of-principal experiment which mainly aims to show the efficiency of the control system. The measuring devices were connected to a data acquisition card via a noise-shielding I/O junction box with a sample time of less than 0.1 [μs]. The real-time codes of the control algorithm were explicitly implemented in computational software. The sampling period was set to 0.1[ms].

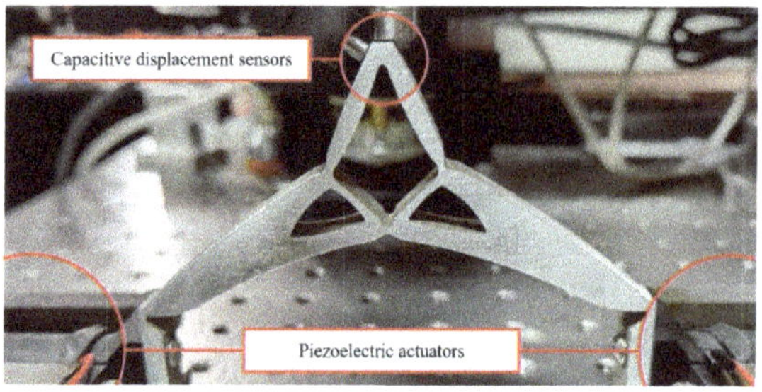

Figure 8. Image showing the experimental setup of the system. Two capacitance displacement sensors were used to measure the coupled output trajectory.

4.2. Experimental Results and Discussion

The values of the parameters of the system described by Equation (35) were estimated using an online least-squares estimator provided by the System Identification Toolbox in Simulink. The systems described by Equations (31) and (34) were discretized in time to make use of the estimators. The parameters to be evaluated, their corresponding regressors, and the reference outputs are given by the following:

$$\varphi\& = \begin{bmatrix} A_0\hat{x}_{in}(t-T_s) - q_o(t-T_s) \\ A_0[\hat{x}_{in}(t) - \hat{x}_{in}(t-2T_s)] - [q_o(t) - q_o(t-2T_s)] \\ -\hat{x}(t-T_s) \\ -[\hat{x}_{in}(t) - \hat{x}_{in}(t-2T_s)] \end{bmatrix}, \tag{54}$$

$$\theta\& = \begin{bmatrix} T_s^2 M^{-1} K \\ \frac{T_s}{2} M^{-1} C \\ T_s^2 M_{in}^{-1} K_{in} \\ \frac{T_s}{2} M_{in}^{-1} C_{in} \end{bmatrix}, \tag{55}$$

$$u_r = \begin{bmatrix} q_o(t) - 2q_o(t-T_S) + q_o(t-2T_s) \\ \hat{x}_{in}(t) - 2\hat{x}_{in}(t-T_S) + \hat{x}_{in}(t-2T_s) - u + M_{in}^{-1}K_{in}x_{in}(t-T_S) \end{bmatrix}, \tag{56}$$

where T_s represents the sampling period of the controller, and was set to 1×10^{-4} [s] in our case. Note that some of the parameters were not evaluated. The masses were weighed on an electronic scale. The amplification ratio was approximated using the maximum as shown in Equation (27). The values of parameters that only appear in the additional terms, such as K_{pea_i}s in K_{in}, were calculated algebraically from the estimated K_{in_i}s and K_{qo_1}. Initial guesses of the model parameters, as well as their final estimations, are provided in Table 4. The state estimator gain and feedback loop gain in the controller design process are also given in Table 4.

Lissajous curves with specified ratios of 3 and 2 were chosen as the desired output trajectories. Specifically, Figure 9a shows the time-domain performance of the designed system tracking a ratio-3 Lissajous curve without compensating for the nonlinearities by a robust controller, as a comparison. This is realized by aborting the robust feedback output in the block diagram of Figure 8. Figure 9b shows the performance of the complete system tracking the same desired ratio-3 Lissajous trajectory. The tracking of a ratio-2 curve is given in Figure 10a; a circular trajectory was also performed, as shown in Figure 10b. A two-dimensional plot for the space-domain trajectories is shown in Figure 11, for all four different experiments.

Table 4. Evaluated system parameters and controller design specifications.

Part	Quantity	Guess	Estimate
CM	Horizontal stiffness $k_{qo1} \times 10^6$ [N/m]	5	3.78
	Vertical stiffness $k_{qo2} \times 10^6$ [N/m]	5	4.03
	Mass M [kg]	0.046	– [a]
PZT set	Amplification ratio α	1.012	–
	PZT1 stiffness $K_{pea1} \times 10^8$ [N/m]	2.5	–
	PZT2 stiffness $K_{pea2} \times 10^8$ [N/m]	2.5	–
	Input plate stiffness $K_{in1} \times 10^5$ [N/m]	1	0.92
	Input plate stiffness $K_{in2} \times 10^5$ [N/m]	1	0.89
	Input plate mass M_{ini} [kg]	0.042	–
State Estimator	Estimator gain k_{e_1} Estimator gain k_{e_2}	1×10^5	
Controller design [b]	Error gain k_{ϵ_1}	2×10^4	
	Error gain k_{ϵ_2}	1.5×10^4	
	Feedback gain k_1	1.2×10^4	
	Feedback gain k_2	1.2×10^3	

[a] The masses in this table were measured on an electronic scale and were assumed to remain constant throughout the experiments. [b] The functionalities of the gains in the controller design can be found in Appendix C.

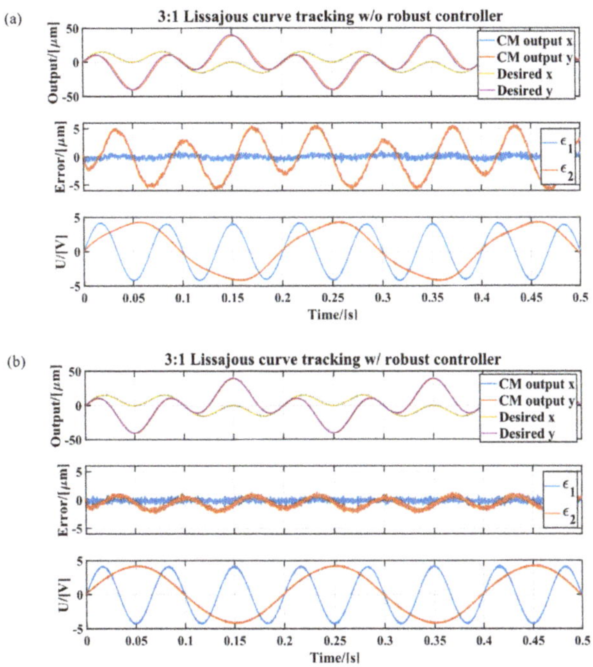

Figure 9. Experiment result showing a comparison between (**a**) Tracking of 3:1 Lissajous curve without the robust controller and (**b**) Tracking of 3:1 Lissajous curve with the robust controller.

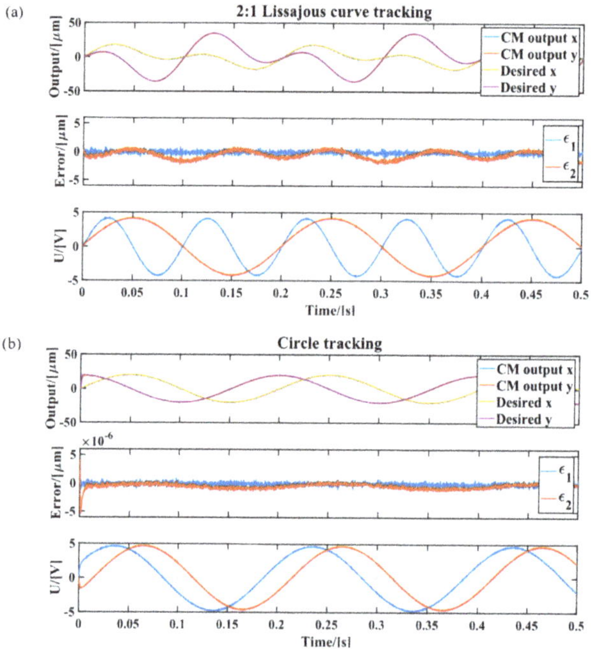

Figure 10. Experiment result showing the CM tracking (**a**) A 2:1 Lissajous curve and (**b**) A circular trajectory.

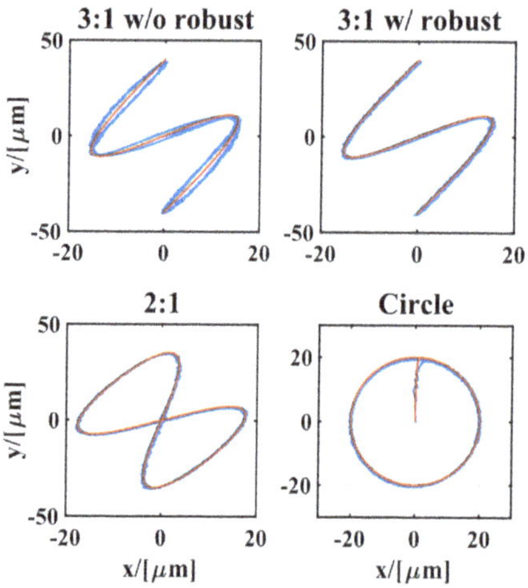

Figure 11. Two-dimensional plot of the results from Figures 9 and 10, showing the actual trajectory of the closed-loop CM.

In general, the results shown in Figures 9 and 10 indicate that despite the errors in the physical parameters, i.e., the vibration at the very beginning of the error signals, the controller achieves boundedness and guaranteed transient performance for error signals. As a comparison, as shown in Figure 9a, the non-robust controller suffers from higher noise levels and a larger overall tracking error amplitude. Further, the maximum linear feedback gain of the non-robust controller is also lower than the robust one due to uncompensated noises and disturbances, resulting in a slower response to fast variating input signals.

On the other hand, however, although the proposed method outperforms the non-robust one to some degree, the deduction for the robust controller is cumbersome and varies in accordance with different system designs. In addition, the nonlinearity in our system is suppressed by a robust controller with knowledge of only a part of the nonlinear deformation denoted by $\delta \boldsymbol{u}$. The whole modeling and controller design process is performed in the corresponding tangent space, which is a major limitation to final performance. Work on larger deformations and more profound nonlinearities remains to be conducted.

5. Conclusions

In this paper, we developed an integrated design and control scheme for a large workspace topology-optimized compliant mechanism. The scheme takes the noisy sensors, system distortions, and nonlinearities of the PEAs into consideration. The system model is based on a simplification of a spring–mass system with fully coupled inputs x_{in}. The controller described in this study was designed without deliberately trying to decouple the input–output relationship, which is usually unavailable. The proposed controller employs a backstepping procedure with a robust part and an output feedback part. The theoretical analysis was verified through experimental studies.

Future work will be conducted on controller design using full finite-element models or reduced-order modal analysis to achieve better tracking performance. The error analysis mentioned in Section 3.2.1 is also a potential area of future research. The development of an adaptive controller is also a potential research topic, since in the current study, dynamic response and convergence efficiency is limited due to ignorance of the online parameter variations.

Author Contributions: Conceptualization, B.J. and W.Z.; Data curation, Y.H.; Formal analysis, Y.H.; Funding acquisition, B.J. and W.Z.; Investigation, W.Z.; Methodology, Y.H.; Project administration, B.J. and W.Z.; Resources, B.J. and W.Z.; Software, Y.H.; Supervision, B.J. and W.Z.; Validation, Y.H. and W.Z.; Visualization, Y.H.; Writing—original draft, Y.H. and B.J.; Writing—review & editing, Y.H. and W.Z. All authors have read and agreed to the published version of the manuscript.

Funding: This work was supported by the National Natural Science Foundation of China (NSFC) (No. 52175439), National Natural Science Foundation of Zhejiang Province (No. LD22E050010), National Key R&D Program of China No. 2021YFB3400300, and the Science Fund for Creative Research Groups of the National Natural Science Foundation of China (No. 51821093).

Institutional Review Board Statement: Not applicable.

Informed Consent Statement: Not applicable.

Data Availability Statement: Not applicable.

Conflicts of Interest: The authors declare no conflict of interest. The funders had no role in the design of the study; in the collection, analyses, or interpretation of data; in the writing of the manuscript, or in the decision to publish the results.

Appendix A. Nonlinear Finite-Element Method Synthesis

The spatial discretization of the design domain, with element alignment, node connectivity, and DOF indexing, is shown in Figure A1a, which is in accordance with [44]. In general, there are two main contributors to the nonlinearity of the system design [40]. The first one is geometric nonlinearity, which appears in the form of higher-order terms $\boldsymbol{H}^T\boldsymbol{H}$ in the Green–Lagrange strain tensor, when we take large deformations in the elements into consideration:

$$E = \frac{1}{2}\left(\boldsymbol{H} + \boldsymbol{H}^T + \boldsymbol{H}^T\boldsymbol{H}\right), \tag{A1}$$

where $\boldsymbol{H} = \nabla\, \overline{\boldsymbol{u}}$, the displacement gradient when the displacement vector $\overline{\boldsymbol{u}}(\boldsymbol{X}, t)$ is introduced. A detailed iso-parametric mapping of the deformation of a finite element, Ω_e, is depicted in the lower part of Figure A1a: The unit square reference configuration, $\Omega_{\square}$, is first transformed into the initial configuration, Ω_e, with the coordinates $\boldsymbol{X}_{el}$, $l = 1,2,3,4$ by the Jacobian $\boldsymbol{J}_e$. The initial configuration, Ω_e, is then transformed into the current configuration, $\overline{\varphi}(\Omega_e)$, with coordinates $\boldsymbol{x}_{el}$, $l = 1,2,3,4$ via a deformation gradient $\boldsymbol{F}_e = \boldsymbol{H}_e + \boldsymbol{I}$, where $\boldsymbol{I}$ is the identity matrix.

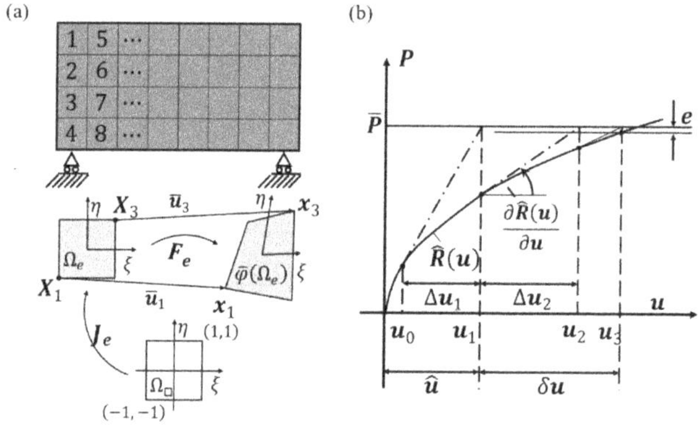

Figure A1. (**a**) Schematic illustration of the design domain discretization and elemental deformation; (**b**) Newton–Raphson method for nonlinear finite element analysis.

The second main contributor to the nonlinearity of the system is the material nonlinearity, which occurs when the relation between the stress and strain is not linear. Material nonlinearities are usually neglected in the finite-element analysis of TO. However, in our case, the use of an additive hyper-elastic element will suppress the numerical instability in the low-stiffness region and is thus an effective way to achieve global convergence [41,63]. The basic idea is to add a soft hyper-elastic material with a strain energy function in accordance with the Yeoh model to the low-density elements that are at risk of instability. The additive stiffness energy function of the eth element Ω_e is of the following form:

$$\widetilde{\Psi}_e(I_1) = \left(1 - \rho_e^p\right)\left(c_{1e}(I_1 - 3) + c_{2e}(I_1 - 3)^2\right), \tag{A2}$$

where $I_1 = \mathrm{tr}(\mathfrak{C}_e)$ is the first invariant of the right Cauchy–Green strain tensor $\mathfrak{C}_e$, ρ_e is the elemental density, p is the penalization factor used in the SIMP approach, and $c_{1e} > 0, c_{2e} > 0$ are the material constants of the additive hyper-elastic material for the eth element. In general, c_{1e} is assumed to be small to sustain a convergent result under little strain, whereas c_{2e} is relatively larger to suppress the instability at larger deformations. These coefficients are updated elementwise in each iteration of the SIMP in accordance with the following [41]:

$$\begin{aligned} c_{1e} &= \frac{\rho_{min}^p E_0}{6} \\ c_{2e}^{(k+1)} &= \begin{cases} c_{e2}^{(k)}\sqrt{\omega_e^{(k)}}, & \text{if } \eta_e^{(k)} \leq 1 \\ c_{e2}^{(k)}\left(\omega_e^{(k)}\right), & \text{if } \eta_e^{(k)} > 1 \end{cases} \\ \eta_e^{(k)} &= \frac{\varepsilon_e^{(k)}}{\varepsilon^\star}, \end{aligned} \tag{A3}$$

where the superscript (k) represents the iteration step of the optimization process, ε_e is the average von Mises strain of the eth element, and $\varepsilon^\star$ is the specified threshold.

The second Piola–Kirchhoff stress $\tilde{\mathfrak{S}}$ of the additive element is the derivative of the strain energy function with respect to the Gauss–Lagrange strain tensor $\boldsymbol{E}$:

$$\tilde{\mathfrak{S}}_e = \frac{\partial\widetilde{\Psi}_e}{\partial \boldsymbol{E}}\& = \frac{2\,\partial\widetilde{\Psi}_e}{\partial \mathfrak{C}} = 2\left(1-\rho_e^p\right)(c_{1e}+2c_{2e}(I_1-3))\frac{\partial I_1}{\partial \mathfrak{C}}. \tag{A4}$$

Notice that the associated nonlinearity of the additive hyper-elastic material only benefits the convergence procedure during the TO iterations and is not included in the actual nonlinear dynamic analysis.

Returning to the spatially discretized design domain with square elements as shown in Figure A1a, the finite-element formulations of the weak form in the initial configuration for the residuum and both boundary loads $\boldsymbol{P}_e^\sigma$ and body force $\boldsymbol{P}_e$ within a single element Ω_e in the SIMP method are obtained as follows [39]:

$$\begin{aligned} \boldsymbol{R}_e(\boldsymbol{u}_e,\rho_e,p)| &= \textstyle\int_{\Omega_e}\boldsymbol{B}(\boldsymbol{u}_e)^T\left(\boldsymbol{S}_e+\tilde{\mathfrak{S}}_e\right)\det\boldsymbol{J}_e\,\mathrm{d}\square, \\ \boldsymbol{P}_e^\sigma| &= \textstyle\int_{\Gamma_r}\boldsymbol{N}^T\bar{\boldsymbol{t}}d\Gamma, \\ \boldsymbol{P}_e| &= \textstyle\int_{\Omega_e}\rho_0\rho_e\bar{\boldsymbol{b}}\mathrm{d}\Omega, \end{aligned} \tag{A5}$$

where $\boldsymbol{u}_e : (X,t)\rightarrow\mathbb{R}^N$ denotes the displacement vector from the original configuration $\boldsymbol{X}_e$ to the current configuration $\boldsymbol{x}_e$. Further, $\boldsymbol{B}$ denotes the strain–displacement matrix, whereas $\boldsymbol{S}_e=\rho_e\boldsymbol{D}\boldsymbol{E}$ is the Piola–Kirchhoff stress of the original elastic material in the SIMP, and $\boldsymbol{N}$ is the shape function used in the evaluation of the deformations. All variables and matrices are evaluated within element Ω_e and are related to the initial configuration. The approximated integration is carried out with Gauss integration due to its efficiency.

The assembly operators for the residual forces and mass are the same and denoted by $\cup_e\bullet$, where $\{e\}_{e\in\mathbb{N}^+}$ is the index set for the interior elements. The boundary loads are assembly operators acting on $\Gamma_r\subseteq\partial\Omega$, denoted by $\cup_r\bullet$, where $\{r\}_{r\in\{e\}}$ is the index set for the boundary elements to be considered. The global matrices are expressed as shown below with the assistance of the assembly operators:

$$\begin{aligned} [\boldsymbol{R}(\boldsymbol{u})]^{2N\times 1} &= \bigcup_{e=1}^{n_e}[\boldsymbol{R}_e(\boldsymbol{u}_e)]^{8\times 1} \\ [\boldsymbol{P}]^{2N\times 1} &= \bigcup_{e=r}^{n_r}[\boldsymbol{P}_r]^{8\times 1}, \end{aligned} \tag{A6}$$

where N is the total number of nodes in the discretized design domain. Therefore, $2N$ is the total number of DOF in the two-dimensional setup for all of the nodes, and n_e and n_r are the number of total elements and number of boundary elements, respectively. A detailed assembly algorithm was followed, as presented in previous research [40,64].

The above derivations lead to a compact nonlinear system of ordinary differential equations of the following form:

$$\boldsymbol{R}(\boldsymbol{u})-\boldsymbol{P}=0. \tag{A7}$$

We usually denote by $\boldsymbol{K}_T=\left.\frac{\partial\boldsymbol{R}}{\partial\boldsymbol{u}}\right|_{\boldsymbol{u}}=\cup_{e=1}^{n_e}\boldsymbol{K}_{eT}$ the assembly of the elementary tangent stiffness matrix $\boldsymbol{K}_{eT}=\left.\frac{\partial\boldsymbol{R}_e(\boldsymbol{u}_e)}{\partial\boldsymbol{u}_e}\right|_{\boldsymbol{u}_e}$ at the state $\boldsymbol{u}(\boldsymbol{u}_e)$, as shown in Figure A1b. The approximated solution $\boldsymbol{u}$ of the nonlinear dynamic system in Equation (A7) is obtained via a modified Newton–Raphson method.

Appendix B. Identifying the Hysteresis Model of the PEAs

Various previous studies have reported on the identification procedure for the PEA coefficients [2,48,56]. In this study, we used recursive least-squares estimation to estimate the model parameters. We rewrote Equation (18) in the form of a state-space representation. Suppose that $\hat{x}_{nom}$ represents the predicted output vector of the model. By selecting the parameter vector $\boldsymbol{\theta}=\begin{bmatrix}\alpha & B_{bw} & \Gamma_{bw} & \delta_{bw}\end{bmatrix}^T$, where $\alpha=d_p-A_{bw}$ is a combined intermediate parameter, and the state variables $\boldsymbol{\varphi} = \begin{bmatrix} \dot{U} & |\dot{U}|h & \dot{U}|h| & -\text{Usgn}\left(\dot{U}\right) \end{bmatrix}^T$, a time derivative of the first formula in Equation (18) yields the following:

$$\hat{\dot{x}}_{nom}(t, \boldsymbol{\theta}) = \boldsymbol{\varphi}(t)^T \boldsymbol{\theta}. \tag{A8}$$

A covariance matrix for the recursive least square estimation process can be given as follows [27,64]:

$$\boldsymbol{Cov}(t) = \left[\sum_{i=1}^{t} \boldsymbol{\varphi}(i) \boldsymbol{W}(t,i) \boldsymbol{\varphi}(i)^T \right]^{-1} \in \mathbb{R}^{p \times p}. \tag{A9}$$

where $\boldsymbol{W}(t,i)$ represents the weighting matrices, in the form of:

$$\boldsymbol{W}(t,n) = \prod_{m=n+1}^{t} \lambda_f(m) \boldsymbol{W}_0. \tag{A10}$$

The $\lambda_f(k)$ in the above expression is the forgetting factor, applied to improve the real-time performance against system disturbances, and $\boldsymbol{W}_0 \in \mathbb{R}^{m \times m}$ is an arbitrary constant weighting matrix. We obtain the following recursive formula for least-squares estimation based on the parameter estimate $\hat{\boldsymbol{\theta}}(t-1)$ and covariance matrix $\boldsymbol{Cov}(t-1)$ obtained in the previous step:

$$\begin{aligned} &\boldsymbol{\Gamma}(t) = \tfrac{1}{\lambda_f(t)} \boldsymbol{Cov}(t-1) \\ &\boldsymbol{Cov}(t) = \boldsymbol{\Gamma}(t) \{ I_p - \varphi(t) [\boldsymbol{W}_0^{-1} + \varphi(t) \boldsymbol{\Gamma}(t) \varphi(t)]^{-1} \varphi(t)^T \boldsymbol{\Gamma}(t) \} \\ &\epsilon^0(t) := \hat{\dot{x}}_{nom}(t) - \boldsymbol{\varphi}(t)^T \hat{\boldsymbol{\theta}}(t-1) \\ &\hat{\boldsymbol{\theta}}(t) = \hat{\boldsymbol{\theta}}(t-1) + \boldsymbol{Cov}(t) \boldsymbol{\varphi}(t) W_0 \epsilon^0(t). \end{aligned} \tag{A11}$$

The intermediate adaptation rate matrix $\boldsymbol{\Gamma}(\mathrm{t})$ is introduced for notational simplicity and efficiency of computation. ϵ^0 is normally called the *a priori* prediction error. The d_p in the combined intermediate variable α is estimated by a simultaneous process with a structure identical to the one given in Equation (A11), where $\varphi(t) = U(t)$ and $h(t)$ is updated via a forward Euler algorithm, in accordance with the first formula in Equation (18). Choose another set of parameter vector $\boldsymbol{\varphi}_{nom} = \begin{bmatrix} d_p & -1 \end{bmatrix}^T$ and state variable vector $\boldsymbol{\theta}_{nom} = \begin{bmatrix} U & h \end{bmatrix}$, the parallel RLSE process can be formulated as follows:

$$\hat{x}_{nom}(t, \boldsymbol{\theta}) = \boldsymbol{\varphi}_{nom}(t)^T \boldsymbol{\theta}_{nom}. \tag{A12}$$

Thus, by fixing the constant -1 in the $\boldsymbol{\varphi}_{nom}$, the value of d_p can be estimated via an integrated process based on U and h. Meanwhile, the value of the parameter $A_{bw} = d_p - \alpha$ can also be calculated at every time step of the identification procedure. Details of the algorithm are given in Algorithm A1, and the estimation results are given in Figure A2 and Table A1.

Algorithm A1 Parallel recursive least square algorithm for online parameter estimation.
Parallel RLSE Algorithm

Initialization:
Make an initial guess of the coefficients, compute corresponding initial values of θ and φ
Main loop:
While PEA is working do
Do process (A11) for regressor Equation (A8)
Update state variable h
Do process (A11) for regressor Equation (A12)
Update d_p and A_{bw}
End while

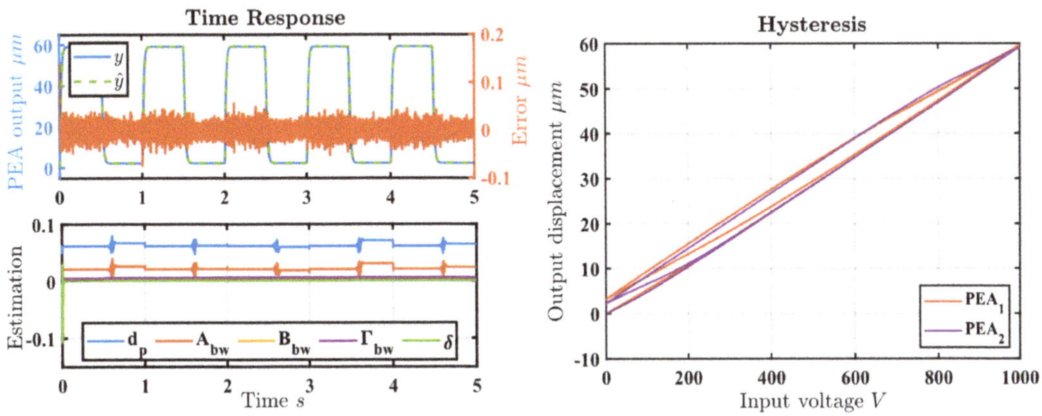

Figure A2. Parameter identification process for one of the PEAs and the corresponding hysteresis curve for both.

Table A1. Estimation results of characteristic coefficients of the PEAs.

Symbol	Initial Guess	Estimates	
		PEA1	PEA2
$\mathbf{d_p}$	0	$\approx 6.2 \times 10^{-2}$	$\approx 6.5 \times 10^{-2}$
$\mathbf{A_{bw}}$	0	$\approx 1.2 \times 10^{-2}$	$\approx 2.5 \times 10^{-2}$
$\mathbf{B_{bw}}$	0	1.9×10^{-3}	1.8×10^{-3}
$\mathbf{\Gamma_{bw}}$	0	1.6×10^{-3}	5.3×10^{-3}
δ	0	5.015×10^{-4}	1.973×10^{-4}

Appendix C. The State-Space Representation of the System Dynamics

Following the steps given in Section 3.2.1, the elementary form of the total system dynamics can be given as follows:

$$
\begin{aligned}
\dot{z}_1 &= z_5 \\
\dot{z}_2 &= z_6 \\
\dot{z}_3 &= z_7 \\
\dot{z}_4 &= z_8 \\
\dot{z}_5 &= \sum_{i=1}^{8} \beta_{5i} z_i \\
\dot{z}_6 &= \sum_{i=1}^{8} \beta_{6i} z_i \\
\dot{z}_7 &= d_{11} u_1 + d_{12} u_2 + \sum_{i=1}^{8} \beta_{7i} z_i \\
\dot{z}_8 &= d_{21} u_2 + d_{22} u_2 + \sum_{i=1}^{8} \beta_{8i} z_i
\end{aligned}
\quad . \qquad \text{(A13)}
$$

This linear differential equation system can be modified into the state-space form as Equation (36), whose coefficient matrices $B = [\beta_{ij}]^{8\times 8}$ and $D = [d_{ij}]^{2\times 2}$ are detailed below:

$$\begin{gathered}
\begin{bmatrix}\beta_{51} & \beta_{52}\\ \beta_{61} & \beta_{62}\end{bmatrix} = \begin{bmatrix}-\frac{2K_{qo1}}{M} & 0\\ 0 & -\frac{K_{qo2}}{M}\end{bmatrix}, \begin{bmatrix}\beta_{53} & \beta_{54}\\ \beta_{63} & \beta_{64}\end{bmatrix} = \begin{bmatrix}\frac{K_{qo1}}{M} & \frac{K_{qo1}}{M}\\ \frac{\alpha K_{qo2}}{M} & -\frac{\alpha K_{qo2}}{M}\end{bmatrix},\\
\begin{bmatrix}\beta_{55} & \beta_{56}\\ \beta_{65} & \beta_{66}\end{bmatrix} = \begin{bmatrix}-\frac{2C_{qo1}}{M} & 0\\ 0 & -\frac{C_{qo2}}{M}\end{bmatrix}, \begin{bmatrix}\beta_{57} & \beta_{58}\\ \beta_{67} & \beta_{68}\end{bmatrix} = \begin{bmatrix}\frac{C_{qo1}}{M} & \frac{C_{qo1}}{M}\\ \frac{\alpha C_{qo2}}{M} & -\frac{\alpha C_{qo2}}{M}\end{bmatrix},\\
\begin{bmatrix}\beta_{71} & \beta_{72}\\ \beta_{81} & \beta_{82}\end{bmatrix} = \begin{bmatrix}\frac{K_{qo1}}{M_{in1}} & 0\\ \frac{K_{qo1}}{M_{in2}} & 0\end{bmatrix},\\
\begin{bmatrix}\beta_{73} & \beta_{74}\\ \beta_{83} & \beta_{84}\end{bmatrix} = \begin{bmatrix}-\frac{K_{in1}+K_{pea1}+K_{qo1}}{M_{in1}} & 0\\ 0 & -\frac{K_{in2}+K_{pea2}+K_{qo1}}{M_{in2}}\end{bmatrix},\\
\begin{bmatrix}\beta_{75} & \beta_{76}\\ \beta_{85} & \beta_{86}\end{bmatrix} = \begin{bmatrix}\frac{C_{qo1}}{M_{in1}} & 0\\ \frac{C_{qo1}}{M_{in2}} & 0\end{bmatrix}, \begin{bmatrix}\beta_{77} & \beta_{78}\\ \beta_{87} & \beta_{88}\end{bmatrix} = \begin{bmatrix}-\frac{C_{in1}+C_{qo1}}{M_{in1}} & 0\\ 0 & -\frac{C_{in2}+C_{qo1}}{M_{in2}}\end{bmatrix},\\
\begin{bmatrix}d_{11} & d_{12}\\ d_{21} & d_{22}\end{bmatrix} = \begin{bmatrix}\frac{K_{pea1}}{M_{in1}} & 0\\ 0 & \frac{K_{pea2}}{M_{in2}}\end{bmatrix}.
\end{gathered} \tag{A14}$$

Appendix D. Deduction of the Backstepping Controller

Recall that $\boldsymbol{a}(\eta) = [a_1, a_2]^T$ represents the ideal control law for the inputs $[x_{in1}, x_{in2}]^T$ according to Equation (A13), the notation that $\boldsymbol{z} = \begin{bmatrix} q_o & \dot{q}_o & x_{in} & \dot{x}_{in}\end{bmatrix}^T$, and the subsystem Equation (43a) in terms of $\boldsymbol{a}$ is given as follows:

$$\begin{aligned}
\dot{\eta}_1 &= \eta_5 := a_1\\
\dot{\eta}_2 &= \eta_6 := a_2\\
\dot{\eta}_3 &= \dot{\xi}_1\\
\dot{\eta}_4 &= \dot{\xi}_2\\
\dot{\eta}_5 &= \sum_i^6 \beta_{5i}\eta_i + \beta_{57}\dot{\xi}_1 + \beta_{58}\dot{\xi}_2 + \Delta_{\eta 1}\\
\dot{\eta}_6 &= \sum_i^6 \beta_{6i}\eta_i + \beta_{67}\dot{\xi}_1 + \beta_{68}\dot{\xi}_2 + \Delta_{\eta 2}
\end{aligned}. \tag{A15}$$

When denoting $\epsilon_i = \eta_i - \eta_{di} = z_i - z_{di}$, $i = 1,2$, the tracking error between the output node $\eta_i = z_i$, and the desired trajectory $\eta_{di} = z_{di}$, a selection of the feedback error $\boldsymbol{s} = \dot{\epsilon} + \boldsymbol{k}_\epsilon \boldsymbol{\epsilon}$ yields the total error dynamic:

$$\dot{s}_1 = -\frac{2K_{qo1}}{M}\eta_1 + \frac{K_{qo1}}{M}a_1 + \frac{K_{qo1}}{M}a_2 + \beta_{57}\dot{\xi}_1 + \beta_{58}\dot{\xi}_2 - \ddot{\eta}_{d1} + k_{\epsilon 1}(\eta_5 - \dot{\eta}_{d1}) + \Delta_{\eta 1}, \tag{A16}$$

$$\dot{s}_2 = -\frac{K_{qo2}}{M}\eta_2 + \alpha\frac{K_{qo2}}{M}a_1 - \alpha\frac{K_{qo2}}{M}a_2 + \beta_{67}\dot{\xi}_1 + \beta_{68}\dot{\xi}_2 - \ddot{\eta}_{d2} + k_{\epsilon 2}(\eta_6 - \dot{\eta}_{d2}) + \Delta_{\eta 2}. \tag{A17}$$

Consider the control Lyapunov function (CLF) Equations (47) and (48):

$$\boldsymbol{V}(\boldsymbol{\eta}) = \frac{1}{2}\begin{bmatrix} s_1^2\\ s_2^2\end{bmatrix}. \tag{A18}$$

The path derivative of the CLF with respect to the solution $\eta(t)$ is as follows:

$$\dot{V}(\eta) = \begin{bmatrix} s_1\dot{s}_1 \\ s_2\dot{s}_2 \end{bmatrix}. \tag{A19}$$

Next, we choose a specific $\boldsymbol{a}$ such that $\dot{V}(\eta) < 0$, satisfying the Lyapunov stability criterion. Substituting Equations (A16) and (A17) into Equation (A19) and considering the inequalities, we have

$$\begin{cases} s_1\left(\frac{K_{qo1}}{M}(a_1 + a_2 - 2\eta_1) - \ddot{\eta}_{d1} + k_{\epsilon 1}(\eta_5 - \dot{\eta}_{d1}) + (\beta_{57}\hat{\dot{\xi}}_1 + \beta_{58}\hat{\dot{\xi}}_2) + \Delta_{\eta 1}\right) < 0 \\ s_2\left(\frac{\alpha K_{qo2}}{M}(a_1 - a_2 - \frac{\eta_2}{\alpha}) - \ddot{\eta}_{d2} + k_{\epsilon 2}(\eta_6 - \dot{\eta}_{d2}) + (\beta_{67}\hat{\dot{\xi}}_1 + \beta_{68}\hat{\dot{\xi}}_2) + \Delta_{\eta 2}\right) < 0 \end{cases}. \tag{A20}$$

The desired controller output $\boldsymbol{a}$ as a combination of $\boldsymbol{a}_m$ and $\boldsymbol{a}_s$ from Equation (49) satisfies the following:

$$\begin{cases} \frac{K_{qo1}}{M}(a_{m1} + a_{m2}) - Z_5\left(\eta_1, \dot{\eta}_1, \hat{\dot{\xi}}_1\right) = -(k_{\eta 1} - 1)s_1 \\ \frac{\alpha K_{qo2}}{M}(a_{m1} - a_{m2}) - Z_6\left(\eta_1, \dot{\eta}_1, \hat{\dot{\xi}}_1\right) = -(k_{\eta 2} - 1)s_2 \end{cases}, \tag{A21}$$

$$\begin{cases} \frac{K_{qo1}}{M}(a_{m1} + a_{m2}) + \Delta_{\eta 1} < \frac{K_{qo1}}{M}(a_{m1} + a_{m2}) + \max|\Delta_{\eta 1}| = -s_1 \\ \frac{\alpha K_{qo2}}{M}(a_{m1} - a_{m2}) + \Delta_{\eta 2} < \frac{\alpha K_{qo2}}{M}(a_{m1} - a_{m2}) + \max|\Delta_{\eta 2}| = -s_2 \end{cases}, \tag{A22}$$

where $k_{\eta i} > 1, i = 1,2$ are the linear feedback gains and can be chosen freely, and

$$\begin{aligned} Z_5\left(\eta_1, \dot{\eta}_1, \dot{\xi}_1\right) &= \frac{2K_{qo1}}{M}\eta_1 + (\ddot{\eta}_{d1} - k_{\epsilon 1}(\eta_5 - \dot{\eta}_{d1})) - \left(\beta_{57}\hat{\dot{\xi}}_1 + \beta_{58}\hat{\dot{\xi}}_2\right), \\ Z_6\left(\eta_2, \dot{\eta}_2, \dot{\xi}_2\right)\& &= \frac{K_{qo2}}{M}\eta_2 + (\ddot{\eta}_{d2} - k_{\epsilon 2}(\eta_6 - \dot{\eta}_{d2})) - \left(\beta_{67}\hat{\dot{\xi}}_1 + \beta_{68}\hat{\dot{\xi}}_2\right). \end{aligned} \tag{A23}$$

By solving Equations (A21) and (A22), we get the following expressions for the ideal control law $\boldsymbol{a}$:

$$\boldsymbol{a}_m = \begin{bmatrix} K_{qo1}/M & K_{qo1}/M \\ \alpha K_{qo2}/M & -\alpha K_{qo2}/M \end{bmatrix}^{-1} \left(\begin{bmatrix} -(k_{\eta 1} - 1)s_1 + Z_5 \\ -(k_{\eta 2} - 1)s_2 + Z_6 \end{bmatrix}\right), \tag{A24}$$

$$\boldsymbol{a}_s = \begin{bmatrix} K_{qo1}/M & K_{qo1}/M \\ \alpha K_{qo2}/M & -\alpha K_{qo2}/M \end{bmatrix}^{-1} \begin{bmatrix} -\max|\Delta_{\eta 1}| - s_1 \\ -\max|\Delta_{\eta 2}| - s_2 \end{bmatrix}. \tag{A25}$$

Let ϵ_a be the deviation from Equation (50):

$$\boldsymbol{\epsilon}_a = \boldsymbol{\xi} - \boldsymbol{a}. \tag{A26}$$

The total system in the new error coordinates $\begin{bmatrix} \boldsymbol{s} & \boldsymbol{\epsilon}_a \end{bmatrix}^T$ is then given by:

$$\begin{aligned} \dot{\boldsymbol{s}} &= -\begin{bmatrix} k_{\eta 1}s_1 \\ k_{\eta 2}s_2 \end{bmatrix} \\ \dot{\boldsymbol{\epsilon}}_a &= \boldsymbol{v} - \begin{bmatrix} K_{qo1}/M & K_{qo1}/M \\ \alpha K_{qo2}/M & -\alpha K_{qo2}/M \end{bmatrix}^{-1} \left(\begin{bmatrix} k_{\eta 1}(k_{\eta 1} - 1)s_1 + \frac{\partial Z_5(\eta)}{\partial \eta} \\ k_{\eta 2}(k_{\eta 2} - 1)s_2 + \frac{\partial Z_6(\eta)}{\partial \eta} \end{bmatrix}\right) + \boldsymbol{\Delta}_{\epsilon}, \end{aligned} \tag{A27}$$

where the error term Δ_ϵ is derived from Equation (43b), such that $\boldsymbol{\Delta}_\epsilon = \boldsymbol{\Delta}_{\xi}$.

The CLF for the whole system in accordance with Equation (51) can be given as follows:

$$V_\alpha(\eta,\xi) = V(\eta) + \frac{1}{2}\begin{bmatrix}\epsilon_{\alpha_1}^2 \\ \epsilon_{\alpha_2}^2\end{bmatrix}. \tag{A28}$$

The path derivative of the above Equation (A28) along the solutions of Equation (43a) can be calculated by substituting the expression of $\dot{s}$ in Equation (A27) into Equation (52), in the following form:

$$\dot{V}_\alpha = \begin{bmatrix}-k_{\eta 1}s_1^2 \\ -k_{\eta 2}s_2^2\end{bmatrix} + \begin{bmatrix}\epsilon_{\alpha 1}\dot{\epsilon}_{\alpha 1} \\ \epsilon_{\alpha 2}\dot{\epsilon}_{\alpha 2}\end{bmatrix}. \tag{A29}$$

The term $\epsilon_\alpha\dot{\epsilon}_\alpha$ can further be written as a linear combination of input v. Denoting $\begin{bmatrix}\theta_1 & \theta_2\end{bmatrix} = \begin{bmatrix}\frac{K_{q01}}{M} & \frac{K_{q02}}{M}\end{bmatrix}$, we have the following total system formulation for Equation (A29):

$$\begin{aligned}\dot{V}_{\alpha 1} &= -k_{\eta 1}s_1^2 + \left(\hat{\dot{\xi}}_1 + \Pi_{11}\right)(v_1 - \Pi_{12}) + \epsilon_{\alpha 1}\Delta_{\epsilon 1} \\ \dot{V}_{\alpha 2} &= -k_{\eta 2}s_2^2 + \left(\hat{\dot{\xi}}_2 + \Pi_{21}\right)(v_2 - \Pi_{22}) + \epsilon_{\alpha 1}\Delta_{\epsilon 2}\end{aligned}, \tag{A30}$$

where

$$\begin{aligned}\Pi_{11} &= \frac{\left(\max|\Delta_{\eta 2}| - Z_6 + k_{\eta 2}s_2\right)\theta_1 + \alpha\left(\max|\Delta_{\eta 1}| - Z_5 + k_{\eta 1}s_1\right)\theta_2}{2\alpha\theta_1\theta_2} \\ \Pi_{12} &= \frac{\left(k_{\eta 2}(k_{\eta 2}-1)s_2 + \frac{\partial Z_6(\eta)}{\partial \eta}\right)\theta_1 + \alpha\left(s_1 k_{\eta 1}(k_{\eta 1}-1) + \frac{\partial Z_5(\eta)}{\partial \eta}\right)\theta_2}{2\alpha\theta_1\theta_2}\end{aligned}, \tag{A31}$$

$$\begin{aligned}\Pi_{21} &= -\frac{\left(\max|\Delta_{\eta 2}| - Z_6 + k_{\eta 2}s_2\right)\theta_1 - \alpha\left(\max|\Delta_{\eta 1}| - Z_5\theta_2 + k_{\eta 1}s_1\right)\theta_2}{2\alpha\theta_1\theta_2} \\ \Pi_{22} &= \frac{\left(k_{\eta 2}(k_{\eta 2}-1)s_2 + \frac{\partial Z_6(\eta)}{\partial \eta}\right)\theta_1 - \alpha\left(k_{\eta 1}(k_{\eta 1}-1)s_1 + \frac{\partial Z_5(\eta)}{\partial \eta}\right)\theta_2}{2\alpha\theta_1\theta_2}\end{aligned}. \tag{A32}$$

The task is to choose inputs v_is such that $\dot{V}_{\alpha i}$, $i = 1,2$ are negative definite, in which case the whole system is Lyapunov stable. Suppose that our controller input is of the following form:

$$v_m + v_s = \begin{bmatrix}v_{m1} + v_{s1} \\ v_{m2} + v_{s2}\end{bmatrix}, \tag{A33}$$

and notice that $-k_{\eta i}s_i^2 \leq 0$ always holds. The values of v_{mi}s are then chosen such that:

$$\begin{cases}\left(\hat{\dot{\xi}}_1 + \Pi_{11}\right)(v_{m1} - \Pi_{12}) = -k_1\epsilon_{a1} \\ \left(\hat{\dot{\xi}}_2 + \Pi_{21}\right)(v_{m2} - \Pi_{22}) = -k_2\epsilon_{a2}\end{cases}, \tag{A34}$$

where $k = \begin{bmatrix}k_1 & k_2\end{bmatrix} > 0$ is the feedback gain, thus yielding:

$$\begin{cases}v_{m1} = -\frac{k_1\epsilon_{a1}}{\hat{\dot{\xi}}_1 + \Pi_{11}} + \Pi_{12} \\ v_{m2} = -\frac{k_2\epsilon_{a2}}{\hat{\dot{\xi}}_2 + \Pi_{21}} + \Pi_{22}\end{cases}. \tag{A35}$$

The system nonlinearity is overcome by the robust controller v_{si}s, satisfying the following relationships:

$$\begin{cases}\left(\hat{\dot{\xi}}_1 + \Pi_{11}\right)v_{s1} + \epsilon_{\alpha 1}\Delta_{\epsilon 1} < (\hat{\dot{\xi}}_1 + \Pi_{11})v_{s1} + \max|\epsilon_{\alpha 1}\Delta_{\epsilon 1}| = -\epsilon_{\alpha 1} \\ \left(\hat{\dot{\xi}}_2 + \Pi_{21}\right)v_{s2} + \epsilon_{\alpha 2}\Delta_{\epsilon 2} < (\hat{\dot{\xi}}_2 + \Pi_{21})v_{s2} + \max|\epsilon_{\alpha 2}\Delta_{\epsilon 2}| = -\epsilon_{\alpha 2}\end{cases}, \tag{A36}$$

and the expression of the v_{s_i}s can be solved as follows:

$$\begin{aligned} v_{s1} &= -\frac{\max|\epsilon_{\alpha 1}\Delta_{\epsilon 1}| + \epsilon_{\alpha 1}}{\dot{\hat{\xi}}_1 + \Pi_{11}} \\ v_{s2} &= -\frac{\max|\epsilon_{\alpha 2}\Delta_{\epsilon 2}| + \epsilon_{\alpha 2}}{\dot{\hat{\xi}}_2 + \Pi_{21}} \end{aligned}. \tag{A37}$$

We can also choose identical feedback gains such that $k_i = k_{mi} = k$. The final design of the controller input, Equation (A33), as a combination of Equations (A35) and (A37), is simplified to:

$$\begin{aligned} v_1 &= -\frac{k_1\epsilon_{a1}}{\dot{\hat{\xi}}_1 + \Pi_{11}} - \frac{\max|\epsilon_{\alpha 1}\Delta_{\epsilon 1}| + \epsilon_{\alpha 1}}{\dot{\hat{\xi}}_1 + \Pi_{11}} + \Pi_{12} \\ v_2 &= -\frac{k_2\epsilon_{a2}}{\dot{\hat{\xi}}_2 + \Pi_{21}} - \frac{\max|\epsilon_{\alpha 2}\Delta_{\epsilon 2}| + \epsilon_{\alpha 2}}{\dot{\hat{\xi}}_2 + \Pi_{21}} + \Pi_{22} \end{aligned}. \tag{A38}$$

Also, all the coefficients are calculated with respect to estimation of the state varia.

References

1. Howell, L.L. Compliant Mechanisms. In *21st Century Kinematics*; Springer: London, UK, 2013; pp. 189–216.
2. Gu, G.Y.; Zhu, L.M.; Su, C.Y.; Ding, H.; Fatikow, S. Modeling and control of piezo-actuated nanopositioning stages: A survey. *IEEE Trans. Autom. Sci. Eng.* **2016**, *13*, 313–332. [CrossRef]
3. Solehuddin, S.; Ridzwan, M.I.Z.; Kadarman, A.H. Methodology of Compliant Mechanisms and its Current Developments in Applications: A Review. *Am. J. Appl. Sci.* **2007**, *4*, 160–167. [CrossRef]
4. Wildman, R.; Gaynor, A. 11—Topology optimization for robotics applications. In *Robotic Systems and Autonomous Platforms*; Walsh, S.M., Strano, M.S., Eds.; Woodhead Publishing: Sawston, UK, 2019; pp. 251–292.
5. Wen, S.; Xu, Q. Design of a Novel Piezoelectric Energy Harvester Based on Integrated Multistage Force Amplification Frame. *IEEE/ASME Trans. Mechatron.* **2019**, *24*, 1228–1237. [CrossRef]
6. Abdelnaby, M.A.; Arafa, M. Energy harvesting using a flextensional compliant mechanism. *J. Intell. Mater. Syst. Struct.* **2016**, *27*, 2707–2718. [CrossRef]
7. Choi, K.-B.; Lee, J.J.; Hata, S. A piezo-driven compliant stage with double mechanical amplification mechanisms arranged in parallel. *Sens. Actuators A Phys.* **2010**, *161*, 173–181. [CrossRef]
8. Zhu, Z.; To, S.; Zhu, W.; Li, Y.; Huang, P. Optimum Design of a Piezo-Actuated Triaxial Compliant Mechanism for Nanocutting. *IEEE Trans. Ind. Electron.* **2018**, *65*, 6362–6371. [CrossRef]
9. Zhu, W.-L.; Zhu, Z.; Shi, Y.; Wang, X.; Guan, K.; Ju, B.-F. Design, modeling, analysis and testing of a novel piezo-actuated XY compliant mechanism for large workspace nano-positioning. *Smart Mater. Struct.* **2016**, *25*, 115033. [CrossRef]
10. Li, L.; Chew, Z.J. 11—Microactuators: Design and technology. In *Smart Sensors and Mems*; Nihtianov, S., Luque, A., Eds.; Woodhead Publishing: Sawston, UK, 2014; pp. 305–348.
11. Lourdes Thomas, T.; Kalpathy Venkiteswaran, V.; Ananthasuresh, G.K.; Misra, S. Surgical Applications of Compliant Mechanisms: A Review. *J. Mech. Robot.* **2021**, *13*, 020801. [CrossRef]
12. Sigmund, O. On the Design of Compliant Mechanisms Using Topology Optimization*. *Mech. Struct. Mach.* **1997**, *25*, 493–524. [CrossRef]
13. Lazarov, B.S.; Schevenels, M.; Sigmund, O. Robust design of large-displacement compliant mechanisms. *Mech. Sci.* **2011**, *2*, 175–182. [CrossRef]
14. Sigmund, O. Manufacturing tolerant topology optimization. *Acta Mech. Sin.* **2009**, *25*, 227–239. [CrossRef]
15. Guo Zhan, L.; Tat Joo, T.; Guilin, Y.; Song Huat, Y.; Sitti, M. A hybrid topological and structural optimization method to design a 3-DOF planar motion compliant mechanism. In Proceedings of the 2013 IEEE/ASME International Conference on Advanced Intelligent Mechatronics, Wollongong, NSW, Australia, 9–12 July 2013; pp. 247–254.
16. Liu, C.-H.; Huang, G.-F.; Chen, T.-L. An Evolutionary Soft-Add Topology Optimization Method for Synthesis of Compliant Mechanisms With Maximum Output Displacement. *J. Mech. Robot.* **2017**, *9*, 054502. [CrossRef]
17. Bendsoe, M.P.; Sigmund, O. *Topology Optimization: Theory, Methods, and Applications*; Springer Science & Business Media: Berlin/Heidelberg, Germany, 2013.
18. Zhu, B.; Zhang, X.; Zhang, H.; Liang, J.; Zang, H.; Li, H.; Wang, R. Design of compliant mechanisms using continuum topology optimization: A review. *Mech. Mach. Theory* **2020**, *143*, 103622. [CrossRef]
19. Lobontiu, N.; Garcia, E. Analytical model of displacement amplification and stiffness optimization for a class of flexure-based compliant mechanisms. *Comput. Struct.* **2003**, *81*, 2797–2810. [CrossRef]
20. Li, Z.; Kota, S. Dynamic Analysis of Compliant Mechanisms. In Proceedings of the 27th Biennial Mechanisms and Robotics Conference, Montreal, QC, Canada, 29 September–2 October 2002; Volume 5, pp. 43–50.
21. Li, N.; Su, H.-J.; Zhang, X.-P. Accuracy Assessment of Pseudo-Rigid-Body Model for Dynamic Analysis of Compliant Mechanisms. *J. Mech. Robot.* **2017**, *9*, 054503. [CrossRef]

22. Wang, W.; Yu, Y. New Approach to the Dynamic Modeling of Compliant Mechanisms. *J. Mech. Robot.* **2010**, *2*, 021003. [CrossRef]
23. Sigmund, O. Design of multiphysics actuators using topology optimization–Part I: One-material structures. *J Comput. Methods Appl. Mech. Eng.* **2001**, *190*, 6577–6604. [CrossRef]
24. Sigmund, O. Design of multiphysics actuators using topology optimization–Part II: Two-material structures. *J Comput. Methods Appl. Mech. Eng.* **2001**, *190*, 6605–6627. [CrossRef]
25. Saxena, A. Topology design of large displacement compliant mechanisms with multiple materials and multiple output ports. *Struct. Multidiscip. Optim.* **2005**, *30*, 477–490. [CrossRef]
26. Zhao, K.; Schmiedeler, J.P. Using Rigid-Body Mechanism Topologies to Design Path Generating Compliant Mechanisms. *J. Mech. Robot.* **2015**, *8*, 014506. [CrossRef]
27. Krstic, M.; Kokotovic, P.V.; Kanellakopoulos, I. *Nonlinear and Adaptive Control Design*; John Wiley & Sons, Inc.: Hoboken, NJ, USA, 1995.
28. Bendsøe, M.P. Optimal shape design as a material distribution problem. *Struct. Optim.* **1989**, *1*, 193–202. [CrossRef]
29. Sigmund, O.; Maute, K. Topology optimization approaches. *Struct. Multidiscip. Optim.* **2013**, *48*, 1031–1055. [CrossRef]
30. Cheng, T.; He, M.; Li, H.; Lu, X.; Zhao, H.; Gao, H. A Novel Trapezoid-Type Stick–Slip Piezoelectric Linear Actuator Using Right Circular Flexure Hinge Mechanism. *IEEE Trans. Ind. Electron.* **2017**, *64*, 5545–5552. [CrossRef]
31. Gu, G.-Y.; Zhu, L.-M.; Su, C.-Y. High-precision control of piezoelectric nanopositioning stages using hysteresis compensator and disturbance observer. *Smart Mater. Struct.* **2014**, *23*, 105007. [CrossRef]
32. Liu, K.; Tovar, A. An efficient 3D topology optimization code written in Matlab. *Struct. Multidiscip. Optim.* **2014**, *50*, 1175–1196. [CrossRef]
33. Zhu, W.-L.; Zhu, Z.; Guo, P.; Ju, B.-F. A novel hybrid actuation mechanism based XY nanopositioning stage with totally decoupled kinematics. *Mech. Syst. Signal Process.* **2018**, *99*, 747–759. [CrossRef]
34. Yong, Y.K.; Aphale, S.S.; Moheimani, S.O.R. Design, identification, and control of a flexure-based XY stage for fast nanoscale positioning. *IEEE Trans. Nanotechnol.* **2009**, *8*, 46–54. [CrossRef]
35. Olhoff, N. Multicriterion structural optimization via bound formulation and mathematical programming. *Struct. Optim.* **1989**, *1*, 11–17. [CrossRef]
36. Radić, M. About a determinant of rectangular 2 × n matrix and its geometric interpretation. *Beiträge Zur Algebra Geom.* **2005**, *46*, 321–349.
37. Capasso, G.; Morlier, J.; Charlotte, M.; Coniglio, S. Stress-based topology optimization of compliant mechanisms using nonlinear mechanics. *Mech. Ind.* **2020**, *21*, 304. [CrossRef]
38. Andreasen, C.S.; Elingaard, M.O.; Aage, N. Level set topology and shape optimization by density methods using cut elements with length scale control. *Struct. Multidiscip. Optim.* **2020**, *62*, 685–707. [CrossRef]
39. Wriggers, P. *Nonlinear Finite Element Methods*; Springer: Berlin/Heidelberg, Germany, 2010.
40. Kim, N.-H. *Introduction to Nonlinear Finite Element Analysis*; Springer: New York, NY, USA, 2018.
41. Liu, L.; Xing, J.; Yang, Q.; Luo, Y. Design of Large-Displacement Compliant Mechanisms by Topology Optimization Incorporating Modified Additive Hyperelasticity Technique. *Math. Probl. Eng.* **2017**, *2017*, 467974. [CrossRef]
42. Michaleris, P.; Tortorelli, D.A.; Vidal, C.A. Tangent operators and design sensitivity formulations for transient non-linear coupled problems with applications to elastoplasticity. *Int. J. Numer. Methods Eng.* **1994**, *37*, 2471–2499. [CrossRef]
43. Dabrowski, M.; Krotkiewski, M.; Schmid, D.W. MILAMIN: MATLAB-based finite element method solver for large problems. *Geochem. Geophys. Geosystems* **2008**, *9*, Q04030. [CrossRef]
44. Andreassen, E.; Clausen, A.; Schevenels, M.; Lazarov, B.S.; Sigmund, O. Efficient topology optimization in MATLAB using 88 lines of code. *Struct. Multidiscip. Optim.* **2010**, *43*, 1–16. [CrossRef]
45. Gu, G.-Y.; Zhu, L.-M. Comparative experiments regarding approaches to feedforward hysteresis compensation for piezoceramic actuators. *Smart Mater. Struct.* **2014**, *23*, 095029. [CrossRef]
46. Cao, Y.; Chen, X.B. A Survey of Modeling and Control Issues for Piezo-electric Actuators. *J. Dyn. Syst. Meas. Control* **2014**, *137*, 014001. [CrossRef]
47. Krejci, P.; Kuhnen, K. Inverse control of systems with hysteresis and creep. *IEE Proc. Control Theory Appl.* **2001**, *148*, 185–192. [CrossRef]
48. Rakotondrabe, M. Bouc–Wen Modeling and Inverse Multiplicative Structure to Compensate Hysteresis Nonlinearity in Piezoelectric Actuators. *IEEE Trans. Autom. Sci. Eng.* **2011**, *8*, 428–431. [CrossRef]
49. Zhu, W.; Wang, D.-h. Non-symmetrical Bouc–Wen model for piezoelectric ceramic actuators. *Sens. Actuators A Phys.* **2012**, *181*, 51–60. [CrossRef]
50. Lin, C.-J.; Lin, P.-T. Tracking control of a biaxial piezo-actuated positioning stage using generalized Duhem model. *J. Comput. Math. Appl.* **2012**, *64*, 766–787. [CrossRef]
51. Zhu, W.; Zhu, Z.; He, Y.; Ehmann, K.F.; Ju, B.; Li, S. Development of a Novel 2-D Vibration-Assisted Compliant Cutting System for Surface Texturing. *IEEE/ASME Trans. Mechatron.* **2017**, *22*, 1796–1806. [CrossRef]
52. Wang, R.; Zhang, X. Preload characteristics identification of the piezoelectric-actuated 1-DOF compliant nanopositioning platform. *Front. Mech. Eng.* **2015**, *10*, 20–36. [CrossRef]
53. Eielsen, A.A.; Vagia, M.; Gravdahl, J.T.; Pettersen, K.Y. Damping and Tracking Control Schemes for Nanopositioning. *IEEE/ASME Trans. Mechatron.* **2014**, *19*, 432–444. [CrossRef]

54. Gu, G.; Zhu, L.; Su, C.; Ding, H. Motion Control of Piezoelectric Positioning Stages: Modeling, Controller Design, and Experimental Evaluation. *IEEE/ASME Trans. Mechatron.* **2013**, *18*, 1459–1471. [CrossRef]
55. Hossein Mousavi, S.; Khayatian, A. Adaptive Control for a Class of Hysteretic Systems. *J. Comput. Nonlinear Dyn.* **2012**, *8*, 011003. [CrossRef]
56. Wei, Z.; Xiang, B.L.; Ting, R.X. Online parameter identification of the asymmetrical Bouc–Wen model for piezoelectric actuators. *Precis. Eng.* **2014**, *38*, 921–927. [CrossRef]
57. Tong, S.; Sui, S.; Li, Y. Fuzzy Adaptive Output Feedback Control of MIMO Nonlinear Systems With Partial Tracking Errors Constrained. *IEEE Trans. Fuzzy Syst.* **2015**, *23*, 729–742. [CrossRef]
58. Nguyen, A.T.; Rath, J.; Guerra, T.M.; Palhares, R.; Zhang, H. Robust Set-Invariance Based Fuzzy Output Tracking Control for Vehicle Autonomous Driving Under Uncertain Lateral Forces and Steering Constraints. *IEEE Trans. Intell. Transp. Syst.* **2020**, *22*, 5849–5860. [CrossRef]
59. Liu, G.R.; Quek, S.S. Chapter 2—Briefing on Mechanics for Solids and Structures. In *The Finite Element Method*, 2nd ed.; Liu, G.R., Quek, S.S., Eds.; Butterworth-Heinemann: Oxford, UK, 2014; pp. 13–41.
60. McGuire, W.; Gallagher, R.H.; Ziemian, R.D. *Matrix Structural Analysis*; Wiley: Hoboken, NJ, USA, 1999.
61. Yao, B.; Xu, L. Output Feedback Adaptive Robust Control of Uncertain Linear Systems with Disturbances. *J. Dyn. Syst. Meas. Control* **2006**, *128*, 938–945. [CrossRef]
62. Li, X.; Bin, Y. Output feedback adaptive robust control of uncertain linear systems with large disturbances. In Proceedings of the 1999 American Control Conference (Cat. No. 99CH36251), San Diego, CA, USA, 2–4 June 1999; Volume 551, pp. 556–560.
63. Luo, Z.; Tong, L. A level set method for shape and topology optimization of large-displacement compliant mechanisms. *Int. J. Numer. Methods Eng.* **2008**, *76*, 862–892. [CrossRef]
64. Yao, B.; Tomizuka, M. Smooth Robust Adaptive Sliding Mode Control of Manipulators with Guaranteed Transient Performance. *J. Dyn. Syst. Meas. Control* **1996**, *118*, 764–775. [CrossRef]

actuators

MDPI

Article

An Inertial Impact Piezoelectric Actuator Designed by the Asymmetric Friction Principle and Achieved by Laser Texturing of the Driving Feet

Wuxiang Sun, Yanwei Liu, Xuan Li, Zhi Xu, Zhaojun Yang and Hu Huang *

Key Laboratory of CNC Equipment Reliability, Ministry of Education, School of Mechanical and Aerospace Engineering, Jilin University, Changchun 130022, China; wxsun18@mails.jlu.edu.cn (W.S.); lyw19@mails.jlu.edu.cn (Y.L.); lixuan21@mails.jlu.edu.cn (X.L.); xuzhi19@mails.jlu.edu.cn (Z.X.); yzj@jlu.edu.cn (Z.Y.)
* Correspondence: huanghu@jlu.edu.cn

Abstract: An asymmetric friction principle is newly proposed for the design of inertial impact piezoelectric actuators. There are two ways to achieve asymmetric frictions: either by tuning the positive pressure or by tuning the friction coefficient. Compared with tuning the positive pressure by an asymmetric structure, the structural parameters can be reduced by employing a symmetric structure and tuning the friction coefficient. In this study, an asymmetric friction inertial impact actuator was developed using a symmetric compliant mechanism (SCM), and the asymmetric frictions were realized by laser texturing of the driving feet at one end of the SCM. Four kinds of microstructures were initially fabricated on the driving feet, and their friction properties were experimentally tested. Accordingly, two kinds of microstructures, namely T_a and T_b microstructures, were selected. Output characteristics of the actuator with these two microstructures were measured and comparatively analyzed. The experimental results indicate that the actuator could achieve stable step motion, and the output characteristics were affected by the fabricated microstructure, as it determined the friction coefficient. The actuator with the T_b microstructure achieved a maximum speed of 2.523 mm/s, a resolution of 188 nm, a vertical loading capacity of 2 N and a horizontal loading capacity of 0.6 N, whereas the actuator with the T_a microstructure had a higher resolution of 74 nm. This study provides a novel idea for the design of asymmetric friction inertial impact actuators by tuning the friction coefficient.

Keywords: asymmetric friction principle; inertial impact actuator; microstructure; laser texturing; friction coefficient

Citation: Sun, W.; Liu, Y.; Li, X.; Xu, Z.; Yang, Z.; Huang, H. An Inertial Impact Piezoelectric Actuator Designed by the Asymmetric Friction Principle and Achieved by Laser Texturing of the Driving Feet. *Actuators* **2022**, *11*, 211. https://doi.org/10.3390/act11080211

Academic Editor: Kenji Uchino

Received: 12 July 2022
Accepted: 29 July 2022
Published: 30 July 2022

1. Introduction

With features of fast response, nanopositioning resolution, miniaturization, no coils and magnets, etc., piezoelectric actuators have been widely used in aerospace [1–3], optical scanning [4–6], biomedical engineering [7–9], precision/ultraprecision manufacturing, etc. [10–12]. Depending on the difference in driving principle, piezoelectric actuators can be divided into the following types: direct-driving actuators [13–15], inchworm actuators [16–18], ultrasonic actuators [19–24], stick-slip actuators (friction inertia type) [25–27] and inertial actuators (inertia impact type) [28–32]. Among them, the inertial impact piezoelectric actuators have the advantages of simple structure and control, micro/nanopositioning resolution and large stroke, so they have attracted wide attention and have been applied in various fields [33,34].

Figure 1a illustrates the working principle of traditional inertial impact piezoelectric actuators; they are normally composed of a main mass, a piezoelectric element and an inertial mass. The main mass is placed on the guiding surface, and the inertial mass does not contact the guiding surface. As the piezoelectric element slowly elongates, the inertial mass moves to the right, and the main mass remains stationary due to the static friction between the main mass and the guiding surface. Then, the piezoelectric element rapidly shrinks, and an inertial impact force is generated. If the generated inertial impact force is greater than the maximum static friction between the main mass and guiding surface, the main mass overcomes the static friction and moves to the right. Based on the traditional impact inertial working principle and two kinds of piezoelectric elements (piezoelectric wafer and piezoelectric stack), various inertial impact piezoelectric actuators have been designed [33,35–37]. However, some issues, such as incompatibility between the speed and resolution, the low frequency bandwidth and loading capacity, the overturning moment and motion instability, generally exist for most such actuators. To improve the output performance, traditional inertial impact piezoelectric actuators usually employ a relatively large inertial mass, which in turn significantly increases the size of the actuator and decreases the frequency bandwidth.

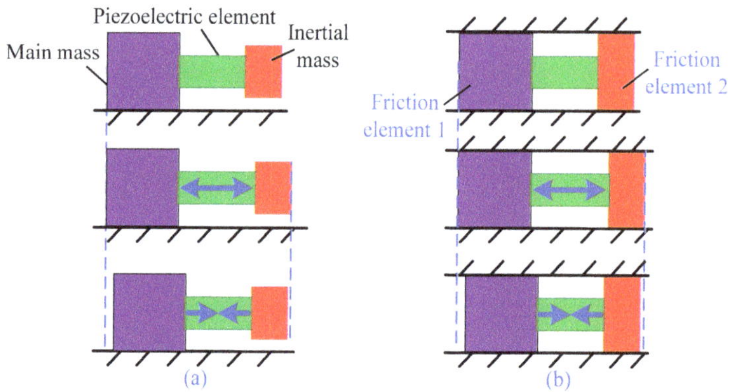

Figure 1. The working principles of (**a**) the traditional inertial impact actuator and (**b**) the asymmetric friction inertial impact actuator.

The main cause of the above problems is the non-contact inertial mass. If the inertial mass is also in contact with the guiding surface, as illustrated in Figure 1b, many of the above problems could be solved. In this case, these two masses in Figure 1a are replaced by two friction elements, and the frictions at the two ends should be different. Accordingly, we proposed an asymmetric friction principle for the design of inertial impact piezoelectric actuators in our previous study [38]. The asymmetric friction working principle is similar to the traditional inertial impact working principle. It includes friction element 1 and friction element 2. Friction element 1 corresponds to the traditional main mass, and friction element 2 corresponds to the inertial mass. The asymmetric frictions are realized by tuning the positive pressure between the two friction elements and the guiding surface. Compared with traditional inertial impact piezoelectric actuators, the actuator designed with the asymmetric friction working principle shows the stable motion, no overturning moment, high-frequency bandwidth and loading capacity. In particular, the contradiction between the speed and resolution is solved. However, in our previous study [38], an asymmetrically compliant mechanism was used as the mover. Its asymmetric structure led to an increase in the structural parameters, making structural design and parameter selection quite difficult. Therefore, it is necessary to explore new methods to achieve asymmetric frictions with a simpler structure.

The friction is mainly determined by two factors: the positive pressure and the friction coefficient. In our previous study [38], we tuned the positive pressure by employing an asymmetric structure; however this method increases the design difficulty. If the friction is tuned by the friction coefficient, a symmetric structure could be used, which would simplify the structure design. Many previous studies [39–41] have indicated that fabrication of microstructures on the surface of materials is an effective method to tune the friction coefficient. Therefore, in this study, we attempted to develop an asymmetric friction inertial impact actuator using a symmetrically compliant mechanism (SCM), and asymmetric frictions were achieved by laser texturing of the driving feet at one end of the SCM. Four kinds of microstructures were initially fabricated on the driving feet, and their friction properties were experimentally tested. Accordingly, two kinds of microstructures, namely T_a and T_b microstructures, were selected. Output characteristics of the actuator with these two microstructures were further measured and comparatively analyzed.

2. Structure of the Developed Actuator

Figure 2a shows the detailed structure of the developed asymmetric friction inertial impact piezoelectric actuator. It includes a symmetrically compliant mechanism (SCM) with laser-textured driving feet at one end, a piezoelectric stack (PES), two parallel guide rails and a base. The overall size of the actuator is 100 mm × 94 mm × 31 mm. These two parallel guide rails are installed on the base by screws. The SCM with Al 7075 has two pairs of arc-shaped driving feet, and it is nested in the two parallel guide rails, working as the mover. To generate the asymmetric frictions, microstructures are fabricated on the surface of the driving feet at one end by laser texturing. The driving feet with microstructures are referred to as the laser-textured end, and the other end without microstructures is referred to as the original end. Figure 1b shows the detailed structure of the arc-shaped driving foot. The main structural parameters are given in Table 1 based on our previous study [38]. To generate positive pressure, the inside width of two parallel guide rails is designed to be 0.3 mm smaller than that of the SCM. When the SCM is nested into the two parallel guide rails, the driving feet are deformed, and preloading forces will emerge between each driving foot and the guide rails. Furthermore, the friction coefficients of the laser-textured driving feet differ from those of the original driving feet, generating asymmetric frictions.

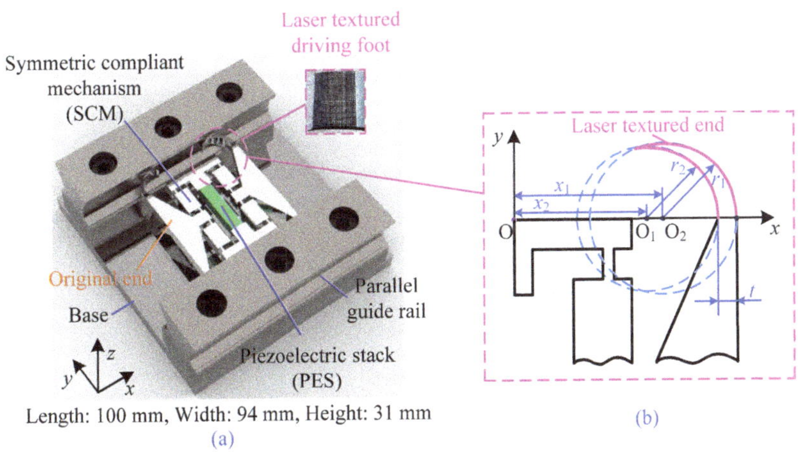

Figure 2. (**a**) Structure of the developed actuator with laser-textured driving feet and (**b**) detailed structure of a quarter of the arc-shaped driving foot.

Figure 3 shows the motion process of the actuator according to its structure. As the SCM has a symmetric structure, the forces and deformations are completely consistent. Therefore, the actuator cannot achieve effective step motion if the surfaces of the driving feet at the two ends are the same, as shown in Figure 3a. However, with microstructures fabricated on the surfaces of the driving feet at one end, the friction coefficients (μ_o and μ_L) of the two ends would be different. When the SCM is deformed with the elongation of the PES, asymmetric frictions occur at the two ends. According to the asymmetric friction motion principle [38], the actuator with laser-textured driving feet can generate displacement (L_0) along the negative-x axis in one motion period, as shown in Figure 3b.

Table 1. The main structural parameters of a quarter of the arc-shaped driving foot.

Parameter	Meaning	Value
r_1	Radius of circle O_1	7.5 mm
r_2	Radius of circle O_2	7 mm
t	Thickness of the driving foot	1.6 mm
x_1	Distance between points O_1 and O	14.5 mm
x_2	Distance between points O_2 and O	13.4 mm

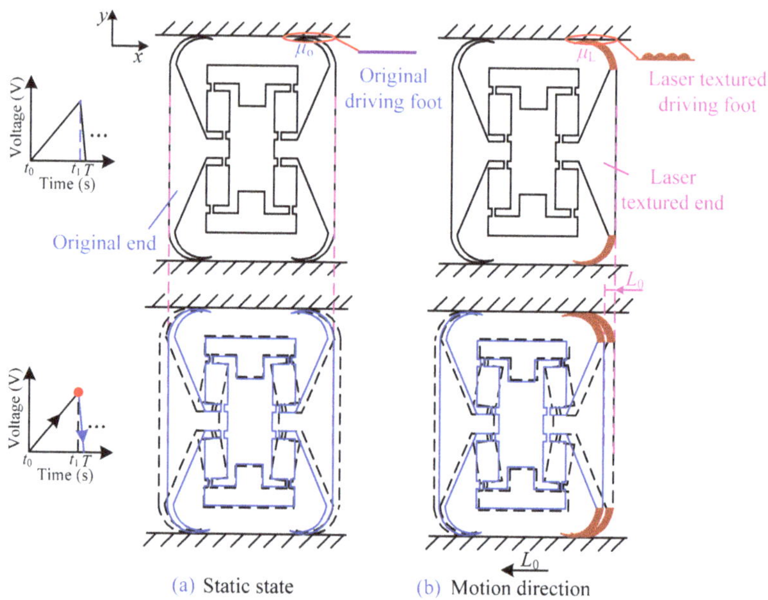

Figure 3. The motion process of the actuator with (**a**) the original driving feet and (**b**) the laser-textured driving feet.

3. Selection of Microstructures for Laser Texturing of the Driving Foot

According to some previous studies [39–41], the non-smooth surface of natural animal bodies, such as convex hulls, ridges and concaves, can effectively tune the surface friction properties. Therefore, four kinds of typical microstructures were selected as the basic microstructures and further experimentally compared and analyzed. Four arc-shaped driving feet were machined by wire electrical discharge machining (WEDM) with Al 7075, as shown in the insert in Figure 4a. Four kinds of microstructures, i.e., striped convex, square convex, circular convex and circular concave, were fabricated on the four arc-shaped driving feet by a fiber nanosecond pulsed laser with a wavelength of 1064 nm, pulse duration of 7 ns, repetition frequency of 600 kHz, scanning speed of 2 mm/s and average power of 15.8 W. An optical microscope (OM, DSX500, Olympus, Tokyo, Japan) was used to observe the four kinds of microstructures of the laser-textured driving feet surfaces; the results are shown in Figure 4.

An experimental system was established to test the friction properties of the four laser-textured driving feet, as shown in Figure 5a, including an IPC (industrial personal computer), a friction test unit, an analog output card (PCI NI6722, National Instruments Corporation, Austin, TX, USA), a voltmeter and a 24 V DC power supply. As the core part of the experimental system, the friction test unit is constituted by an X positioning stage, two lifting platforms, sensor I, signal amplifier I, sensor II, signal amplifier II, a rope, a polished disc with a mandrel, a driving foot, a base, a bearing and a transition plate. To reduce the effect of environmental vibration, all experiments were performed on a vibration-isolated optical table.

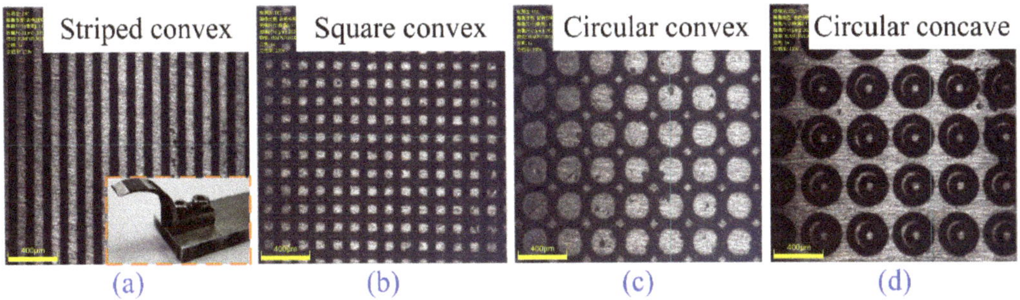

Figure 4. Four kinds of microstructures fabricated on the driving feet surfaces by laser texturing: (**a**) striped convex, (**b**) square convex, (**c**) circular convex and (**d**) circular concave.

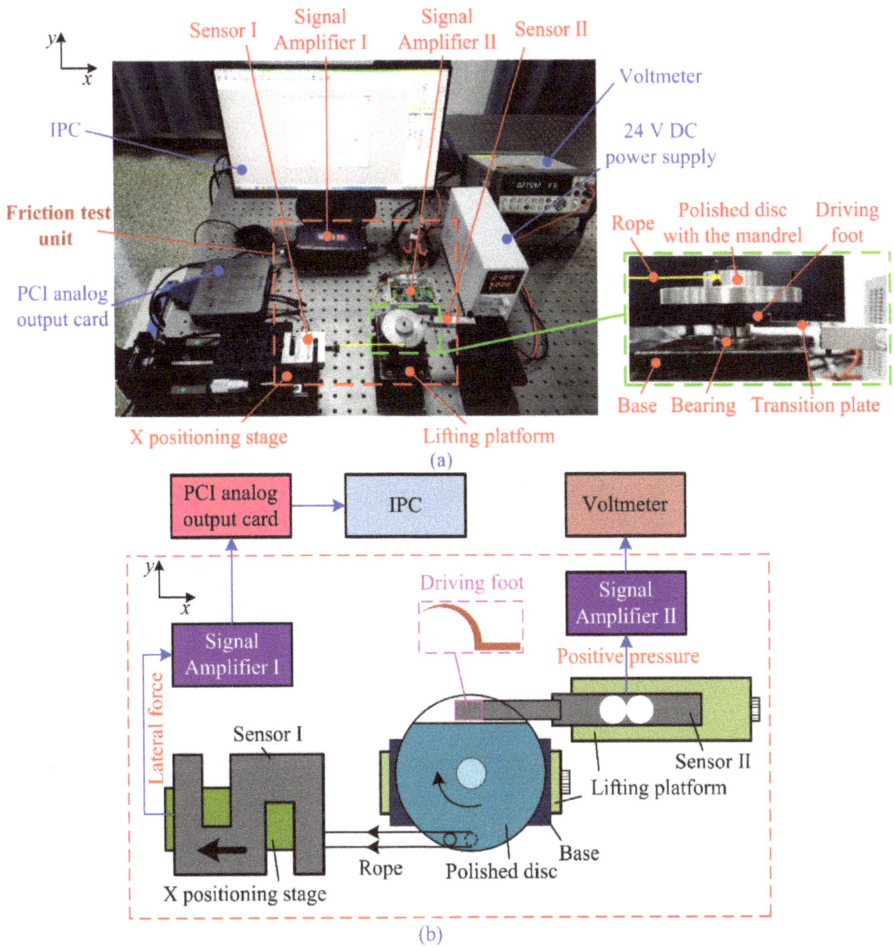

Figure 5. (**a**) The established experimental system for testing the friction properties of the four laser-textured driving feet and (**b**) a schematic diagram illustrating the measuring principle.

The measuring principle corresponding to the established experimental system is illustrated in Figure 5b. Sensor I and sensor II were used to measure the lateral force and positive pressure between the driving foot and polished disc, respectively. The standard weights were used to calibrate the voltage–force relationship of sensor I and sensor II. To measure the positive pressure between the driving foot and the polished disc, the driving foot was installed below the tangential position of the polished disc. The base with a polished disc and sensor II were fixed on two lifting platforms. The positive pressure was adjusted by two lifting platforms. The detected signal (positive pressure) of sensor II was amplified by signal amplifier II and recorded by the voltmeter. To test the lateral force, sensor I was fixed to the X positioning stage and controlled by software on the IPC. Sensor I was connected with the polished disc by a rope. Finally, when the positive pressure is constant, the X positioning stage with sensor I moves along the negative-x axis, driving the rope to pull the polished disc to rotate clockwise. During the test process, the detected signal (the lateral force) of sensor I was amplified by signal amplifier I and transmitted to the PCI analog output card, for further processing by the IPC. In addition, although the measured frictions include the friction between the bearing and the mandrel, all tests were completed using the same experimental system. Therefore, the measured lateral force can be approximately regarded as the maximum static friction.

Given the aforementioned experimental system and measuring principle, the maximum static friction (f_o) of the original driving foot was tested first, with a result of 0.47 N under positive pressure of 1.92 N, as shown in Figure 6. Then, the maximum static frictions (f_L) of the four laser-textured driving feet were tested under the same positive pressure, with results of 0.57 N, 0.6 N, 0.57 N and 0.48 N, respectively, as shown in Figure 7. Accordingly, the friction coefficient (μ_o) of the original driving foot is 0.24, and the friction coefficients (μ_L) of the four laser-textured driving feet are 0.30, 0.31, 0.30 and 0.25, respectively. Compared with the original driving foot, the friction coefficients of the four laser-textured driving feet are increased. According to the asymmetric friction principles [38], if the difference in friction coefficient between the original end and the laser-textured end is larger, the actuator would have better output characteristics. Therefore, two kinds of microstructures (striped convex and square convex) were selected for subsequent experiments, named T_a and T_b microstructures, respectively, for convenience. To test the output characteristics of the actuator with these two microstructures, T_a and T_b microstructures were fabricated on the surfaces of the driving feet at one end by laser texturing.

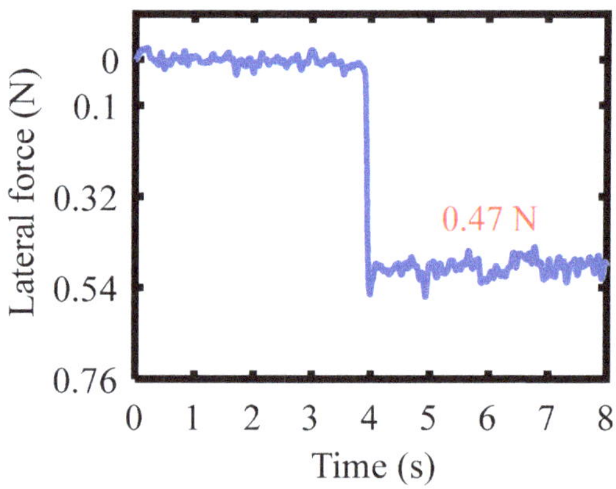

Figure 6. Lateral force–time curves of the original driving foot.

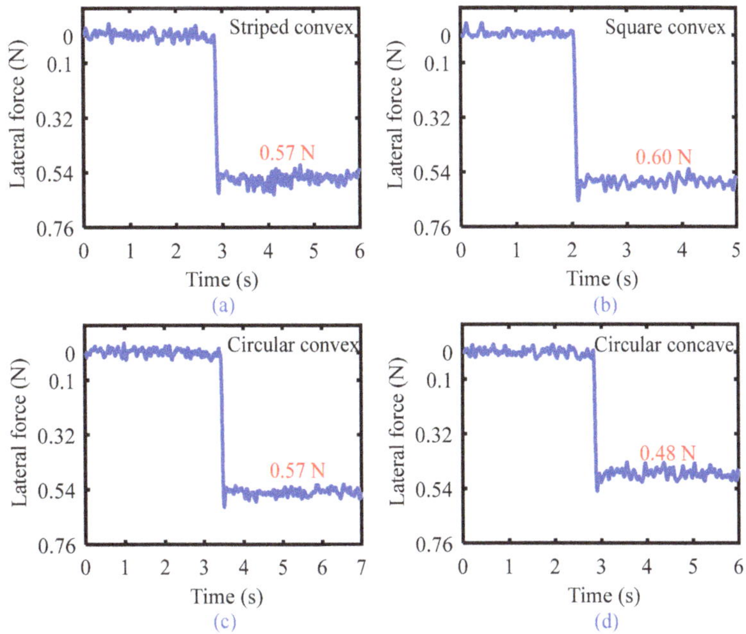

Figure 7. Lateral force–time curves of the four laser-textured driving feet: (**a**) striped convex, (**b**) square convex, (**c**) circular convex and (**d**) circular concave.

4. Experiments and Output Characteristics of the Actuator

To verify that the actuator could also achieve the stable motion by tuning the friction coefficients between the SCM and two parallel guide rails, a prototype was fabricated, and its output characteristics were tested using the experimental system, as shown in Figure 8. The signal generator (DG4062, RIGOL Technologies, Suzhou, China) produces a sawtooth-type driving voltage signal. Then, the signal is enlarged 15 times by the signal amplifier (E01.A3, Harbin Core Tomorrow Science & Technology Co., Ltd., Harbin, China) and applied to the PES (5 mm × 5 mm × 20 mm, AE0505D16DF, TOKIN, Japan; nominal displacement output: 17.4 ± 2.0 µm at 150 V) to drive the prototype. The motion displacement of the prototype was tested by a laser displacement sensor (ILD2300-2, Micro-Epsilon, Ortenburg, Germany), and the collected data were further processed by the IPC. The corresponding height between the prototype and laser displacement sensor was adjusted by the lifting platforms.

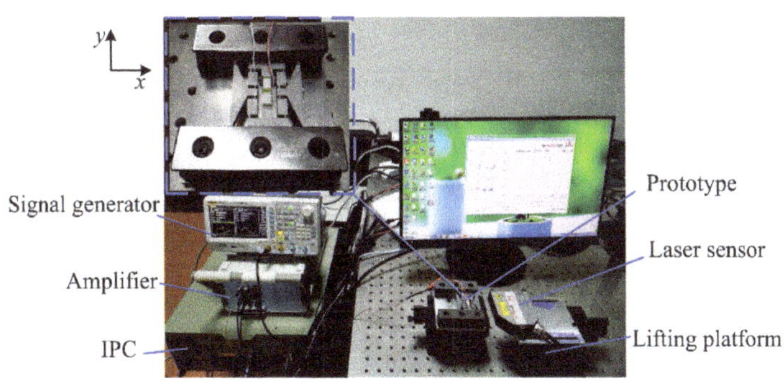

Figure 8. The established experimental system.

4.1. Output Characteristics with Various Working Gaps

The width of the two parallel guide rails, defined as the working gap, could significantly affect the output characteristics of the actuator. Taking the T_b microstructure as an example, Figure 9 shows the output displacement characteristics of the actuator with different working gaps when the driving voltage and frequency are 100 V and 10 Hz, respectively. A working gap of 58.6 mm was selected as the initial value based on our previous study [38]. However, with this working gap, the output displacement within 1 s is only 1.79 μm, possibly because the thickness of the driving foot is the same as that of the thick end in the previous study, resulting in high frictions between the SCM and the two parallel guide rails. The high frictions cause the output displacement to decrease. Therefore, it is necessary to increase the working gap. When the working gap is 58.7 mm, the actuator can achieve stable motion, as shown in the insert in Figure 9. When the working gap increases to 58.8 mm, the output displacement of the actuator can reach about 1 mm within 0.8 s. The reason for the large displacement is that the preload force between the SCM and guide rails is relatively small with a working gap of 58.8 mm, which can generate the small frictions. However, with this working gap, the output force of the actuator is relatively low. Therefore, to obtain better comprehensive output characteristics for the actuator, the subsequent experiments were performed with a working gap of 58.7 mm.

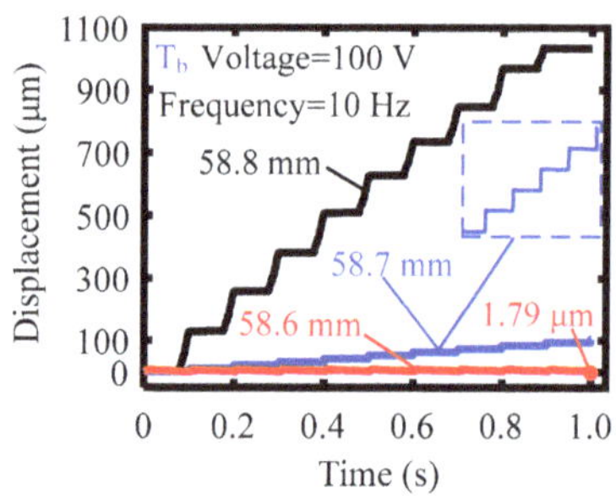

Figure 9. Output displacement of the actuator obtained with various working gaps. The driving voltage and frequency are 100 V and 10 Hz, respectively.

4.2. Output Characteristics with Various Driving Voltages and Frequencies

The output characteristics of the actuator with the T_a and T_b microstructures were further tested with various driving voltages (40 to 120 V) and driving frequencies (1 to 10 Hz) and a working gap of 58.7 mm. The results are presented in Figure 10. Figure 10a shows the accumulated displacement of the actuator within 1 s with a T_a microstructure, a fixed driving voltage of 100 V and various driving frequencies (1 to 10 Hz). Although the displacement increases with increased driving frequency, the evolution of the speed according to frequency fluctuates considerable, as shown in the insert in Figure 10a. The relatively small difference in friction coefficient between the T_a microstructure surface and the original surface results in an unstable motion. On the other hand, due to this small difference in the friction coefficient, the contact state between the SCM and guide rails is easily affected by fabrication and assembly errors, resulting in motion instability. Compared with the T_a microstructure, the difference in friction coefficient between the T_b microstructure surface and the original surface is relatively large. Therefore, the actuator with the T_b microstructure has better output characteristics, as shown in Figure 10c and the insert in Figure 10c. For example, when the driving frequency increases from 1 to 10 Hz, the speed–frequency curve obtained with the T_b microstructure shows a linear increasing tendency, indicating stable motion. In addition, under 100 V and 10 Hz, the output displacement obtained with the T_b microstructure reaches 104.8 μm within 1 s, compared to only 50.1 μm with the T_a microstructure. Therefore, a relatively high speed can be achieved by the T_b microstructure.

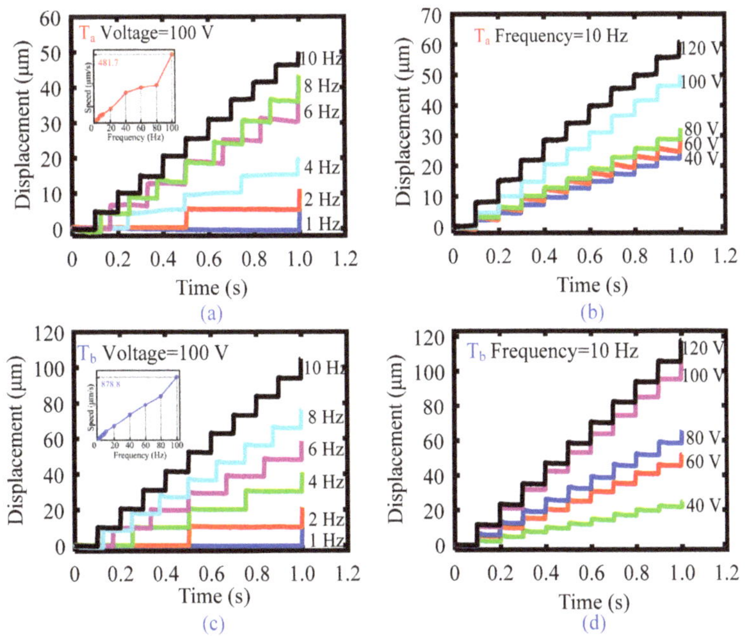

Figure 10. Output displacement change over time obtained under various driving voltages (40 to 120 V) and driving frequencies (1 to 10 Hz): (**a**,**b**) output characteristics of the actuator with the T_a microstructure, (**c**,**d**) output characteristics of the actuator with the T_b microstructure.

Figure 10b,d shows the displacement–time curves of the actuator obtained with T_a and T_b microstructures with a fixed driving frequency of 10 Hz and various driving voltages (40 to 120 V). For both T_a and T_b microstructures, the actuator can achieve stable motion under different driving voltages, but the actuator with the T_b microstructure has obvious advantages in terms of accumulated output displacement within 1 s. For example, under 120 V and 10 Hz, the accumulated output displacement is 117.5 μm for the T_b microstructure compared to only 60.6 μm for the T_a microstructure. Therefore, under various driving voltages and frequencies, relatively high motion stability and speed can be achieved for the actuator by employing the T_b microstructure.

4.3. Loading Capacity and Resolution

Loading capacity is an indispensable output characteristic for practical applications of actuators. Therefore, the vertical and horizontal loading capacities of the actuator with were tested with T_a and T_b microstructures under 100 V and 10 Hz. Figure 11a,b shows the experimental methods and results. To test the vertical loading capacity, a standard weight was placed directly on the SCM. When the vertical load is in the range of 0 to 2 N, the speed of the actuator with the T_b microstructure is still higher than that of the actuator with the T_a microstructure. For tests of horizontal loading capacity, a pulley was used to convert the standard weight to the horizontal force and applied to the SCM. The speeds of the actuator with the T_a and T_b microstructures reduce to 4.85 μm/s and 31.91 μm/s when the horizontal load is 0.4 N and 0.6 N, respectively. In comparison, the actuator with the T_b microstructure has a better loading capacity than that with the T_a microstructure.

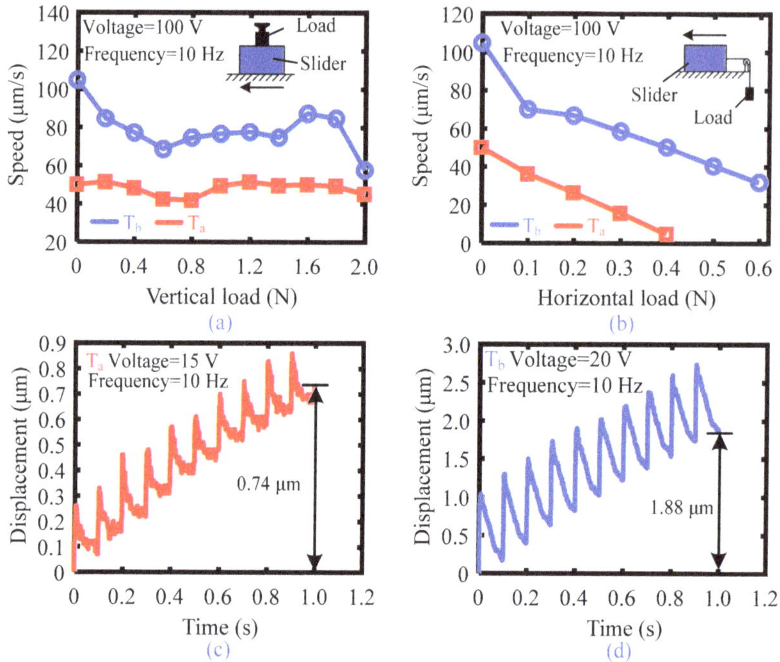

Figure 11. Motion speed change with (**a**) a vertical load and (**b**) a horizontal load. (**c**,**d**) Resolution testing results of the actuator with the T_a and T_b microstructures without an external load.

For piezoelectric actuators, the minimum stable stepping displacement, i.e., the resolution, is another important parameter. To measure the resolution, the driving frequency was kept at 10 Hz, and the driving voltage was gradually increased from zero without an external load. The minimum stable stepping motions of the actuator with the T_a and T_b microstructures are achieved when the voltages are increased to 15 V and 20 V, respectively. Figure 11c,d shows the output displacements of the actuator in ten steps obtained with the T_a and T_b microstructures. The accumulated displacements are 0.74 μm and 1.88 μm, so the resolution of the actuator with the T_a and T_b microstructures can be derived as 74 nm and 188 nm, respectively. Although the actuator with the T_b microstructure exhibits stable motion and relatively high loading capacity, its resolution is lower than that of the actuator with the T_a microstructure. Therefore, for some practical applications that require a higher resolution, the T_a microstructure should be selected

4.4. *Output Characteristics of the Actuator with the T_b Microstructure*

According to the above comparison, the actuator with the T_b microstructure exhibits relatively high motion stability, speed and loading capacity, so its output characteristics were further tested. Figure 12a presents the speed change according to driving frequency (1 to 450 Hz) under a fixed driving voltage of 100 V. The actuator with the T_b microstructure reaches a maximum speed of 2.523 mm/s under 100 V and 300 Hz.

Both forward and reverse motions of the actuator are required in practical applications. By simply changing the driving voltage waveform to that shown in the insert in Figure 12b, the actuator can realize the reverse motion along the positive *x* axis. The reverse motion speed is faster than the forward motion speed, as explained in our previous study [38].

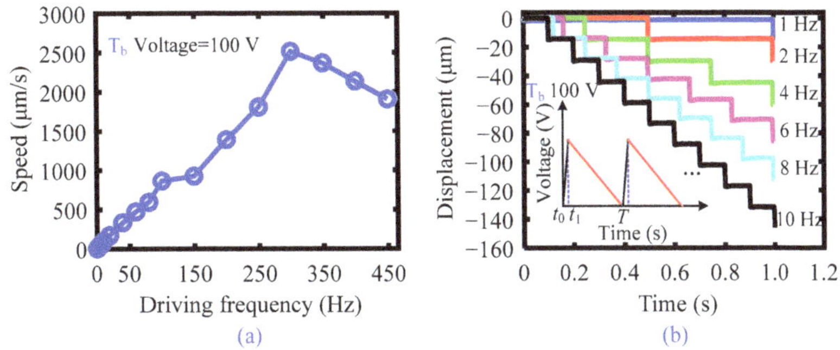

Figure 12. (**a**) Speed change according to driving frequency (1 to 450 Hz) under a fixed driving voltage of 100 V. (**b**) Reverse stepping characteristics of the actuator in the frequency range of 1 to 10 Hz.

4.5. Comparison and Discussion

Table 2 shows a comparison of output characteristics between previously reported inertial impact piezoelectric actuators and the actuator with the T_b microstructure developed in the present study, including the frequency bandwidth, maximum speed, resolution, and horizontal and vertical loading capacities. The actuator presented in this study has a high-frequency bandwidth and motion speed. The main aim of [37] was to improve the frequency bandwidth of the inertial impact actuator so that the actuator could move stably in a large frequency range. It is well known that inertial impact piezoelectric actuators generally have a low-frequency bandwidth due to their intrinsic structure (the main mass and inertial mass), as listed in Table 2 (Refs. [29,33,42,43]). Compared to these actuators, the actuator designed according to the newly proposed asymmetric friction principle presents with an improved frequency bandwidth and motion speed.

Table 2. Comparison of output characteristics between actuators reported in previous studies and the present study.

Reference	**[29]**	**[33]**	**[42]**	**[43]**	**[37]**	**[38]**	**This Work (T_b)**
Frequency bandwidth (Hz)	45	35	13	11	275	390	300
Maximum speed (mm/s)	/	0.44	0.2	0.03	1.218	7.311	2.523
Resolution (nm)	600	30	2000	20	214	221	188/T_a-74
Horizontal load (N)	/	/	/	/	0.6	1.4	0.6
Vertical load (N)	4.25	0.06	0.9	8	17	20	2

Although the actuator designed in our previous study in [38] achieved improved output characteristics by employing an asymmetric structure, this resulted in an increase in the structural parameters, causing difficulty with respect to structural design and optimization. Alternatively, the actuator in this study employs a symmetric structure and achieves asymmetric friction driving by tuning the friction coefficient. Although some output characteristics are weakened compared to those reported in [38], the structure is simplified, and the output characteristics are still quite competitive compared to traditional inertial impact actuators.

5. Conclusions

In this study, we developed an asymmetric friction inertial impact actuator using a symmetrically compliant mechanism (SCM), and asymmetric frictions were achieved by laser texturing of the driving feet at one end of the SCM. Four kinds of microstructures were initially fabricated on the driving feet, and their friction properties were experimentally tested. Accordingly, two kinds of microstructures (striped convex and square convex, i.e., T_a and T_b microstructures) with relatively large friction coefficients were selected for subsequent experiments. Output characteristics of the actuator with these two microstructures were measured and comparatively analyzed. The experimental results showed that the actuator with the T_b microstructure had better output characteristics in terms of motion stability, speed and loading capacity. It achieved a maximum speed of 2.523 mm/s, a resolution of 188 nm, a vertical loading capacity of 2 N and a horizontal loading capacity of 0.6 N, whereas the actuator with the T_a microstructure had a higher positioning resolution of 74 nm. According to comparison with previously reported inertial impact piezoelectric actuators, the output characteristics of the actuator developed herein are quite competitive.

By tuning the friction coefficient, a symmetric structure can be used to simplify the structural design of asymmetric friction inertial impact actuators. The topography of the microstructure affects the output characteristics of the actuator. Therefore, various output characteristics can be obtained by selecting different microstructures.

Author Contributions: Conceptualization, H.H.; methodology, W.S.; validation, W.S., Y.L. and X.L.; investigation, Z.X. and Z.Y.; resources, H.H.; writing—original draft preparation, W.S. and Y.L.; writing—review and editing, H.H.; supervision, H.H.; project administration, H.H.; funding acquisition, H.H. All authors have read and agreed to the published version of the manuscript.

Funding: This work was supported by the National Natural Science Foundation of China (Grant No. 52075221), the Graduate Innovation Fund of Jilin University (Grant No. 101832020CX100) and the Fundamental Research Funds for the Central Universities (2019–2022).

Institutional Review Board Statement: Not applicable.

Informed Consent Statement: Not applicable.

Data Availability Statement: The data that support the findings of this study are available from the corresponding authors upon reasonable request.

Conflicts of Interest: The authors declare no conflict of interest.

References

1. Elahi, H.; Eugeni, M.; Gaudenzi, P.; Qayyum, F.; Swati, R.F.; Khan, H.M. Response of piezoelectric materials on thermomechanical shockingand electrical shocking for aerospace applications. *Microsyst. Technol.* **2018**, *24*, 3791–3798. [CrossRef]
2. Li, M.; Yuan, J.; Guan, D.; Chen, W. Application of piezoelectric fiber composite actuator to aircraft wing for aerodynamic performance improvement. *Sci. China Technol. Sci.* **2011**, *54*, 395–402. [CrossRef]
3. Qing, X.; Li, W.; Wang, Y.; Sun, H. Piezoelectric Transducer-Based Structural Health Monitoring for Aircraft Applications. *Sensors* **2019**, *19*, 545. [CrossRef] [PubMed]
4. Holmström, S.T.S.; Baran, U.; Urey, H. Laser Scanners A Review. *J. Microelectromechanical Syst.* **2014**, *23*, 259–274. [CrossRef]
5. Koh, K.H.; Kobayashi, T.; Hsiao, F.L.; Lee, C. Characterization of piezoelectric PZT beam actuators for driving 2D scanningmicromirrors. *Sens. Actuators A Phys.* **2010**, *162*, 336–347. [CrossRef]
6. Wang, Z. Piezopotential gated nanowire devices Piezotronics and piezo-phototronics. *Nano Today* **2010**, *5*, 540–552. [CrossRef]
7. Chorsi, M.T.; Curry, E.J.; Chorsi, H.T.; Das, R.; Baroody, J.; Purohit, P.K.; Ilies, H.; Nguyen, T.D. Piezoelectric Biomaterials for Sensors and Actuators. *Adv. Mater.* **2019**, *31*, 1802084. [CrossRef] [PubMed]
8. Yuan, H.; Lei, T.; Qin, Y.; He, J.H.; Yang, R. Design and application of piezoelectric biomaterials. *J. Phys. D Appl. Phys.* **2019**, *52*, 194002. [CrossRef]
9. Zheng, Q.; Shi, B.; Li, Z.; Wang, Z. Recent Progress on Piezoelectric and Triboelectric Energy Harvesters in Biomedical Systems. *Adv. Mater.* **2017**, *4*, 1700029. [CrossRef]
10. Ji, F.; Li, T.; Yu, S.; Wu, Z.; Zhang, L. Propulsion Gait Analysis and Fluidic Trapping of Swinging Flexible Nanomotors. *ACS Nano* **2021**, *15*, 5118–5128. [CrossRef]
11. Li, T.; Zhang, A.; Shao, G.; Wei, M.; Guo, B.; Zhang, G.; Li, L.; Wang, W. Janus Microdimer Surface Walkers Propelled by Oscillating Magnetic Fields. *Adv. Funct. Mater.* **2018**, *28*, 1706066. [CrossRef]
12. Wang, H.; Yu, S.; Liao, J.; Qing, X.; Sun, D.; Ji, F.; Song, W.; Wang, L.; Li, T. A Robot Platform for Highly Efficient Pollutant Purification. *Front. Bioeng. Biotechnol.* **2022**, *10*, 903219. [CrossRef] [PubMed]
13. Guan, C.; Jiao, Z. A piezoelectric direct-drive servo valve with a novel multi-body contacting spool-driving mechanism: Design, modelling and experiment. *P. I. Mech. Eng. C-J. Mec.* **2014**, *228*, 169–185. [CrossRef]
14. Mohith, S.; Upadhya, A.R.; Navin, K.P.; Kulkarni, S.M.; Rao, M. Recent trends in piezoelectric actuators for precision motion and their applications: A review. *Smart Mater. Struct.* **2021**, *30*, 013002. [CrossRef]
15. Wang, L.; Chen, W.; Liu, J.; Deng, J.; Liu, Y. A review of recent studies on non-resonant piezoelectric actuators. *Mech. Syst. Signal Process.* **2019**, *133*, 106254. [CrossRef]

16. Gao, Y.; Wen, J.; Ma, J.; Zhang, Y.; Wang, R.; Hu, Y.; Li, J. A self-adapting linear inchworm piezoelectric actuator based on a permanent magnets clamping structure. *Mech. Syst. Signal Process.* **2019**, *132*, 429–440. [CrossRef]
17. Li, J.; Zhao, H.; Shao, M.; Zhou, X.; Huang, H.; Fan, Z. Design and experiment performances of an inchworm type rotary actuator. *Rev. Sci. Instrum.* **2014**, *85*, 085004. [CrossRef]
18. Shao, S.; Song, S.; Shao, Y.; Xu, M. Long-range piezoelectric actuator with large load capacity using inchworm and stick-slip driving principles. *Precis. Eng.* **2022**, *75*, 167–179. [CrossRef]
19. Deng, J.; Liu, Y.; Chen, W.; Yu, H. A XY Transporting and Nanopositioning Piezoelectric Robot Operated by Leg Rowing Mechanism. *IEEE-ASME Trans. Mechatron.* **2019**, *24*, 207–217. [CrossRef]
20. Liu, R.; Wen, Z.; Cao, T.; Lu, C.; Wang, B.; Wu, D.; Li, X. A precision positioning rotary stage driven by multilayer piezoelectric stacks. *Precis. Eng.* **2022**, *76*, 226–236. [CrossRef]
21. Liu, Y.; Wang, L.; Gu, Z.; Quan, Q.; Deng, J. Development of a Two-Dimensional Linear Piezoelectric Stepping Platform Using Longitudinal-Bending Hybrid Actuators. *IEEE Trans. Ind. Electron.* **2019**, *66*, 3030–3040. [CrossRef]
22. Delibas, B.; Koc, B. L1B2 Piezo Motor Using D33 Effect. In Proceedings of the ACTUATOR 2018: 16th International Conference on New Actuators, Bremen, Germany, 25–27 June 2018; pp. 1–4.
23. Feng, Y.; Chang, X.; Liu, H.; Hu, Y.; Li, T.; Li, L. Multi-response biocompatible Janus micromotor for ultrasonic imaging contrast enhancement. *Appl. Mater. Today* **2021**, *23*, 101026. [CrossRef]
24. Tian, X.; Liu, Y.; Deng, J.; Wang, L.; Chen, W. A review on piezoelectric ultrasonic motors for the past decade: Classification, operating principle, performance, and future work perspectives. *Sens. Actuators A Phys.* **2020**, *306*, 111971. [CrossRef]
25. Tang, J.; Wei, J.; Wang, Y.; Xu, Z.; Huang, H. A Novel Rotation-Structure Based Stick-Slip Piezoelectric Actuator with High Consistency in Forward and Reverse Motions. *Actuators* **2021**, *10*, 189. [CrossRef]
26. Yang, X.; Tang, J.; Guo, W.; Huang, H.; Fan, H.; Liu, J.; Li, T. Design and Analysis of a Stepping Piezoelectric Actuator Free of Backward Motion. *Actuators* **2021**, *10*, 200. [CrossRef]
27. Nguyen, X.H.; Mau, T.H.; Meyer, I.; Dang, B.L.; Pham, H.P. Improvements of Piezo-Actuated Stick–Slip Micro-Drives: Modeling and Driving Waveform. *Coatings* **2018**, *8*, 62. [CrossRef]
28. Shao, Y.; Shao, S.; Xu, M.; Song, S.; Tian, Z. An inertial piezoelectric actuator with miniaturized structure and improved load capacity. *Smart Mater. Struct.* **2019**, *28*, 055023. [CrossRef]
29. Zhang, E.; Hu, Y.; Bao, H.; Li, J.; Ma, J.; Wen, J. A linear inertial piezoelectric actuator using a single bimorph vibrator. *Smart Mater. Struct.* **2019**, *28*, 115020. [CrossRef]
30. Hunstig, M. Piezoelectric Inertia Motors—A Critical Review of History, Concepts, Design, Applications, and Perspectives. *Actuators* **2017**, *6*, 7. [CrossRef]
31. Higuchi, T.; Watanabe, M.; Kudou, K. Precise positioner utilizing rapid deformations of a piezoelectric element. *J. Jpn. Soc. Precis. Eng.* **1988**, *54*, 2107–2112. [CrossRef]
32. Okamoto, Y.; Yoshida, R.; Sueyoshi, H. The development of a smooth impact drive mechanism (SIDM) using a piezoelectric element. *Konica Minolta Technol. Rep.* **2004**, *1*, 23–26.
33. Hu, Y.; Wang, R.; Wen, J. A Low-Frequency Structure-Control-Type Inertial Actuator Using Miniaturized Bimorph Piezoelectric Vibrators. *IEEE Trans. Ind. Electron.* **2019**, *66*, 6179–6188. [CrossRef]
34. Zhang, Z.M.; An, Q.; Li, J.W.; Zhang, W.J. Piezoelectric friction–inertia actuator-a critical review and future perspective. *Int. J. Adv. Manuf. Technol.* **2012**, *62*, 669–685. [CrossRef]
35. Cao, Y.; Xu, Z.; You, L.; Wu, Y.; Huang, H. An inertial piezoelectric actuator with small structure but large loading capacity. *Rev. Sci. Instrum.* **2021**, *92*, 085004. [CrossRef] [PubMed]
36. Hu, Y.; Lin, S.; Ma, J.; Zhang, Y.; Li, J.; Wen, J. Piezoelectric inertial rotary actuator operating in two-step motion mode for eliminating backward motion. *Appl. Phys. Lett.* **2020**, *117*, 031902. [CrossRef]
37. Xu, Z.; Yang, Z.; Wang, K.; Li, X.; Dong, J.; Huang, H. A bionic inertial piezoelectric actuator with improved frequency bandwidth. *Mech. Syst. Signal Process.* **2021**, *156*, 107620. [CrossRef]
38. Sun, W.; Xu, Z.; Wang, K.; Li, X.; Tang, J.; Yang, Z.; Huang, H. An impact inertial piezoelectric actuator designed by means of the asymmetric friction. *IEEE Trans. Ind. Electron.* **2022**, 1. [CrossRef]
39. Bushnell, D.M. Drag Reduction in Nature. *Annu. Rev. Fluid Mech.* **1991**, *23*, 65–79. [CrossRef]
40. Ren, L. Progress in the bionic study on anti-adhesion and resistance reduction of terrain machines. *Sci. China Ser. E Technol. Sci.* **2009**, *52*, 273–284. [CrossRef]
41. Ren, L.; Liang, Y. Biological couplings: Classification and characteristic rules. *Sci. China Ser. E Technol. Sci.* **2009**, *52*, 2791–2800. [CrossRef]
42. Shen, D.; Wen, J.; Ma, J.; Hu, Y.; Wang, R.; Li, J. A novel linear inertial piezoelectric actuator based on asymmetric clamping materials. *Sens. Actuators A Phys.* **2020**, *303*, 111746. [CrossRef]
43. Wang, R.; Hu, Y.; Shen, D.; Ma, J.; Wen, J. Design and Experimental Performance of a Novel Piezoelectric Inertial Actuator for Magnetorheological Fluid Control Using Permanent Magnet. *IEEE Access* **2019**, *7*, 43573–43580. [CrossRef]

actuators MDPI

Article

Research on the Influence of Friction Pairs on the Output Characteristics of the Piezoelectric Ultrasonic Actuator

Jie Deng *,†, Jianfei Cheng †, Yuntian Guan, He Li, Fei Lu and Weishan Chen *

State Key Laboratory of Robotics and System, Harbin Institute of Technology, Harbin 150001, China; jf_cheng@hit.edu.cn (J.C.); gyt_hit@163.com (Y.G.); 17b908020@stu.hit.edu.cn (H.L.); 19b908032@stu.hit.edu.cn (F.L.)

* Correspondence: dengjie21@hit.edu.cn (J.D.); cws@hit.edu.cn (W.C.)

† These authors contributed equally to this work.

Abstract: The piezoelectric ultrasonic actuator is driven by the friction coupling between the stator and mover. Its friction pairs are very important, but there are few studies on the long-term output stability. Therefore, zirconia (ZrO_2) is selected as a stator material to form friction pairs with four different wear-resistant materials: silicon nitride (Si_3N_4), ZrO_2, bearing steel (GCr_{15}) and polyether ether ketone (PEEK). Experiments show that the friction pair composed of ZrO_2-Si_3N_4 is the best, and the attenuation percentage of the speed from the initial state to the end state in the speed rising stage of 50 m is 3.66%. A linear piezoelectric platform is developed based on the best friction pair; a maximum speed of 426.2 mm/s and a resolution of 85 nm are achieved.

Keywords: piezoelectric ultrasonic actuator; friction pairs; stable output characteristics; long service life

Citation: Deng, J.; Cheng, J.; Guan, Y.; Li, H.; Lu, F.; Chen, W. Research on the Influence of Friction Pairs on the Output Characteristics of the Piezoelectric Ultrasonic Actuator. *Actuators* **2022**, *11*, 212. https://doi.org/10.3390/act11080212

Academic Editor: Kenji Uchino

Received: 18 July 2022
Accepted: 28 July 2022
Published: 30 July 2022

1. Introduction

The piezoelectric actuator has the advantages of a flexible structure design, no electromagnetic interference, a fast response speed and no reducer [1–6]. Therefore, it has attracted more and more attention from researchers and is widely used in aerospace, military, robotics, medical and other fields [7–11]. According to the difference of actuation modes, piezoelectric actuators can be divided into four actuation modes: inertia [12,13], direct drive [14], inchworm [15] and ultrasonic [16–18]. With the advantages of a high output speed, a strong load capacity, a simple structure, a flexible design and an unlimited stroke, the piezoelectric actuator has been well used in precision linear motion platforms [19–22]. The output force of the PUA is transmitted by the friction coupling between the stator and the mover. Therefore, the selection of friction materials has a great impact on the driving effect and the service life of the piezoelectric actuator [23].

In order to reduce the wear between the stator and the mover, a lot of research is carried out. The selection of friction materials not only needs to meet the high friction coefficient but also needs to have excellent wear resistance [24]. Friction pairs usually contain three combinations: organic materials and organic materials, organic materials and inorganic materials and inorganic materials and inorganic materials. In terms of the friction characteristics between organic materials and organic materials, Gu et al. studied the dry sliding friction and wear of polytetrafluoroethylene (PTFE)/Kevlar composites under different vacuums, loads, sliding speeds and temperature performances; the results show that the friction coefficient of the PTFE/Kevlar composite material under vacuum conditions is lower than that of normal temperature conditions, and it has good wear resistance under low and medium vacuum conditions and poor wear resistance under high vacuum conditions. Under vacuum conditions, the friction coefficient and wear rate decrease with the increases in the sliding speed, load and temperature within a certain range [25]. So, the application of the friction pair of organic materials in vacuum is greatly affected by the degree of vacuum, load and temperature.

In terms of the friction characteristics between organic materials and inorganic materials, Liu et al. studied the influence of the PTFE composite and polyimide (PI) composite as mover friction materials on the energy conversion efficiency and mechanical output performance of PUA with phosphorous bronze as the stator friction material. The experimental results show that the PI composite as a friction material can significantly improve the output performance of PUA compared with the PTFE composite. The maximum energy conversion efficiency of the PUA is increased from 24.27% to 41.32% by replacing the PTFE-based friction material with the PI composite material [26]. Song et al. used graphite, molybdenum disulfide or PTFE-filled aramid fiber/PI composite materials as friction materials, which were formed friction pairs with phosphor bronze, respectively, and each friction pair was used to carry out the reciprocating friction and wear systematic research. A comparative study of the friction reduction and wear resistance of polyimide composites shows that graphite exhibits the best lubricity with a lower friction coefficient and wear rate [27]. Qu et al. glued polyphenylene (EKonol) composites to the stator of a traveling-wave PUA, which was combined with rotors of grade 45 steel, copper, stainless steel and aluminum, respectively, and the influence of the drive mode and friction pair combination on the tribological properties of the motor contact interface was analyzed. The experimental results show that, when polyphenylene (EKonol) is combined with a copper rotor, the contact layer can obtain a greater coefficient of friction, a higher output torque and a better wear performance, which can meet the actual operation needs of a traveling ware ultrasonic motor [28]. Therefore, the friction pair composed of suitable organic materials and inorganic materials has a positive effect on improving the energy conversion efficiency, output torque, friction stability and wear resistance of the actuator.

In the research on friction pairs composed of inorganic materials and inorganic materials, Zhang et al. selected ZrO_2, alumina (Al_2O_3), silicon carbide (SiC) and silicon nitride (Si_3N_4) ceramics as the contact material of the stator, which is matched with the ZrO_2, Al_2O_3, SiC and Si_3N_4 ceramic of movers, respectively, and the output performance, friction, wear characteristics and performance life of the standing wave linear PUA in the contact interface are studied, respectively. The experimental results show that the combination of SiC-ZrO_2, Si_3N_4-ZrO_2 and ZrO_2-ZrO_2 stator and rotor friction pairs has a better working reliability and durability [29]. Adachi et al. used Al_2O_3 as a stator and mover materials, respectively, to form friction pairs. The research results show that a smooth wear surface can be formed by two processes of micro-wear and small-wear particles filling surface depression [30]. Olofsson et al. carried out disc and ball tests with Al_2O_3 as the friction material under different humidity conditions, and the results show that the wear characteristics of the brake disc remained stable between 14,000 and 200,000 RPM under all humidity conditions. For the water lubrication test, the width of the wear track increases significantly between 800 and 200,000 RPM but does not increase significantly in the dryer test. The friction coefficient is the highest under dry conditions and the lowest under water lubrication conditions [31]. Under high temperature and high-pressure conditions, friction sintering of the wear particles is possible when Al_2O_3 is in friction with itself. Under dry conditions, the friction coefficient is at its maximum when the wear is at its minimum. Therefore, the friction pairs composed of inorganic and inorganic materials and the selection of suitable friction pairs and lubrication conditions have an important impact on the stability and life of the actuator.

Although the research work of friction pairs has been carried out for a long time and a lot of research work has been done, the previous research work mainly aims to evaluate the output characteristics. The research focused on the efficiency and wear characteristics under different environments and different combinations of friction pairs. So, this work uses PUA to study the output speed, output force and long-term running speed stability of different friction pair materials and selects the friction pair with the best output characteristics to develop a linear motion platform, which will provide an idea for improving the long-term running stability of the ultrasonic motor. The main content of the work is as follows. The experimental platform of the friction pair is constructed, and the speed, output force and speed stability of different friction pairs are illustrated in Section 2. The experimental results were discussed, and the material of the friction pair with better characteristics was selected in Section 3. The linear motion platform was developed, and platform characteristics were tested in Section 4.

2. Experimental Platform Construction and Friction Pair Performance Test

The basic configuration and principle of the actuator is introduced first, and then the friction pair testing system built by the actuator is presented. ZrO_2 is used as the stator material in the experiment, which constitutes the friction pairs with silicon nitride (Si_3N_4), ZrO_2, bearing steel (GCr_{15}) and polyether ether ketone (PEEK), respectively. The speed, output force and motion stability of different friction pairs were tested, respectively.

2.1. Configuration of the Actuator and Actuating Principle

A. Actuator configuration

Considering the working stability and output characteristics of PUA, the sandwich PUA with four zones of the d_{33} working mode was selected as the basic configuration in this work. As shown in Figure 1a, the actuator is composed of a luffing lever made of 2A12, six pieces of the four-zone ceramic, six pieces of the electrode plate, a pedestal and a connecting bolt. Two sets of sinusoidal signals with a phase difference of 90°/270° were selected to apply to the ceramic in the horizontal and vertical directions, respectively. Under this driving condition, the motion trajectory of the driving foot is a positive ellipse when it is actuated. The motion trajectory under the excitation signal is shown in Figure 1b, which can ensure that the motion trajectory of the point on the actuator driving foot is a positive ellipse. The main structural parameters of the piezoelectric ultrasonic actuator are shown in Figure 1c. Its overall dimensions in the X, Y and Z directions are 30 mm, 30 mm and 34.2 mm, respectively, and its mass is 0.074 kg.

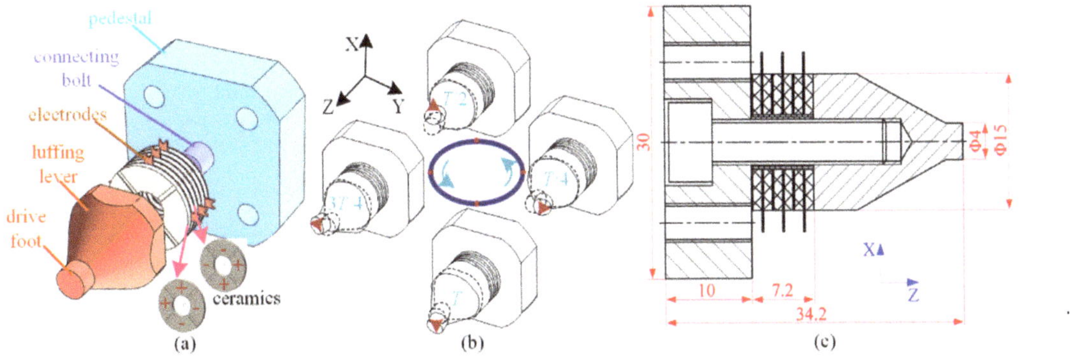

Figure 1. Actuator configuration and driving signal. (**a**) Basic structure. (**b**) The movement process of the stator. (**c**) The dimensions of the stator.

B. Actuating principle

The motion process of the PUA in a sinusoidal AC excitation signal cycle and the working process of the PUA in one cycle (T) is illustrated by the motion states of the PUA at four special time points: $T/4$, $T/2$, $3T/4$ and T.

(1) $t = T/4$: The excitation voltage applied to the two vertical zones of the piezoelectric ceramics reaches the maximum positive value, and the bending displacement of the PUA along the negative direction of the Y axis reaches the maximum to press down the actuator.

(2) $t = T/2$: The excitation voltage applied to the horizontal two regions of the piezoelectric ceramic reaches the maximum value, and the bending displacement of the

PUA reaches the maximum along the positive direction of the X-axis. The driving foot moves along the positive X-axis of the actuator between $t = T/4$ and $t = T/2$.

(3) $t = 3T/4$: The excitation voltage applied to the two vertical zones of the piezoelectric ceramics reaches the maximum negative value, the bending displacement of the PUA along the positive direction of the Y axis reaches the maximum and the driving foot leaves the actuator. The driving foot does not contact the mover between $t = T/2$ and $t = 3T/4$.

(4) $t = T$: The excitation voltage applied to the left and right zones of the piezoelectric ceramics reaches the maximum negative value, and the bending displacement of the PUA along the negative direction of the X axis reaches the maximum. The driving foot does not contact the mover between $t = 3T/4$ and $t = T$.

This is the working process of the PUA in an excitation signal cycle. The PUA pushes the mover in the positive direction of the X-axis during each excitation signal cycle. The mover can realize the movement of a large stroke by the continuous accumulation of multiple cycles. The reverse driving process of the mover can be realized when the phase difference of two sinusoidal AC excitation signals is changed to 270°.

2.2. Friction Pair Performance Test

A. Configuration friction pair performance test

The experimental device shown in Figure 2 was used to test the mechanical output performance and long-term operation output speed stability of the PUA. Because the main wear mechanism of ceramic materials is mechanical fracture, ZrO_2 materials have a higher fracture toughness and less wear [20]. In the experiment, ZrO_2 was used as the stator material to form friction pairs with four different wear-resistant materials: Si_3N_4, ZrO_2, GCr_{15} and PEEK. The material parameters are shown in Table 1. The experimental device took the bearing as the mover, the PUA drives the bearing to move, the lifting frame is connected with the force sensor, the force sensor is connected with the mover through the connecting device, the height of the lifting frame is controlled by the hand wheel, the accurate preload value between the stator and the mover can be obtained through the force sensor and the mover is pasted with reflective paper. The laser emitted by the tachometer returns to the tachometer after being reflected by the reflective paper. The speed value of the mover can be obtained by calculating the time interval of the reflected laser.

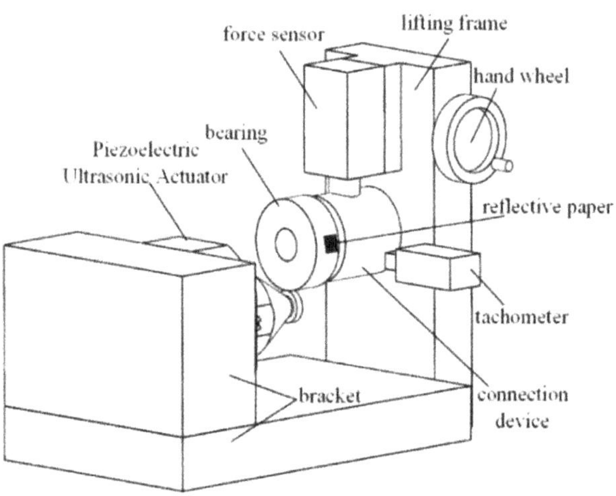

Figure 2. Friction pair test experimental device.

Table 1. Basic parameters of friction pair materials.

Material Parameters	Si_3N_4	ZrO_2	GCr_{15}	PEEK	2A12
Density ρ (g/cm^3)	3.2	6	7.8	1.32	2.81
Elastic modulus E (GPa)	320	210	208	3.8	72
Poisson's ratio	0.26	0.3	0.3	0.4	0.33

B. Experiment development and data processing

The initial preload value was set to 1 N, the commercial ultrasonic power supply (Model: QD-8D, Harbin Jiangjun precision testing technology Co., Ltd., Harbin, China) was used to provide the excitation signal, the excitation voltage was increased from 40 $V_{p\text{-}p}$ to 200 $V_{p\text{-}p}$ and the relationship between the output speed of the actuator and the excitation voltage is shown in Figure 3a. The experimental results show that the output speed of the PUA increases approximately linearly with the increase in excitation voltage. The output speeds of the friction pair materials from high to low are ZrO_2-Si_3N_4, ZrO_2-ZrO_2, ZrO_2-GCr_{15} and ZrO_2-PEEK. When the excitation voltage is 200 $V_{p\text{-}p}$, the maximum output speed of the ZrO_2-Si_3N_4 friction pair can reach 470.66 r/min. By contrast, the minimum output speed of the ZrO_2-PEEK friction pair is 230.95 r/min. The output speed of the ZrO_2-Si_3N_4 friction pair is about 2.04 times that of the ZrO_2-PEEK friction pair, and the ZrO_2-PEEK friction pair cannot work when the excitation voltage is less than 130 $V_{p\text{-}p}$.

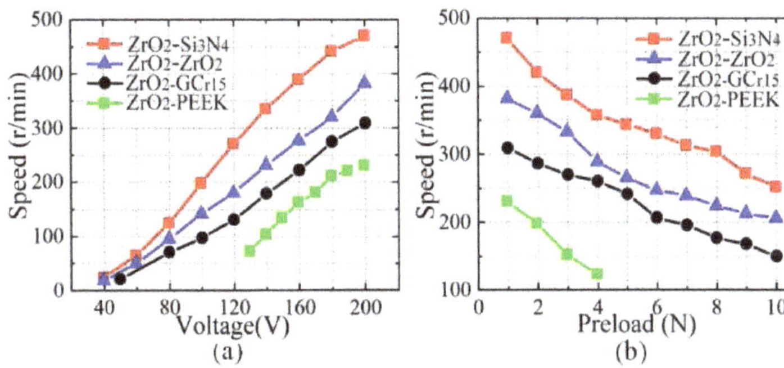

Figure 3. Characteristics of speed. (**a**) Relationship between the velocity and excitation voltage characteristics of the PUA. (**b**) Relationship between the output and preload characteristics of the PUA.

The excitation voltage was 200 $V_{p\text{-}p}$, and the preload between the stator and the mover was increased from 1 N to 10 N. The relationship between the output speed of the actuator and the preload is shown in Figure 3b. The experimental results show that the output speed of the PUA decreases approximately linearly with the increase in the preload. The output speeds of the friction pair materials from high to low are ZrO_2-Si_3N_4, ZrO_2-ZrO_2, ZrO_2-GCr_{15} and ZrO_2-PEEK. When the friction pair material is ZrO_2-PEEK, the PUA cannot work when the preload is greater than 4 N.

The preload is 8 N, the value of the excitation voltage increases from 100 $V_{p\text{-}p}$ to 200 $V_{p\text{-}p}$ and the relationship between the actuator output force and excitation voltage is shown in Figure 4a. The experimental results show that the output force of the PUA increases approximately linearly with the increase in the excitation voltage. The friction pair materials with the overall output force from high to low are ZrO_2-Si_3N_4, ZrO_2-ZrO_2 and ZrO_2-GCr_{15}. When the excitation voltage is 200 $V_{p\text{-}p}$, the maximum output force of the ZrO_2-Si_3N_4 friction pair can reach 1.32 N. Under the same conditions, the minimum output force of the ZrO_2-GCr_{15} friction pair is 0.54 N, and the output force of the ZrO_2-Si_3N_4 friction pair is about 2.44 times that.

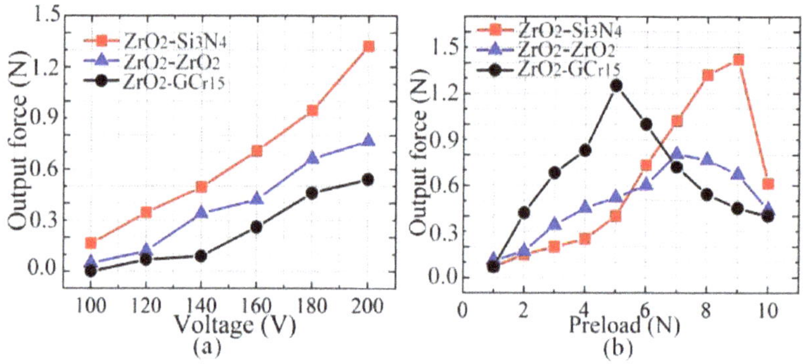

Figure 4. Characteristics of the output force. (**a**) Relationship between the output force and excitation voltage. (**b**) Relationship between the output force and preload.

The value of the excitation voltage is 200 $V_{p\text{-}p}$, and the preload increases from 1 N to 10 N. The relationship between the actuator output force and the preload is shown in Figure 4b. The experimental results show that the output force of the PUA first increases and then decreases with the increase in the preload. At a lower preload, the output forces of the friction pair materials from high to low are ZrO_2-GCr_{15}, ZrO_2-ZrO_2 and ZrO_2-Si_3N_4. At a high preload, the high output forces of the friction pair materials from high to low are ZrO_2-Si_3N_4, ZrO_2-ZrO_2 and ZrO_2-GCr_{15}, respectively. When the preload is 5 N, 7 N and 9 N, the output forces of the ZrO_2-GCr_{15}, ZrO_2-ZrO_2 and ZrO_2-Si_3N_4 friction pairs reach the maximum, which are 1.25 N, 0.805 N and 1.42 N, respectively. When the preload force is greater than 6.5 N, the output force of ZrO_2-Si_3N_4 is always greater than that of the other friction pairs. Therefore, the output force characteristics of the ZrO_2-Si_3N_4 friction pair combination are better.

The stability of the output speed of the PUA under different material combinations of friction pairs is the key factor to evaluate each friction pair. Under the experimental conditions that the peak value of the excitation voltage was 200 $V_{p\text{-}p}$ and the preload was 2 N, the PUA drove the bearing to run continuously for 500 m, and the variation law of its output speed with the operating distance was measured, as shown in Figure 5. Firstly, the change in the output speed of the PUA running for a long distance was analyzed. It can be seen from the experimental results that, under the four different friction pair material combinations, the output speed of the PUA experienced a process of rising at the beginning and then fell. The specific output speed changes of the four friction pairs are shown in Table 2.

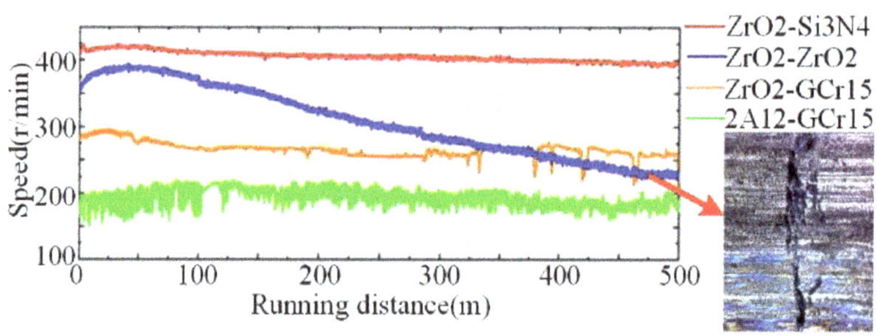

Figure 5. Output speed stability experiment under long-term operation (500 m).

Table 2. Variation in the output speed of different friction pairs.

Friction Pairs	Running Distance (m)	Initial Velocity (r/min)	Peak Velocity (r/min)	End Speed (r/min)
ZrO_2-Si_3N_4	50	410	420	395
ZrO_2-ZrO_2	40	355	390	228
ZrO_2-GCr_{15}	25	285	295	258
2A12-Cr_{15}	130	198	218	190

The running distances of the four kinds of friction pairs of ZrO_2-Si_3N_4, ZrO_2-ZrO_2, ZrO_2-GCr_{15} and 2A12-GCr_{15} in the rising stage are 50 m, 40 m, 25 m and 130 m, respectively; the initial speeds are 410 mm/s, 355 mm/s, 285 mm/s and 198 mm/s, respectively. The rising peak speeds are 420 mm/s, 390 mm/s, 295 mm/s and 218 mm/s, respectively. The end speeds are 395 mm/s, 228 mm/s, 258 mm/s and 190 mm/s, respectively. Through calculation, the attenuation percentages of the output speed of the four friction pairs from the initial state to the end state are 3.66%, 35.77%, 9.47% and 4.04%, respectively. The speed attenuation degree of the ZrO_2-Si_3N_4 friction pair is the lowest, which is followed by ZrO_2-GCr_{15} and 2A12-GCr_{15}. ZrO_2-ZrO_2 is the same kind of friction material, the wear of which is the most severe, and the output speed of the PUA attenuation is the fastest.

The discreteness of the output velocity of the PUA for long-distance operation was analyzed. It can be seen from Figure 6 that the velocity dispersions of the ZrO_2-Si_3N_4 and ZrO_2-GCr_{15} friction pairs are small, and the velocity dispersion of 2A12-GCr_{15} is the largest. However, the output speed of the ZrO_2-GCr_{15} combination fluctuates violently in the second half of the whole operation cycle of the PUA, which is caused by cracks on the surface of the abrasive particles and mover. The discrete type of the output speed of the PUA is accurately analyzed by a statistical method. A histogram of the output velocity distribution of the four different friction pairs is shown in Figure 6. It can be seen intuitively that the output speed distributions of ZrO_2-Si_3N_4, ZrO_2-GCr_{15} and 2A12-GCr_{15} are approximately normal, the output speed distribution of ZrO_2-ZrO_2 is not obvious and the speed stability is not good.

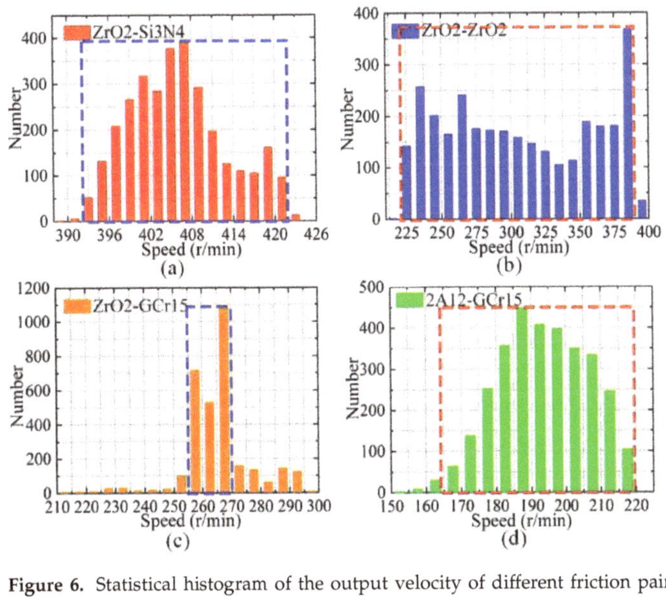

Figure 6. Statistical histogram of the output velocity of different friction pairs: (**a**) ZrO_2-Si_3N_4; (**b**) ZrO_2-ZrO_2; (**c**) ZrO_2-GCr_{15}; (**d**) 2A12-GCr_{15}.

In order to further quantitatively evaluate the stability of the output speeds of the four friction pairs, the part of the concentrated distribution of the output speeds in the dotted box in Figure 6 was selected for speed distribution analysis, and the mean value (V_m), standard deviation (SD), corrected sum of squares (CSS), coefficient of variation (CV), skewness (S) and kurtosis (K) of their output speeds were calculated, respectively. The calculation formula is as follows:

$$\begin{cases} CV = \frac{SD}{\overline{x}} \\ V_m = \frac{x_1 + x_2 ... + x_n}{n} \\ SD = \sqrt{\frac{(x_1-\overline{x})^2 + (x_2-\overline{x})^2 ... + (x_n-\overline{x})^2}{n}} \\ CSS = (x_1-\overline{x})^2 + (x_2-\overline{x})^2 \ldots + (x_n-\overline{x})^2 \\ S = \frac{\sqrt{n(n-1)}}{n-2} \left[\frac{\frac{1}{n}\sum_{i=1}^{n}(x_i-\overline{x})^3}{\left(\frac{1}{n}\sum_{i=1}^{n}(x_i-\overline{x})^2\right)^{\frac{3}{2}}} \right] \\ K = \frac{(n+1)n}{(n-1)(n-2)(n-3)} \sum_{i=1}^{n} \left(\frac{(x_i-\overline{x})^4}{s} \right) - \frac{3(n-1)^2}{(n-2)(n-3)} \end{cases} \quad (1)$$

where x_i is the speed sample, n is the number of samples and $\overline{x}$ is the speed average.

The calculation results are shown in Table 3.

Table 3. Discrete analysis parameters of the output speed of actuators with different friction pairs.

Friction Pairs	V_m	SD	CSS	CV	S	K
ZrO_2-Si_3N_4	406.06	7.01	1.53×10^5	0.01726	0.36572	−0.50952
ZrO_2-ZrO_2	307.39	53.29	88.5×10^5	0.17336	0.08296	−1.36762
ZrO_2-GCr_{15}	265.50	11.08	3.83×10^5	0.04173	−0.03711	2.60159

The mean velocity (V_m) reflects the output performance of different friction pairs. The mean velocities (V_m) of ZrO_2-Si_3N_4, ZrO_2-ZrO_2, ZrO_2-GCr_{15} and 2A12-GCr_{15} are 406.06 r/min, 307.39 r/min, 265.50 r/min and 193.46 r/min respectively. The output performance of ZrO_2-Si_3N_4 is the best, followed by those of ZrO_2-ZrO_2 and ZrO_2-GCr_{15}, and that of 2A12-GCr_{15} is the worst. The standard deviation (SD), corrected sum of squares (CSS) and coefficient of variation (CV) reflect the dispersion of speed. The order of the three evaluation indexes from small to large is ZrO_2-Si_3N_4, ZrO_2-GCr_{15} and 2A12-GCr_{15}, ZrO_2-ZrO_2, which fully shows that the discrete type of the ZrO_2-Si_3N_4 output speed is the smallest, that of ZrO_2-GCr_{15}, 2A12-GCr_{15} is the second smallest and the ZrO_2-ZrO_2 output speed is the largest. Skewness (S) and kurtosis (K) reflect the distribution of the output speed of the PUA, and skewness (S) reflects the symmetry of the operating speed. The skewness (S) values of ZrO_2-Si_3N_4 and ZrO_2-ZrO_2 are positive, indicating that their speed is concentrated on the left side of the mean value, and the skewness (S) values of ZrO_2-GCr_{15} and 2A12-GCr_{15} are negative; it shows that the velocity is concentrated on the right side of the mean value. Kurtosis (K) reflects the aggregation degree of the velocity in the mean center. The kurtosis (K) value of ZrO_2-GCr_{15} is the largest, and the velocity aggregation degree is the best. However, in the second half of the operation, the output velocity fluctuates violently due to the generation of abrasive particles and mover surface cracks.

3. Discussion and Friction Material Selection

According to Hertz contact theory [32], when the stator of the PUA is in contact with the mover, the depth h of the driving foot pressing into the mover can be calculated as follows:

$$h = \frac{F}{\pi l} \times \frac{1-v^2}{E}\left(2\ln\frac{2R}{\sqrt{\frac{F}{\pi l}\times\frac{R}{E}}} - 1\right) \quad (2)$$

$$\frac{1}{E} = \left(\frac{1-v_1^2}{E_1} + \frac{1-v_2^2}{E_2}\right) \quad (3)$$

where F is the preload between the driving foot and the mover, l is the contact thickness between the driving foot and the mover (mm), R is the radius of the driving foot (mm), h is the driving foot pressuring depth into the mover, v is the equivalent Poisson's ratio of the stator, E is the equivalent modulus of the friction pair composed of the actuator and the stator, E_1 is the elastic modulus of the stator and E_2 is the elastic modulus of the mover.

The output force of the PUA was tested by combining different friction pair materials, and the variation law of the output force with the excitation voltage and preload was measured, respectively, the equivalent elastic modulus E and friction coefficient of the four friction pair materials are shown in Table 4.

Table 4. Equivalent elastic modulus and friction coefficient of different friction pair combinations.

Friction Pair Material	ZrO_2-Si_3N_4	ZrO_2-ZrO_2	ZrO_2-GCr_{15}	ZrO_2-PEEK
Equivalent elastic modulus E (GPa)	126.8	105	104.5	3.73
Friction coefficient	0.13	0.11	0.13	0.14

Assuming F = 100 N, l = 2 mm, R = 4 mm and v = 0.3, the relationship between the depth h of the driving foot pressing into the mover and the equivalent elastic modulus E of the stator and mover is obtained, as shown in Figure 7a. It can be seen that the depth h of the driving foot pressing into the mover gradually decreases with the increase in the equivalent elastic modulus E, and the depth h of the driving foot pressing into the mover decreases rapidly when the equivalent elastic modulus E is small. When the equivalent elastic modulus E is large, the depth h of the driving foot into the mover decreases slowly.

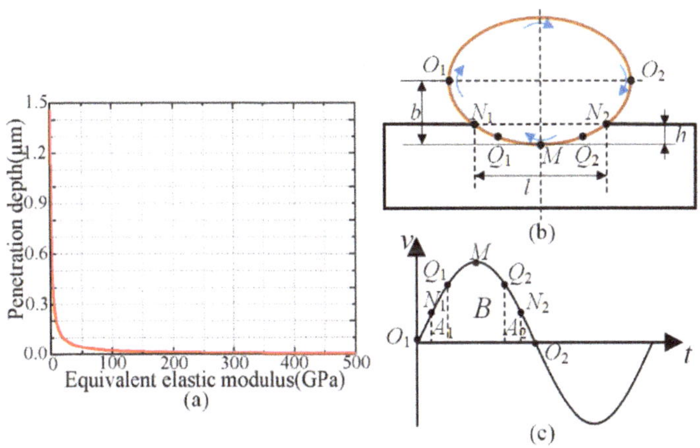

Figure 7. Micro-mechanism analysis. (**a**) Relationship between pressuring depth and equivalent elastic modulus. (**b**) Stator and mover micro-drive model. (**c**) Horizontal movement speed of the driving point.

The power transmission process between the stator and the mover of the PUA is analyzed. As shown in Figure 7b,c, during the elliptical motion of the point on the driving foot of the PUA, its horizontal velocity can be expressed as follows:

$$v_x = d_x{}' = A\omega \sin(\omega t) \tag{4}$$

where d_x represents the displacement in the horizontal direction, A is the velocity amplitude, ω is the frequency of the speed and its velocity is a sinusoidal curve. When the driving foot is at point O_1 in the initial state, the horizontal speed is 0, and when it moves to point M, the horizontal speed reaches the maximum. Then, in the process of driving the foot from point M to point O_2, the speed gradually decreases to 0. The stator experiences the process from contact to separation with the mover. Point N_1 is the contact point, point N_2 is the separation point and points Q_1 and Q_2 are the points where the horizontal speed of the stator is equal to the speed of the mover, which is called the isokinetic point. When the stator moves from point N_1 to point Q_1, the stator speed is less than the mover speed, and the friction does negative work to the mover. When the stator moves from point Q_1 to point Q_2, the speed of the stator is greater than that of the mover, and the friction does positive work to the mover. When the stator moves from point Q_2 to point N_2, the stator speed is less than the mover speed, and the friction does negative work to the mover. When the positive work and negative work done by the friction on the mover are equal, the output speed of the mover is stable.

For different friction materials, the pressuring depth h between the stator and actuator is closely related to its elastic modulus. The pressuring depth h between the stator and actuator is smaller when the equivalent elastic modulus E is larger, so the position of the contact point between the stator and actuator moves backward. In this case, if the location of the uniform point remains at the previous position, the positive work done by the friction force on the mover remains unchanged, while the negative work decreases, which does not meet the law of energy conservation. Therefore, the isokinetic point must be increased to reduce the positive work and increase the negative work to achieve an energy balance. Therefore, the friction pair combination with the larger equivalent elastic modulus E has a higher output speed.

When the equivalent elastic modulus E of the friction pair material is smaller, the pressuring depth h between the stator and the mover is larger, which leads to a larger effective contact area between the stator and the mover, and the friction force is increased to a certain extent, thus increasing the output force of the PUA. However, when the pressuring depth is too large and exceeds a certain critical value, the elliptical motion trajectory of the stator will "fall" into the mover, which reduces the power transmission efficiency and the output force of the PUA. By comprehensively comparing the mechanical output performance and the stability of the output speed of the PUA under different friction pair material combinations, it can be found that the ZrO_2-Si_3N_4 material combination has a better performance in terms of output speed and speed stability than that of other friction pair materials. The output force of the ZrO_2-Si_3N_4 combination is better under a high preload, but the performance under a low preload is not as good as that of other friction pair combinations. The piezoelectric ultrasonic actuated linear platform is developed in this work, which is rarely used in applications where the output force is required, so the combination of the ZrO_2-Si_3N_4 friction pair is selected. The piezoelectric linear driving platform developed in this work has a low demand for the output force in application, so the ZRO2-Si3N4 friction pair combination is selected.

4. Results

In order to measure the mechanical output characteristics of the PUA linear motion platform, an experimental test system was built. The overall dimensions of the platform in the X, Y and Z directions are 80 mm, 51.5 mm and 66 mm, respectively, as shown in Figures 8 and 9a. The driving power supply of the system adopted the commercial QD-8D ultrasonic power supply. The magnetic grid displacement sensor (Model: MSK

200/1-0085, SIKO GmbH, Buchenbach, Germany) and the data acquisition card (DAQ card) (Model: PCI 2394, Beijing Art Technology Development Co., Ltd., Beijing, China) were used to measure the motion displacement of the linear platform and collect the test data, respectively. The collected data were displayed and processed on the PC; the weight block was placed on the platform to measure the load speed output characteristics of the linear platform.

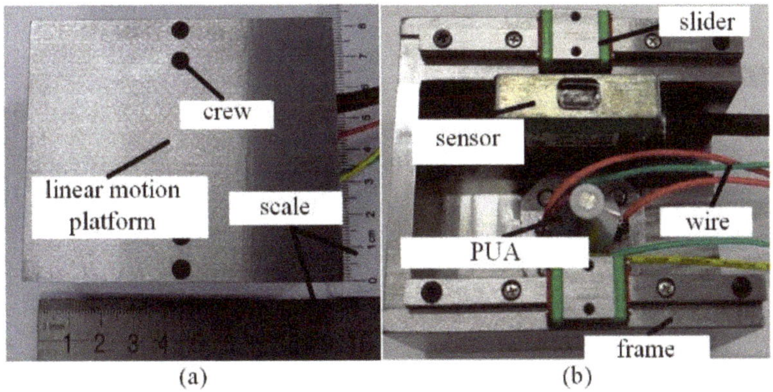

Figure 8. Linear platform system. (**a**) Appearance top view. (**b**) Internal structure.

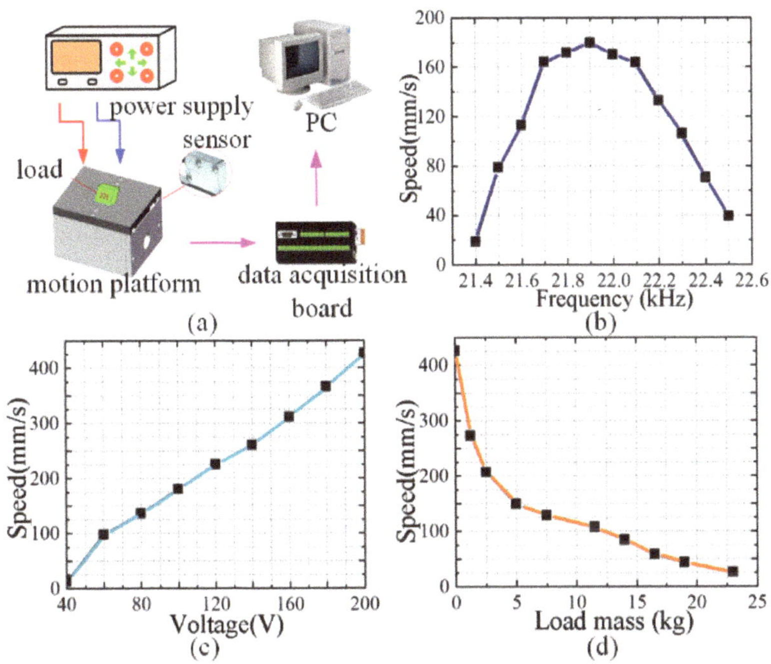

Figure 9. Application test system and experimental results. (**a**) Experimental test system for the mechanical output characteristics of the linear motion platform. (**b**) Relationship between the speed and frequency. (**c**) Relationship between the speed and excitation voltage. (**d**) Relationship between the speed and load mass.

The speed and frequency characteristics of the linear motion platform were tested to determine the best working frequency point of the linear motion platform by the experimental platform. During the experiment, the excitation voltage was 100 $V_{p\text{-}p}$, the phase difference of the two excitation signals was 90°, the preload between the driving foot and the mover was 18 N and the frequency of the excitation signal increases from 21.4 kHz to 22.5 kHz. As shown in Figure 9b, the experimental results show that, with the increase in the excitation frequency, the motion speed of the linear platform increases at the beginning and then decreases. When the excitation frequency is 21.9 kHz, the maximum output speed of the linear platform is 179.8 mm/s, indicating that 21.9 kHz is the best working frequency point of the linear motion platform.

The speed and excitation voltage characteristics of the linear platform were tested with the measured optimal frequency point. The experimental results are shown in Figure 9c. It can be seen from the experimental results that the output speed of the linear platform increases significantly with the increase in the excitation voltage, which is approximately linear with the excitation voltage. When the excitation voltage reaches 200 $V_{p\text{-}p}$, the output speed can reach 426.2 mm/s. When the excitation voltage is less than 40 $V_{p\text{-}p}$, the output speed is 0, which shows that, when the excitation voltage is less than 40 $V_{p\text{-}p}$, the foot end displacement of the PUA is not enough to overcome the maximum static friction between the actuator and the mover.

The speed and load characteristics of the platform were tested under an excitation voltage of 200 $V_{p\text{-}p}$, a frequency of 21.9 kHz and a phase difference of 90°. The experimental results are shown in Figure 9d. The experimental results show that the output speed decreases with the increase in the load mass. When the load mass is less than 5 kg, the output speed decreases faster, and when the load mass is greater than 5 kg, the output speed decreases slowly. When the load mass is 23 kg, the output speed of the platform is 25.75 mm/s, which performs a strong load capacity.

The output displacement resolution of the linear platform was tested under the excitation of the pulsed AC sinusoidal excitation signal, as shown in Figure 10a. The excitation voltage was 100 $V_{p\text{-}p}$, the excitation signal frequency fr was 21.9 kHz, the frequency fs of the step cycle was 100 Hz and the preload between the driving foot and the mover was 18 N. The output displacement of the linear platform was obtained by changing the number of sinusoidal signals in a step cycle to 20, 30, 40, 50 and 60, respectively, as shown in Figure 10b. The experimental results show that, under the excitation of the pulsed AC sinusoidal excitation signal, the linear platform outputs a stepping motion, and the number n of sinusoidal signals in a step cycle significantly affects the stepping displacement of the linear platform. The measurement resolution of the magnetic grid displacement sensor is 0.5 μm. When the number of signals is less than 5, the magnetic grid displacement sensor can no longer meet the measurement requirements; it was replaced by capacitive displacement sensors (PI D-E20.050, German PI Co., Ltd., Karlsruhe, Germany) with a range of 50 μm and a dynamic resolution of 1nm. The experimental results show that, when there is only one sinusoidal signal, the linear platform can reach a motion step of about 85 nm, that is, the PUA can achieve a nano-level high resolution in the resonant working state when there is only one sinusoidal signal. The output displacement of the linear platform is shown in Figure 10c.

The comparisons between the proposed ultrasonic motor and some existing ultrasonic motors are listed in detail in Table 5. Izuhara et al. proposed a miniature linear ultrasonic motor with a size of 2.6 × 2.6 × 2.2 mm^3 and a maximum output speed of 140 mm/s at 150 Vp-p. The motor has a compact structure but a small load capacity [33]. Smithmaitrie et al. proposed a dual transducer ultrasonic motor with a motor size of 85 × 6 × 1 mm^3 and a maximum output speed of 176.2 mm/s under 60 Vp-p. The structure of the dual transducer is complex, and the machining accuracy is complex, so it is difficult to ensure the consistency of the actuator [34]. Shi et al. proposed a precision linear ultrasonic motor with a size of 55 × 26 × 14 mm^3, and the speed reaches 159 mm/s under a 400 Vp-p driving voltage [35]. The maximum speed of the motor proposed in this paper is 179.8 mm/s under 100 Vp-p, the load mass can reach 23 kg and the resolution in the resonance mode is 85 nm. These performances are significantly better than that of the several motors listed in Table 5.

Therefore, the proposed motor has a wide application potential in the fields of high-speed, large-load and precision drive applications.

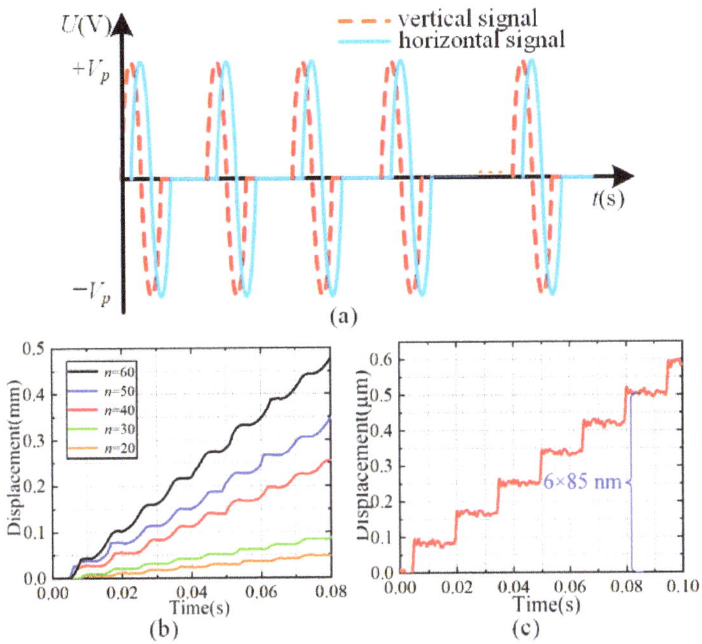

Figure 10. Excitation signal and output displacement characteristics. (**a**) Sinusoidal alternating current pulse signal. (**b**) Output displacement under the pulse signal. (**c**) Output displacement under one signal.

Table 5. Comparison between the proposed ultrasonic motor and some existing works.

Parameters	This Work	The Motors in [33]	The Motors in [34]	The Motors in [35]
Total size (mm)	30 × 30 × 34.2	2.6 × 2.6 × 2.2	85 × 6 × 1	55 × 26 × 14
Voltage (Vp-p)	100	150	60	400
Speed (mm/s)	179.8	140	176.2	159
Load mass (kg)	23	/	/	/
Resolution (μm)	0.085	/	/	/

5. Conclusions

In this study, ZrO_2 was used as the driving foot surface material of the PUA to form a friction pair with Si_3N_4, ZrO_2, GCr_{15} and PEEK, respectively. The variation rules of the output speed with the excitation voltage and preload were tested, respectively, by four friction pairs. The output speeds of the friction pair composed of ZrO_2-Si_3N_4 is the best under the same excitation voltage or preload. Then, the variation rules of the output force with the excitation voltage and preload were tested, respectively. The output force of ZrO_2-Si_3N_4 is always the largest under the same excitation voltage. When the preload force is greater than 6.5 N, the output force of ZrO_2-Si_3N_4 is always greater than that of other friction pairs. Under an excitation signal of 200 V_{p-p} and a preload of 2 N, different combinations of friction pairs are used to drive the mover for 500 m to test the output stability. In the long-term test of the friction pair composed of ZrO_2-Si_3N_4, the attenuation percentage of the speed from the initial state to the end state in the speed rising stage of 50 m is 3.66%. According to the experimental results, the combination of ZrO_2-Si_3N_4

has the best output characteristics and was thus selected and applied to the linear motion platform. The maximum output speed of the platform can reach 426.2 mm/s; it can achieve a high resolution of about 85 nm and has good load characteristics, which are important to push the technological development of high-end manufacturing equipment. This work also provides an idea for improving the output capacity and long-term operation stability of ultrasonic motors and provides a possibility for expanding the application range of ultrasonic motors.

Author Contributions: Conceptualization, designing, writing—original draft preparation: J.D. and J.C.; experimentation, simulation, analysis: Y.G.; writing—review and editing, supervision: H.L., F.L. and W.C. All authors have read and agreed to the published version of the manuscript.

Funding: This research is funded by the National Natural Science Foundation of China, the China Postdoctoral Science Foundation and the Postdoctoral Science Foundation of Heilongjiang Province; the grant numbers are 5210051275, 2021M690830 and LBH-Z21018, respectively.

Institutional Review Board Statement: Not applicable.

Informed Consent Statement: Not applicable.

Data Availability Statement: Not applicable.

Conflicts of Interest: The authors declare no conflict of interest.

References

1. Shi, Y.L.; Lou, C.S.; Zhang, J. Investigation on a linear piezoelectric actuator based on stick-slip/scan excitation. *Actuators* **2021**, *10*, 39. [CrossRef]
2. Tang, J.Z.; Wei, J.S.; Wang, Y.M.; Xu, Z.; Huang, H. A novel rotation-structure based stick-slip piezoelectric actuator with high consistency in forward and reverse motions. *Actuators* **2021**, *10*, 189. [CrossRef]
3. Yu, H.P.; Liu, Y.X.; Tian, X.Q.; Zhang, S.J.; Liu, J.K. A precise rotary positioner driven by piezoelectric bimorphs: Design, analysis and experimental evaluation. *Sens. Actuators A Phys.* **2020**, *313*, 112197. [CrossRef]
4. Wang, F.J.; Shi, B.C.; Tian, Y.L.; Huo, Z.C.; Zhao, X.Y.; Zhang, D.W. Design of a novel dual-axis micromanipulator with an asymmetric compliant structure. *IEEE/ASME Trans. Mechatron.* **2013**, *24*, 656–665. [CrossRef]
5. Deng, J.; Liu, Y.X.; Li, J.; Zhang, S.J.; Li, K. Displacement linearity improving method of stepping piezoelectric platform based on leg wagging mechanism. *IEEE Trans. Ind. Electron.* **2022**, *69*, 6429–6432. [CrossRef]
6. Deng, J.; Liu, S.H.; Liu, Y.X.; Wang, L.; Gao, X.; Li, K. A 2-DOF needle insertion device using inertial piezoelectric actuator. *IEEE Trans. Ind. Electron.* **2022**, *69*, 3918–3927. [CrossRef]
7. Gao, X.; Deng, J.; Zhang, S.J.; Li, J.; Liu, Y.X. A compact 2-DOF micro/nano manipulator using single miniature piezoelectric tube actuator. *IEEE Trans. Ind. Electron.* **2022**, *69*, 3928–3937. [CrossRef]
8. Ghenna, S.; Giraud, F.; Giraud-Audine, C.; Amberg, M. Vector control of piezoelectric transducers and ultrasonic actuators. *IEEE Trans. Ind. Electron.* **2018**, *65*, 4880–4888. [CrossRef]
9. Liu, Y.X.; Li, J.; Deng, J.; Zhang, S.J.; Chen, W.S.; Xie, H.; Zhao, J. Arthropod-metamerism-inspired resonant piezoelectric millirobot. *Adv. Intell. Syst.* **2021**, *3*, 2100015. [CrossRef]
10. Du, P.F.; Liu, Y.X.; Chen, W.S.; Zhang, S.J.; Deng, J. Fast and precise control for the vibration amplitude of an ultrasonic transducer based on fuzzy PID control. *IEEE Trans. Ultrason. Ferroelect. Freq. Control* **2012**, *68*, 2766–2774. [CrossRef]
11. Ma, X.F.; Liu, Y.X.; Liu, J.K.; Deng, J. Crabbot: A pole-climbing robot driven by piezoelectric stack. *IEEE Trans. Robot.* **2021**, *38*, 765–778.
12. Cheng, J.F.; Deng, J.; Liu, Y.X.; Zhang, S.J.; Lu, F.; Tian, X.Q.; Chen, W.S. Step consistency active control method for inertial piezoelectric actuator using embedded strain gauges. *Rev. Sci. Instrum.* **2021**, *92*, 25005. [CrossRef] [PubMed]
13. Wang, L.; Wang, H.R.; Cheng, T.H. Design and performance of a compact stick-slip type piezoelectric actuator based on right triangle flexible stator. *Smart Mater. Struct.* **2022**, *31*, 055013. [CrossRef]
14. Kim, J.H.; Kim, S.H.; Kwak, Y.K. Development and optimization of 3-D bridge-type hinge mechanisms. *Sens. Actuators A Phys.* **2004**, *116*, 530–538. [CrossRef]
15. Li, J.P.; Zhao, H.W.; Qu, X.T.; Qu, H.; Zhou, X.Q.; Fan, Z.Q.; Ma, Z.C.; Fu, H.S. Development of a compact 2-DOF precision piezoelectric positioning platform based on inchworm principle. *Sens. Actuators A Phys.* **2015**, *222*, 87–95. [CrossRef]
16. Wang, Y.; Huang, W. A piezoelectric motor with two projections using two orthogonal flexural vibration modes. *Sens. Actuator A Phys.* **2016**, *250*, 170–176. [CrossRef]
17. Makarem, S.; Delibas, B.; Koc, B. Data-driven tuning of PID controlled piezoelectric ultrasonic motor. *Actuators* **2021**, *10*, 148. [CrossRef]
18. Li, H.; Tian, X.Q.; Shen, Z.H.; Li, K.; Liu, Y. A low-speed linear stage based on vibration trajectory control of a bending hybrid piezoelectric ultrasonic motor. *Mech. Syst. Signal Process.* **2019**, *132*, 523–534. [CrossRef]

19. Liu, Y.X.; Wang, L.; Gu, Z.Z.; Quan, Q.Q.; Deng, J. Development of a two-dimensional linear piezoelectric stepping platform using longitudinal-bending hybrid actuators. *IEEE Trans. Ind. Electron.* **2019**, *66*, 3030–3040. [CrossRef]
20. Liu, Y.X.; Yan, J.P.; Wang, L.; Chen, W.S. A two-DOF ultrasonic motor using a longitudinal-bending hybrid sandwich transducer. *IEEE Trans. Ind. Electron.* **2019**, *66*, 3041–3050. [CrossRef]
21. Wallaschek, J. Contact mechanics of piezoelectric ultrasonic motors. *Smart Mater. Struct.* **1998**, *7*, 369–381. [CrossRef]
22. Gao, J.; Altintas, Y. Development of a three-degree-of-freedom ultrasonic vibration tool holder for milling and drilling. *IEEE/ASME Trans. Mechatron.* **2012**, *24*, 1238–1247. [CrossRef]
23. Qiu, W.; Mizuno, Y.; Nakamura, K. Tribological performance of ceramics in lubricated ultrasonic motors. *Wear* **2016**, *352–353*, 188–195. [CrossRef]
24. Qu, J.J.; Zhang, Y.H.; Tian, X.; Li, J.B. Wear behavior of filled polymers for ultrasonic motor in vacuum environments. *Wear* **2015**, *322*, 108–116. [CrossRef]
25. Gu, D.P.; Duan, C.S.; Fan, B.L.; Chen, S.W.; Yang, Y.L. Tribological properties of hybrid PTFE/Kevlar fabric composite in vacuum. *Tribol. Int.* **2016**, *103*, 423–431. [CrossRef]
26. Liu, X.L.; Qiu, J.H.; Zhao, G. Improved energy conversion efficiency of the ultrasonic motor with surface texture. *Ind. Lubr. Tribol.* **2018**, *70*, 1729–1736. [CrossRef]
27. Song, J.F.; Zhao, G.; Ding, Q.J.; Qiu, J.H. Reciprocating friction and wear of polyimide composites filled with solid lubricants. *J. Polym. Eng.* **2018**, *38*, 363–370. [CrossRef]
28. Sun, F.Y.; Qu, J.J. Tribological properties of Ekonol composites friction material in traveling ware ultrasonic motor. *Adv. Mat. Res.* **2012**, *322*, 404–407. [CrossRef]
29. Zhang, Y.H.; Fu, Y.H.; Hua, X.J.; Quan, L.; Qu, J.J. Characteristics and attenuation mechanism of linear standing-wave piezoelectric motors with ceramics-mated friction couples. *Tribol. Int.* **2021**, *153*, 106580. [CrossRef]
30. Adachi, K.; Kato, K. Formation of smooth wear surfaces on alumina ceramics by embedding and tribo-sintering of fine wear particles. *Wear* **2000**, *245*, 84–91. [CrossRef]
31. Olofsson, J.; Johansson, S.; Jacobson, S. Influence from humidity on the alumina friction drive system of an ultrasonic motor. *Tribol. Int.* **2009**, *42*, 1467–1477. [CrossRef]
32. Gourgiotis, P.A.; Zisis, T.; Giannakopoulos, A.E.; Georgiadis, H.G. The hertz contact problem in couple-stress elasticity. *Int. J. Solids Struct.* **2019**, *168*, 228–237. [CrossRef]
33. Izuhara, S.; Mashimo, T. Design and evaluation of a micro linear ultrasonic motor. *Sens. Actuator A Phys.* **2018**, *278*, 60–66. [CrossRef]
34. Smithmaitrie, P.; Suybangdum, P.; Laoratanakul, P.; Muensit, N. Design and performance testing of an ultrasonic linear motor with dual piezoelectric actuators. *IEEE Trans. Ultrason. Ferroelect. Freq. Control* **2012**, *59*, 1033–1042. [CrossRef] [PubMed]
35. Shi, Y.L.; Zhao, C.S. Simple new ultrasonic piezoelectric actuator for precision linear positioning. *J. Electroceram.* **2012**, *28*, 233–239. [CrossRef]

actuators

Article

Design, Modeling, Testing, and Control of a Novel Fully Flexure-Based Displacement Reduction Mechanism Driven by Voice Coil Motor

Yunzhuang Chen and Leijie Lai *

School of Mechanical and Automotive Engineering, Shanghai University of Engineering Science, Shanghai 201620, China
* Correspondence: lailj@sues.edu.cn; Tel.: +86-134-8247-4766

Abstract: This paper presents a flexure-based displacement reduction mechanism driven by a voice coil motor to improve the motion resolution and eliminate the hysteresis nonlinearity of the traditional piezo-actuated micropositioning/nanopositioning stages. The mechanism is composed of three groups of compound bridge-type displacement reduction mechanisms, which adopt distributed-compliance rectangular beams to reduce the concentration of stress and improve the dynamic performance of the mechanism. The symmetrical distribution of the structure can eliminate the parasitic displacement of the mechanism and avoid the bending moment and lateral stress applied to the voice coil motor. Firstly, the analytical model of the mechanism is obtained by the stiffness matrix method. The theoretical displacement reduction ratio, input stiffness, and natural frequency of the displacement reduction mechanism are obtained by solving the analytical model. Then, through the static analysis and modal analysis of the mechanism with the Ansys software, the accuracy of the analytical model is verified, and the experimental prototype is also constructed for performance tests. The results show that the maximum stroke of the mechanism is 197.43 μm with motion resolution of 40 nm. The natural frequency is 291 Hz, and the input stiffness is 28.50 N/mm. Finally, the trajectory tracking experiment is carried out to verify the positioning performance of the mechanism. The experimental results show that the designed feedback controller has good stability, and the introduction of the feedforward controller and disturbance observer can greatly reduce the tracking errors.

Keywords: displacement reduction mechanism; voice coil motor; micropositioning/nanopositioning; compound bridge-type mechanism

Citation: Chen, Y.; Lai, L. Design, Modeling, Testing, and Control of a Novel Fully Flexure-Based Displacement Reduction Mechanism Driven by Voice Coil Motor. *Actuators* **2022**, *11*, 228. https://doi.org/10.3390/act11080228

Academic Editor: Kenji Uchino

Received: 17 July 2022
Accepted: 5 August 2022
Published: 8 August 2022

1. Introduction

Microdisplacement actuators such as piezoelectric actuators (PZTs) and voice coil motors (VCMs) are widely used in the fields of micromanipulation/nanomanipulation, scanning probe microscopy, and ultra-precision machining owing to their high stiffness, high resolution, fast response, and nanopositioning accuracy [1–4]. However, the PZT has only a small stroke of tens of microns and nonlinear characteristics such as hysteresis and creep [5,6]. Therefore, in order to achieve larger output displacement, flexure-based displacement amplification mechanisms are often used to amplify the stroke of the PZT [6]. As a kind of electromagnetic actuator, the VCM has been increasingly used in the large-stroke micropositioning/nanopositioning stages [7]. For instance, micropositioning/nanopositioning stages with millimeter or even centimeter ranges were reported in [8–14]. However, with the increase of the travel range, the motion resolution and accuracy of the micropositioning stage actuated by the VCM will also decline. Therefore, opposite the amplified PZTs, the motion resolution of the long-range electromagnetic actuators such as the VCM can be improved by employing a displacement reduction mechanism, which can also solve the hysteresis problem of the PZTs.

In the previous literature, many works have systematically studied the design and modeling of the flexure-based displacement amplification mechanism [15–18], but few the reduction mechanism. Theoretically, the displacement amplification mechanism can achieve the purpose of displacement reduction by exchanging the input and output positions. However, there will be several problems after exchanging. Firstly, the specifications to be considered in designing the amplification mechanism are obviously different from those in designing the reduction mechanism [19–21]. Different from the displacement amplification mechanism driven by the PZT, the displacement reduction mechanism changes the macrodisplacement generated by the VCM into microdisplacement. Therefore, the stiffness design and the macro-/micro-scale transmission process of the displacement reduction mechanism should be deeply studied. Several flexure-based displacement reducers have been designed in the past few years. For instance, Wu and Chen proposed a bridge differential displacement reducer with a reduction ratio of 100 by combining two bridge-type flexure mechanisms [22]. However, this kind of displacement reducer with a high reduction ratio is generally suitable for the PZT. The large reduction ratio achieved by combining multiple reducers will also lead to energy transmission inefficiency [19–21].

Therefore, the focus of this paper is to propose a novel fully flexure-based displacement reduction mechanism. Through the displacement reduction mechanism driven by the large-stroke VCM, the millimeter-level output displacement of the motor is transmitted to the end-effector to achieve micron-level output. The mechanism is composed of three groups of compound bridge-type displacement reduction mechanisms. The distributed-compliance rectangular beams used in the mechanism not only reduce the concentration of stress, but also improve the dynamic performance of the mechanism. The symmetrical distribution structure can eliminate the parasitic displacement of the mechanism along the X direction and avoid the bending moments and lateral stress applied to the VCM. In terms of stroke range, through reasonable structural design, the displacement reduction mechanism driven by the VCM has a stroke equivalent to the conventional displacement amplification mechanism driven by the PZT. More importantly, the design avoids various nonlinear characteristics such as hysteresis and creep caused by the PZT in principle. The rest of this paper is organized as follows: In Section 3, based on the stiffness matrix method, the analytical model of the mechanism is established. Section 4 verifies the good static performance of the mechanism through finite element analysis and experiment tests. In Section 5, the position control is carried out to verify the good trajectory tracking performance of the mechanism. Finally, the conclusion is given in Section 6. The main contribution and innovation points of this work are that the displacement reduction mechanism is successfully used to reduce the displacement of the VCM to improve its motion resolution and effectively eliminate the hysteresis nonlinearity of the traditional piezo-actuated nanopositioning stage.

2. Design of Displacement Reduction Mechanism

The displacement reduction mechanism proposed in this paper is a distributed-compliance flexure mechanism, and its structure and working principle are shown in Figure 1. The mechanism is composed of three groups of compound bridge-type displacement reduction mechanisms. The two groups of bridge-type mechanisms on both sides are secondary reduction mechanisms, and the one in the middle is a primary reduction mechanism. After the two-stage displacement reduction of the mechanism, the millimeter-level input displacement of Rigid Body 3 produced by the VCM can be significantly reduced and transmitted to Rigid Body 4 to achieve micron-level output displacement. Instead of the commonly used concentrated notch flexure hinge in the traditional flexure mechanism, the proposed mechanism adopts 12 distributed-compliance rectangular beams, which not only reduce the concentration of stress, but also improve the dynamic performance of the mechanism. This is because, in the deformation process, the stress generated on the notch flexure hinge will become very large, which will make the flexure hinge very fragile in this state. Moreover, the traditional lumped-compliance bridge-type mechanism, using the notch flexure hinges, will introduce rigid bodies with large mass in its arms, which has great weakness for the resonant frequency of the mechanism. Therefore, the bending deformation of rectangular beams in the proposed mechanism makes the stress distribution of the whole flexure mechanism relatively uniform, rather than concentrated in specific areas. This characteristic ensures that the mechanism has better performance and longer service life in dynamic applications. Moreover, the mechanism adopts a symmetrical distribution structure to reduce the input displacement with two stages, which can also eliminate the parasitic displacement of the mechanism in the X direction, avoiding the bending moments and lateral stress applied to the VCM, and effectively eliminate the friction between the mover and the stator in the VCM. In addition, it is worth mentioning that the design can avoid various nonlinear characteristics such as hysteresis and creep caused by the PZT in principle and improve the stability and positioning accuracy.

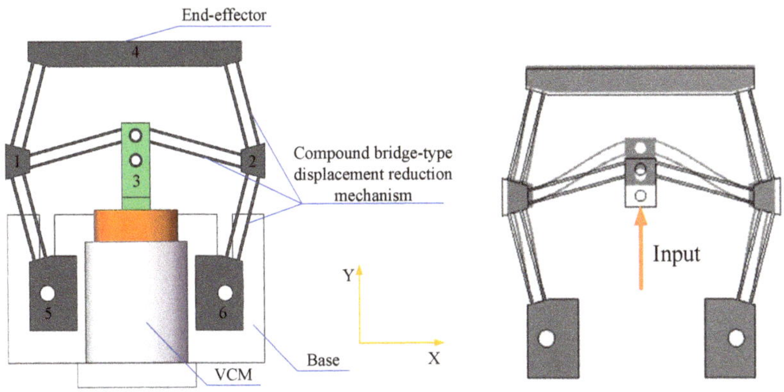

Figure 1. Flexure-based displacement reduction mechanism. (**a**) Structure of the mechanism. (**b**) Working principle of the mechanism.

3. Analytical Model

As shown in Figure 1a, the displacement reduction mechanism is mainly composed of six rigid bodies, of which Rigid Bodies 5 and 6 are used as frames, and the other four are movable rigid bodies. To obtain the dynamic equation of the mechanism, the three degrees of freedom (3-DOFs) $\mathbf{q}_i = \begin{bmatrix} x_i & y_i & \theta_{zi} \end{bmatrix}$ of each movable rigid body are taken as the generalized coordinates of the mechanism.

$$\mathbf{q} = \begin{bmatrix} \mathbf{q}_1 & \mathbf{q}_2 & \cdots & \mathbf{q}_4 \end{bmatrix}^T \tag{1}$$

Therefore, the multi-DOF vibration differential equation of the displacement reduction mechanism is expressed in matrix form as

$$\mathbf{M}\ddot{\mathbf{q}} + \mathbf{K}\mathbf{q} = \mathbf{F} \tag{2}$$

$$\mathbf{M} = \mathrm{diag}(\mathbf{M}_1 \quad \mathbf{M}_2 \quad \mathbf{M}_3 \quad \mathbf{M}_4) \tag{3}$$

The stiffness matrix **K** is

$$\mathbf{K} = \begin{bmatrix} \tilde{K}_{11} & -\tilde{K}_{12} & -\tilde{K}_{13} & -\tilde{K}_{14} \\ -\tilde{K}_{21} & \tilde{K}_{22} & -\tilde{K}_{23} & -\tilde{K}_{24} \\ -\tilde{K}_{31} & -\tilde{K}_{32} & \tilde{K}_{33} & -\tilde{K}_{34} \\ -\tilde{K}_{41} & -\tilde{K}_{42} & -\tilde{K}_{43} & \tilde{K}_{44} \end{bmatrix} \tag{4}$$

The external force vector **F** is

$$\mathbf{F} = \begin{bmatrix} \mathbf{F}_1 & \mathbf{F}_2 & \mathbf{F}_3 & \mathbf{F}_4 \end{bmatrix}^T \tag{5}$$

In Equation (3), the form of the ith element of the mass matrix $\mathbf{M}$ is $\mathbf{M}_i = diag(\begin{matrix} m_{xi} & m_{yi} & J_{zi} \end{matrix})$, where (m_{xi}, m_{yi}) are the mass components of the ith rigid body in the X and Y directions, respectively. J_{zi} is the mass moment of inertia of the ith rigid body in the Z-axis direction. In Equation (5), the ith element of external force vector $\mathbf{F}$ is $\mathbf{F}_i = diag(\begin{matrix} \mathbf{F}_{xi} & \mathbf{F}_{yi} & \mathbf{M}_{zi} \end{matrix})$. When force vector **F** is applied, the relationship between the applied force and displacement is obtained by the following formula:

$$\mathbf{q} = \mathbf{K}^{-1}\mathbf{F} \tag{6}$$

Based on the observation method in vibration theory, each element in the stiffness matrix of **K** can be obtained. This method can be described as the main diagonal element $\tilde{K}_{ii}$ of the stiffness matrix being the sum of the stiffness of the elastic elements connected with the ith rigid body, and the non-diagonal element $\tilde{K}_{ij}$ is the sum of the stiffness of the elastic elements connected with the ith and jth rigid bodies. Therefore, in Equation (4), the elements of the equivalent stiffness matrix are

$$\tilde{K}_{ii} = \sum_{j=1}^{n_i} \mathbf{J}_{ij}\mathbf{K}_{ij}{\mathbf{J}_{ij}}^T \tag{7}$$

$$\tilde{K}_{ij} = \sum_{k=1}^{n_{ij}} \mathbf{J}_{ijk}\mathbf{K}_{ijk}{\mathbf{J}_{jik}}^T, \tilde{K}_{ji} = \tilde{K}_{ij}^T \tag{8}$$

$$\mathbf{J}_{ij} = \begin{bmatrix} 1 & 0 & 0 \\ 0 & 1 & 0 \\ -r_{y_{ij}} & r_{x_{ij}} & 1 \end{bmatrix} \begin{bmatrix} \cos\theta_{ij} & -\sin\theta_{ij} & 0 \\ \sin\theta_{ij} & \cos\theta_{ij} & 0 \\ 0 & 0 & 1 \end{bmatrix} \tag{9}$$

$$\mathbf{H}_{ijk} = \begin{bmatrix} 1 & 0 & 0 \\ 0 & 1 & 0 \\ -h_{y_{ijk}} & h_{x_{ijk}} & 1 \end{bmatrix} \tag{10}$$

$$\begin{aligned} \mathbf{J}_{ijk} &= \begin{bmatrix} 1 & 0 & 0 \\ 0 & 1 & 0 \\ -r_{y_{ijk}} & r_{x_{ijk}} & 1 \end{bmatrix} \begin{bmatrix} \cos\theta_{ijk} & -\sin\theta_{ijk} & 0 \\ \sin\theta_{ijk} & \cos\theta_{ijk} & 0 \\ 0 & 0 & 1 \end{bmatrix}, \\ \mathbf{J}_{jik} &= \begin{bmatrix} 1 & 0 & 0 \\ 0 & 1 & 0 \\ -r_{y_{jik}} & r_{x_{ijk}} & 1 \end{bmatrix} \begin{bmatrix} \cos\theta_{jik} & -\sin\theta_{jik} & 0 \\ \sin\theta_{jik} & \cos\theta_{jik} & 0 \\ 0 & 0 & 1 \end{bmatrix} \end{aligned} \tag{11}$$

In Equation (9), $[r_{x_{ij}} r_{y_{ij}}]$ represents the coordinates of the vector $\overrightarrow{O_iO_{ij}}$ in the $O_i - XYZ$ coordinate system. θ_{ij} represents the rotation angle of $O_{ij} - XYZ$ relative to the $O_i - XYZ$ coordinate system, as shown in Figure 2a. The expression method of Equation (11) is similar to that of Equation (9), as shown in Figure 2b. In addition, the stiffness matrices K_{ij} and K_{ijk} of the rectangular beam can be obtained by taking the inverse of the compliance matrix and expressed as

$$\mathbf{C}_{ij} = \begin{bmatrix} \frac{l}{Ebt} & 0 & 0 \\ 0 & \frac{4l^3}{Ebt} + \frac{3l}{2Gbt} & \frac{6l}{Ebt^3} \\ 0 & \frac{6l^2}{Ebt^3} & \frac{12l}{Ebt^3} \end{bmatrix} \tag{12}$$

$$\mathbf{K}_{ij} = (\mathbf{C}_{ij})^{-1} \tag{13}$$

where E is the elastic modulus of the material and G is the shear modulus of the material.

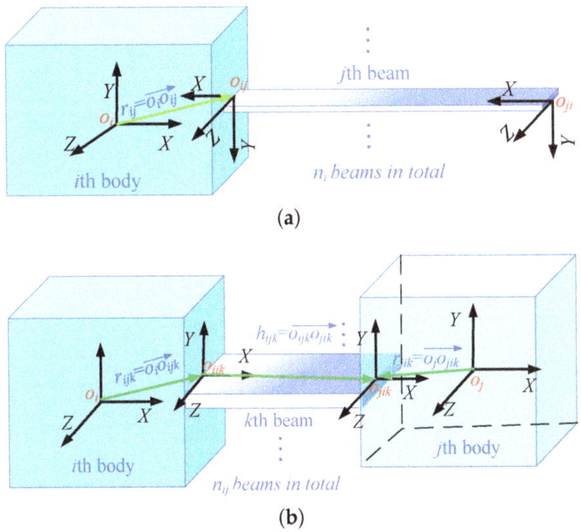

Figure 2. (**a**) The *i*th rigid body connected with the *j*th flexure beam. (**b**) The *i*th rigid body and the *j*th rigid body are connected by the *k*th flexure beam.

According to the theory of vibration, the characteristic equation of the multi-DOF differential equation is obtained by solving Equation (3):

$$\left|\lambda\mathbf{I} - \mathbf{M}^{-1}\mathbf{K}\right| = 0 \tag{14}$$

The natural frequencies of the mechanism are

$$f_i = \frac{1}{2\pi}\sqrt{\lambda_i}, \mathrm{i} = 1, 2, \ldots, 12. \tag{15}$$

Based on the analytical model of the mechanism, when the VCM applies force along the Y direction to Rigid Body 3, the motion and pose of the end-effector can be calculated by Equation (6). Therefore, the displacement reduction ratio and input stiffness of the mechanism can be expressed as

$$\begin{aligned} &\mathbf{C} = \mathbf{K}^{-1} \\ &R_{red} = u_{out}/u_{in} = c_{11,2}/c_{2,2} \\ &K_{in} = 1/c_{2,2} \end{aligned} \tag{16}$$

Based on the analytical model, the structural parameters of the displacement reduction mechanism can be determined under certain constraints. In order to obtain the desired natural frequency and output displacement of the mechanism along the working direction, the size parameters of the mechanism need to be carefully adjusted, as shown in Figure 3. An appropriate safety level must be maintained, and the designing of the structures should be conducted in compliance with optimization procedures. Generally, the structural safety level can be measured with the application of a reliability index [23,24]. The maximum stress produced at the root of the stretched beam shall be less than the allowable stress. According to the known maximum continuous output force of the selected VCM (40 N) and the desired output displacement (200 μm), the structural parameters are carefully adjusted through the trial-and-error method with the assistance of the analytical model. The design parameters of the mechanism are shown in Table 1.

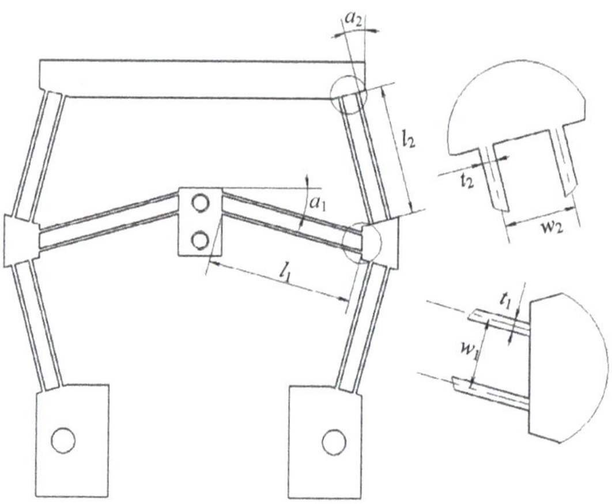

Figure 3. Design parameters of the mechanism.

Table 1. Main parameters of the mechanism.

l_1 40 mm	b_1 10 mm	t_1 0.8 mm	a_1 15°	w_1 4.8 mm
l_2 35 mm	b_2 10 mm	t_2 1 mm	a_2 15°	w_2 5 mm

4. FEA and Experimental Verification

4.1. FEA Verification

In this section, the Ansys Workbench software is used for static and modal analysis of the mechanism to obtain its performances, including the displacement reduction ratio, input stiffness, and natural frequency. The three-dimensional (3D) model of the mechanism was established by the Solidworks software and imported into Ansys Workbench for static analysis. Aluminum alloy was selected as the model material, and its elastic modulus and Poisson ratio were 70 Gpa and 0.33, respectively. As shown in Figure 4, fixed constraints were applied to the two bolt holes at the bottom of the mechanism, and the VCM provides continuous driving force F = 40 N to act on the mechanism along the Y direction. Thus, the end-effector of the mechanism produces the reduced output displacement along the same direction. The static von Mises stress and the total deformation of the mechanism are shown in Figure 5.

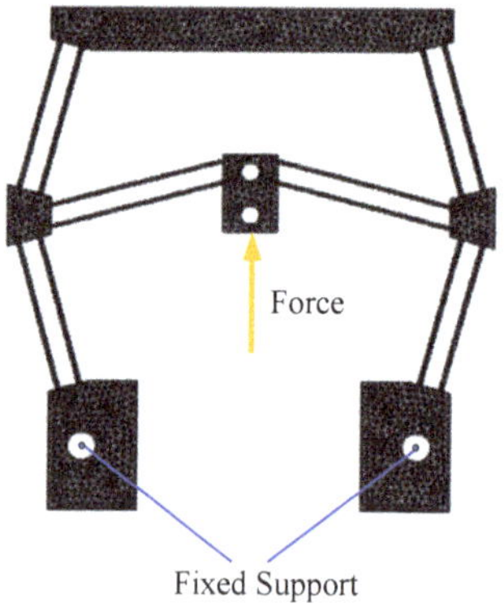

Figure 4. Finite element model of displacement reduction mechanism.

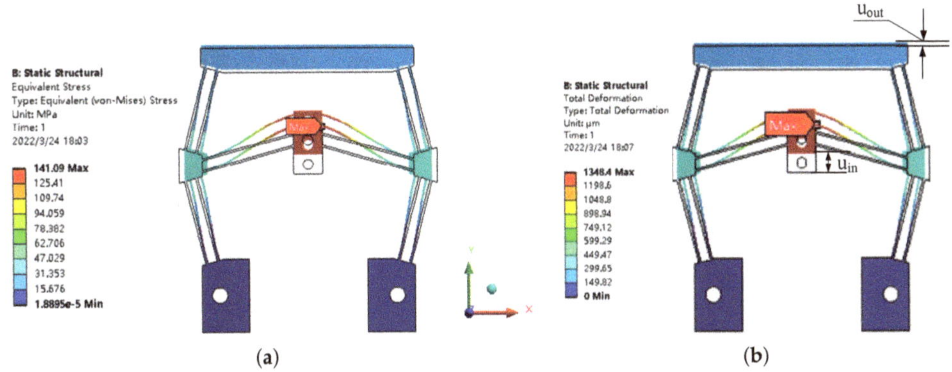

Figure 5. Static deformation and von Mises stress analyzed by FEA. (**a**) von Mises stress. (**b**) Static deformation.

The displacement reduction ratio and input stiffness are obtained by FEA:

$$R_{red} = \frac{u_{out}}{u_{in}}, K_{in} = \frac{F}{u_{in}} \tag{17}$$

where u_{in} and u_{out} are the maximum input and output displacements calculated by the FEA results. F is the continuous thrust of the VCM in FEA.

It can be seen from the FEA results that the maximum input displacement of the mechanism is 1348.4 μm and the maximum output displacement is 184.35 μm. The displacement reduction ratio calculated by Equation (17) is 0.1367, and the input stiffness is 29.67 N/mm, which is basically consistent with the results of 0.1335 and 29.28 N/mm calculated by the analytical model, and the corresponding errors are 2.34% and 1.31%, respectively. Moreover, the maximum von Mises stress obtained from the FEA solution is 141.09 Mpa, which is far lower than the allowable stress of aluminum alloy 7075 (the yield strength is about 480 Mpa), indicating that the displacement reduction mechanism is always in the state of elastic deformation in the whole movement process. At the same time, to analyze the load capacity of the mechanism, a force of 40 N is applied to the end-effector along the working direction in the FEA model. It can be obtained that the displacement of the end-effector under the force is 24.35 μm, and then, the output stiffness of the mechanism is 164 N/mm, about 5.6-times the input stiffness, showing the high load capacity of the mechanism.

In addition, the first four vibration modes of the reduced mechanism are also obtained through FEA, as shown in Figure 6, and the natural frequencies corresponding to the first four modes of the mechanism are listed in Table 2. The results show that the first mode corresponds to the regular left–right swing of the four movable rigid bodies in the XOY plane. The second mode corresponds to the rotation around the fixed constraint in the ZOX plane. The third mode corresponds to the working mode of the mechanism, and the fourth mode corresponds to the irregular left–right swing of the four movable rigid bodies in the XOY plane, while the fourth-order natural frequency is much higher than the first three modes. In practical application, although the displacement reduction mechanism is assembled with the VCM, the stiffness of the VCM is infinite, and its mover is light in weight, which has little impact on the dynamic performance of the mechanism.

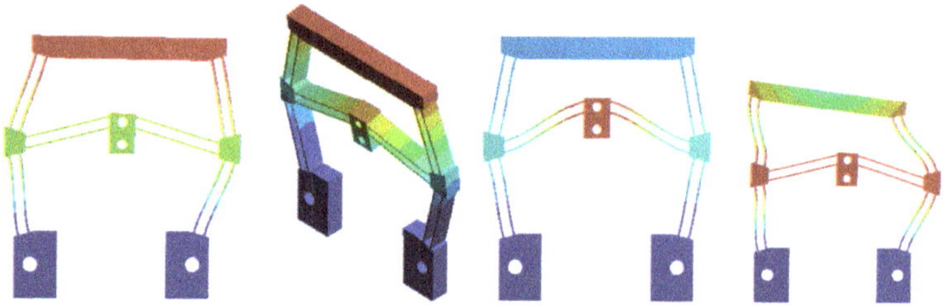

Figure 6. The first four mode shapes of the mechanism. (**a**) First mode, (**b**) 2nd mode, (**c**) 3rd mode, and (**d**) 4th mode.

Table 2. First 4 modal frequencies of the mechanism.

Mode	Nature Frequencies (Hz)
1	152.87
2	276.58
3	310.64
4	571.69

4.2. Performance Testing

To evaluate the performance of the mechanism and verify the accuracy of the analytical model, experimental setups were established as shown in Figure 7. The mechanism was processed by fine milling and wire cutting, and aluminum alloy 7075 was selected as the material of the mechanism to improve its structural strength. In the experimental test, the VCM (VCAR0070-0419-00A) was selected as the driver of the mechanism to generate the input force and displacement. The motor can provide continuous thrust of 40 N (corresponding drive voltage of 5.6 V). The force constant of the VCM was 17.7 N/A, and the maximum allowable current was 4 A. A capacitive displacement sensor (NMT.C1) with a resolution of 2.5 nm and a measuring range of 200 μm was used to measure the output displacement of the mechanism. An inductance micrometer (DGG-8Z) with a resolution 0.01 μm and a measuring range of 2 mm was used to measure the input displacement of the mechanism. Two displacement sensors are shown in Figure 8. In addition, the output voltage (0∼10 V) of the displacement sensor is collected by the data acquisition card (PCI-6251) equipped with 16-bit A/D and D/A converters, and the output voltage (−10 V∼10 V) of the data acquisition card is linearly converted into the driving current (−4 A∼+4 A) through the current linear amplifier (TA115) to drive the VCM.

Firstly, when the motor thrust gradually increased, the input and output displacements of the mechanism were detected simultaneously by two displacement sensors to obtain the maximum stroke and reduction ratio of the mechanism. The experimental setup is shown in Figure 8. The experimental test results of the mechanism are shown in Figure 9. Obviously, the maximum stroke of the mechanism is 197.43 μm when the driving voltage is about 5.6 V, corresponding to the driving current of 2.26 A and the driving force of 40 N. The good linear relationships between the input and output displacements and the control voltage show that the displacement reduction mechanism driven by the VCM does not have various nonlinear characteristics such as hysteresis and creep and has good positioning stability and accuracy. Meanwhile, according to the relationship between the input and output displacement, the reduction ratio of the mechanism is 0.1390. In Figure 10, the displacement reduction ratios calculated by the analytical model, FEA model, and experimental prototype are compared, and the results are in good agreement, indicating that the analytical model has high modeling accuracy.

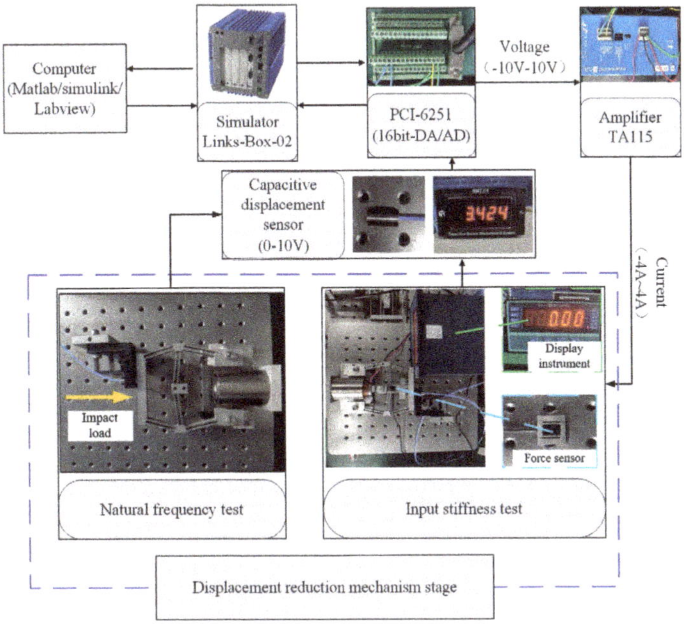

Figure 7. Displacement reduction mechanism and experimental test system.

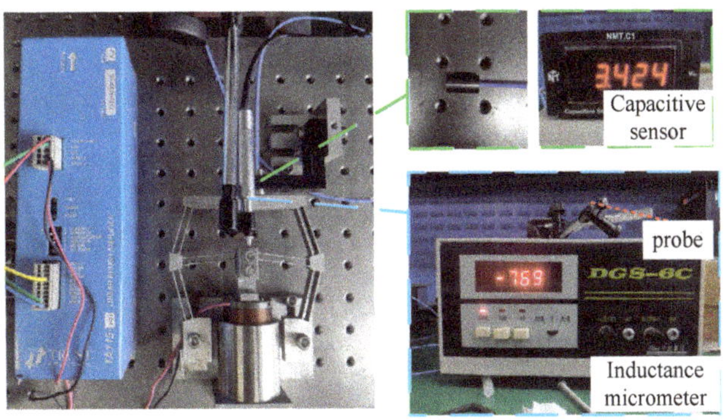

Figure 8. The experimental setup for the reduction ratio test.

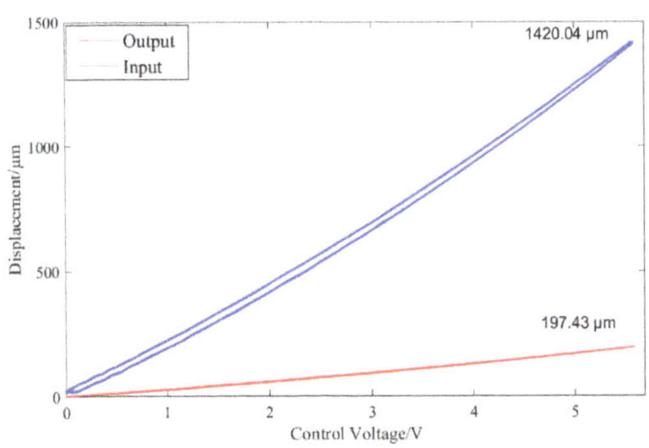

Figure 9. The relationship between the input and output displacements and the control voltage.

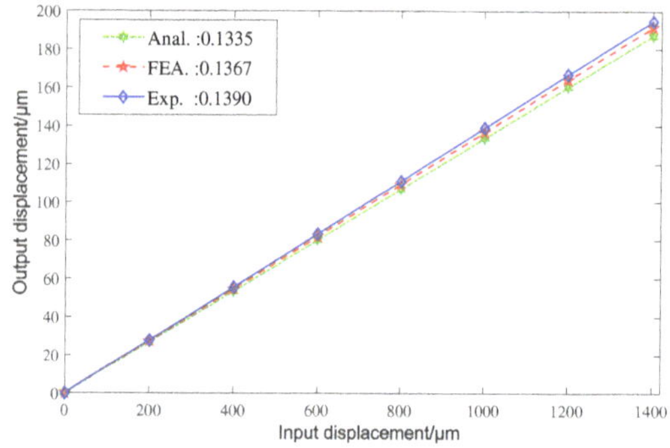

Figure 10. Comparison of displacement reduction ratio.

To obtain the input stiffness of the mechanism, the inductive micrometer (DGC-8Z) and the force sensor (JLBS-MD of Jinnuo company) were combined at the input end of mechanism. The experimental setup is shown in Figure 7. The force sensor was placed between the VCM and the input rigid body. When the VCM pushes and pulls the input rigid body, the force sensor can read its output force, and the inductance micrometer is used to measure the input displacement. According to the relationship between the force and displacement, the input stiffness of the mechanism obtained from the testing result is 28.50 N/mm. The comparison of the input stiffness is given in three ways, as shown in Figure 11, and the results show good consistency.

To obtain the natural frequency along the working direction of the mechanism, the mechanism was separated from the VCM, and then, the impact load was applied to the end-effector of the mechanism along its working direction to make it vibrate freely. At the same time, the capacitive sensor was used to measure the vibration displacement of the mechanism to obtain the dynamic characteristics of the whole system. The data acquisition card collects the input time-domain vibration signals, which are then processed by the fast Fourier transform (FFT) to convert them into frequency-domain signals through the Labview software. The experimental test result of the mechanism is shown in Figure 12. It can be seen that the natural frequency of the mechanism is 291 Hz. The maximum error of the natural frequency among the analytical model, FEA model, and testing results is 9.84%. This difference is mainly attributed to the neglect of the influence of the flexure beams' vibration kinetic energy on the dynamic performance in the analytical model.

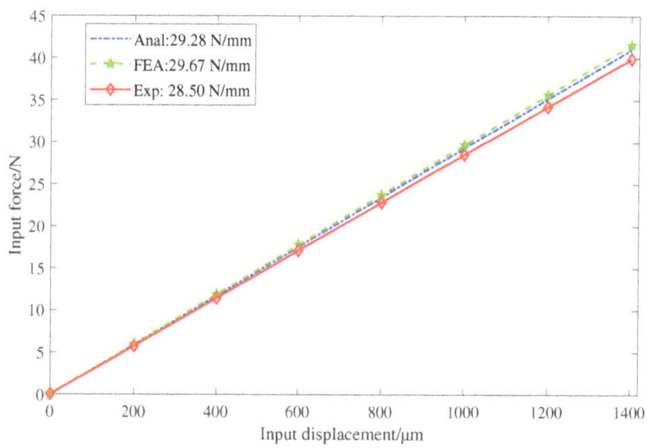

Figure 11. Comparison of input stiffness.

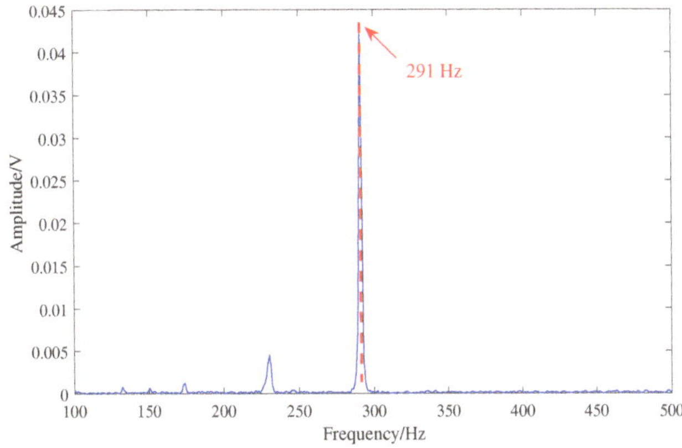

Figure 12. Natural frequency test results.

Table 3 compares the reduction ratio, input stiffness K_{in} (N/mm), and natural frequency f_m (Hz) of the displacement reduction mechanism, which were obtained from the analytical model, FEA, and experimental test. Taking the experimental results as a reference, the relative errors of the analytical model (Error1) and FEA (Error2) can be obtained, respectively. As can be seen from Table 3, the differences among the results of the analytical model, FEA, and the experimental test are less than 10%, indicating that the three results are in good agreement and the analytical model has high calculation accuracy.

Table 3. Comparison results among the analytical model, FEA, and experimental test.

Specifications	Anal.	FEA	Experiment	Error1 (%)	Error2 (%)
R_{red}	0.1335	0.1367	0.1390	3.96	1.65
K_{in} (N/mm)	29.28	29.67	28.50	2.74	4.11
f_m (Hz)	319.62	310.64	291	9.84	6.75

5. Controller Design

In this section, the tracking performance of the mechanism is evaluated. Firstly, the dynamic parameters of the system are identified, and the nominal transfer function of the system is obtained. Then, a two-DOF controller consisting of feedback controller $G_c(s)$ and a dynamic inverse feedforward controller $C_{ff}(s)$ are used to control the position of the mechanism. The feedback controller is obtained by allocating the zero-pole distribution of the system using the root locus method. Finally, to meet the requirements of micropositioning/nanopositioning accuracy, an anti-interference controller based on a disturbance observer (DOB) was introduced to effectively suppress the interference. The control system diagram is shown in Figure 13.

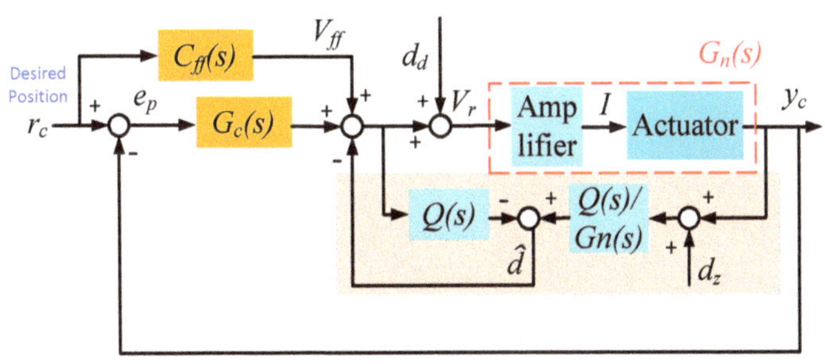

Figure 13. Control system diagram.

5.1. Dynamic Model Identification

To obtain the dynamic model of the system, a semi-physical real-time simulation control system was established by using the PCI-6251 data acquisition card, Links-Box-02 embedded real-time simulator, and MATLAB/Simulink real-time simulation system software packages. The experimental setup of the control system is shown in Figure 14. The chirp signal with a sweep voltage amplitude of 0.1 V was selected as the driving input, and the frequency was changed linearly from 0.1 Hz to 1.0 kHz. The output displacement was collected by the capacitive contactless displacement sensor (NMT.C1). Using the system identification toolbox in MATLAB, the system nominal transfer function is obtained as

$$G_n(s) = \frac{8.472 \times 10^4}{s^2 + 141.4s + 1.301 \times 10^5} \tag{18}$$

The input and output time-domain signals collected in the experiment were processed by the FFT, as shown in Figure 15. The frequency response of the experimental measurement results is plotted and compared with the Bode diagram of the identified transfer function. According to the formant peak, the natural frequency of the system is about 347 rad/s, which is different from the natural frequency of the flexure mechanism itself due to the electrical and control system of the VCM. At the same time, the identified system model fits well with the experimental results, which can accurately describe the dynamic characteristics of the whole system.

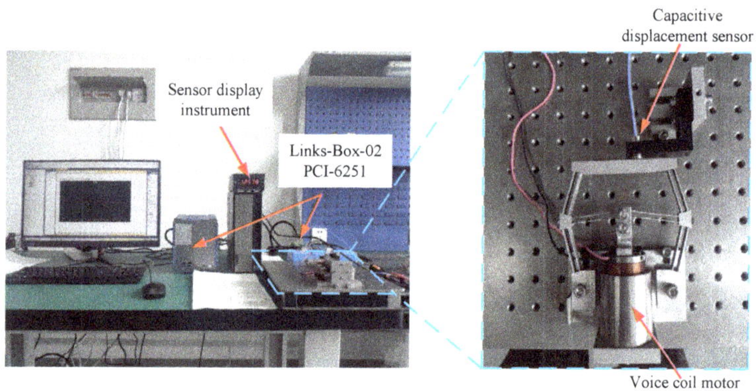

Figure 14. Experimental setup for position control.

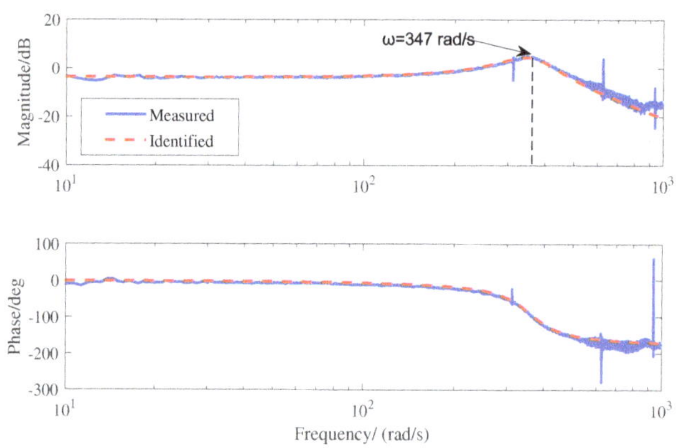

Figure 15. Measured frequency response and Bode plot.

5.2. Design Process of Controller

In the system model of the flexure mechanism, there are two poles very close to the virtual axis, and the resonant mode has a low damping ratio of about 0.2. Mechanical vibration will be easily excited when moving at high speed. The low gain margin caused by low damping resonant mode causes large phase error, which affects the stability and the rapid response of the system. Therefore, in this paper, a high-performance feedback controller that can effectively suppress stage resonance and high-frequency noise disturbance was designed by using zero-pole placement and the root locus correction method. First, a new pole was placed at the origin to achieve zero steady-state error. In the system test, a pair of conjugate poles near the virtual axis may cause large-amplitude vibrations and make the system difficult to stabilize. To offset the influence of these poles, a pair of conjugate zeros was placed near them. Finally, to improve the response speed of the system, two poles were placed far away from the virtual axis. After correctly designing the poles, zeros, and gains of the controller, the system behavior was obtained, as shown in Figure 16. Finally, the transfer function of the designed controller is obtained as

$$G_c(s) = \frac{9437.1 \times (s^2 + 140.8s + 1.293 \times 10^5)}{s \times (s^2 + 4014s + 4.208 \times 10^6)} \tag{19}$$

After the feedback controller is added, the open loop Bode diagram of the system is shown in Figure 16b. The amplitude margin and phase margin of the system after correction are 26.2 dB and 79.1°, respectively. Obviously, the controlled system is stable with a large margin.

Due to the integral element of the feedback controller, the stage system is a type 1 system, which cannot track with zero steady-state error when it is a ramp signal. Feedforward compensation based on the system dynamics inverse is widely used in the position control of complex dynamic systems. A feedforward controller $C_{ff}(s)$ is designed as

$$C_{ff}(s) = G_n{}^{-1}(s) \tag{20}$$

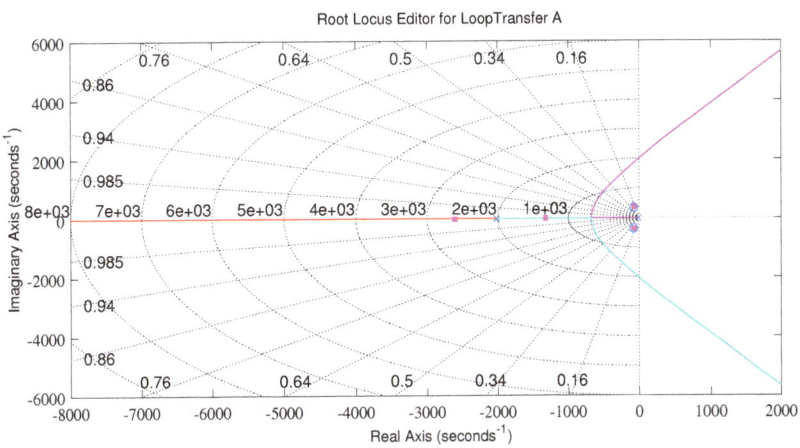

(**a**)

Figure 16. *Cont.*

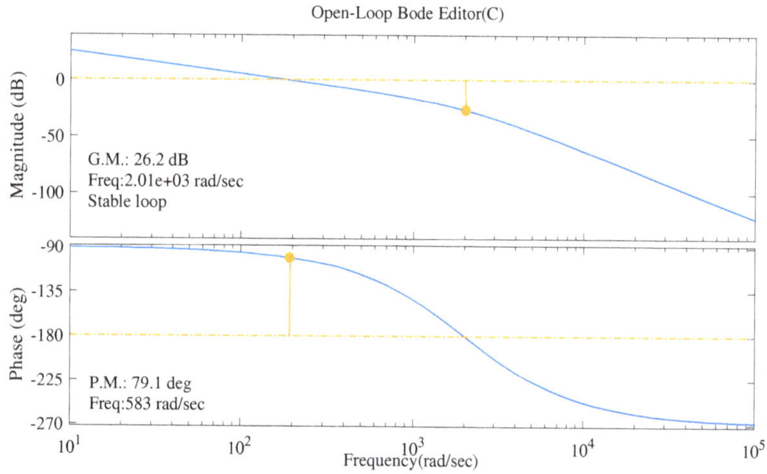

(**b**)

Figure 16. The design results of the controller of the mechanism. (**a**) Pole-zero assignment and root locus correction. (**b**) Open-loop Bode diagram of the system.

To test the motion resolution of the mechanism in the working direction, a staircase signal with a height of 125 nm was applied to the VCM, and the position tracking test was carried out by using the compound control methods of feedback control and dynamic inverse feedforward. A capacitive displacement sensor (NMT.C1) with a resolution of 2.5 nm and a measurement range of 200 μm was used to measure the output displacement of the mechanism. The output displacement curve in Figure 17 shows an about 40 nm peak-to-peak amplitude, meaning that the motion resolution of the mechanism can reach 40 nm in the working direction.

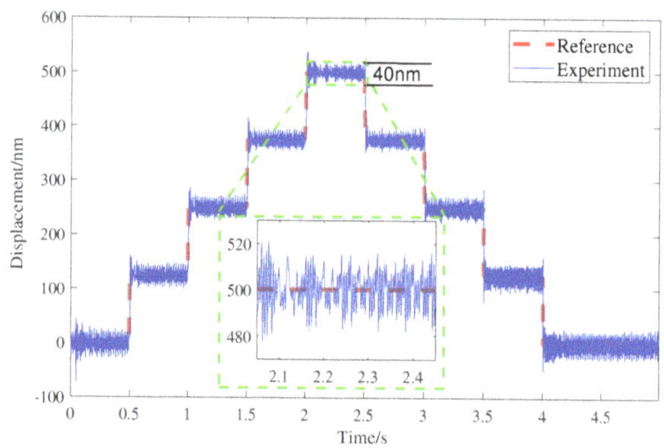

Figure 17. Experiment results for staircase signal input.

5.3. Disturbance Rejection

One problem of the VCM using Lorentz force to drive the low-stiffness-flexure mechanism is the low resonant frequency, which will be easily excited by disturbance and affect the positioning resolution of the system. In the positioning process of the mechanism, the main disturbances are the noise of current amplifier and DA converter, the VCM mover vibration, as well as the vibration of the isolation platform. The disturbance observer can unify the disturbances such as friction and model the uncertainty of the controlled object into equivalent disturbances, then estimate the equivalent disturbance by using the nominal inverse model of the controlled object and the input control value. For instance, Deng and Yao designed an extended state observer, which can estimate the unmeasurable velocity signal for the control of electrohydraulic servomechanisms [25]. Due to the simple structure and convenient parameter adjustment, the DOB has been widely used as a part of controller compensating for disturbances [26–29]. As shown in Figure 13, to obtain micro-/nano-level tracking performance, the disturbance observer (DOB) is introduced into the closed-loop negative feedback control loop to observe the equivalent disturbance $\hat{d}$, which can theoretically achieve the complete suppression of the disturbance. By designing low-pass filter $Q(s)$, the low-frequency disturbance can be observed effectively, and the high-frequency noise signal can be filtered effectively. $Q(s)$ is expressed as

$$Q(s) = \frac{\sum_{k=0}^{M} \alpha_k (\tau s)^k}{(\tau s + 1)^N} \tag{21}$$

where $\alpha_k = \frac{N!}{(N-k)!k!}$ is the coefficient, N is the order of the denominator, M is the order of the numerator, and N-M is the relative order. The value of the parameter determines the bandwidth of $Q(s)$. Considering the ability of the disturbance observer to suppress external disturbance and its sensitivity to measurement noise, τ = 0.001 was selected. Due to the order of $Q(s)$, the filter should not be too high, so N = 3 and M = 1 were selected.

To test the trajectory tracking effect of the designed control system, a sinusoidal signal with an amplitude of 2 V (corresponding to 40 µm displacement) and a frequency of 5 Hz was selected as the desired trajectory. The tracking performances of three controllers, Feedback, Feedback + Feedforward, and Feedback + Feedforward + DOB, were compared. The tracking results are shown in Figure 18. It can be seen from Figure 18b that Feedback control has a relatively large phase error compared with the other two control methods.

The trajectory tracking errors of the three control systems are shown in Figure 18c. The maximum tracking error peak-to-trough value (P-V) for the Feedback control is approximately ±4.98 µm (12.38%). After adding Feedforward control, the error is reduced to ±2.60 µm (6.5%). When DOB anti-interference compensation is further added, the maximum tracking error of the designed control system is reduced to ±1.40 µm (3.5%), which is only 0.28-times that of Feedback control. The maximum tracking error (MTE) and root-mean-squared tracking error (RMSTE) of trajectory tracking are shown in Table 4. The MTE and RMSTE of the latter two control methods are less than 3 µm, indicating that the proposed reduction mechanism can achieve good tracking performance.

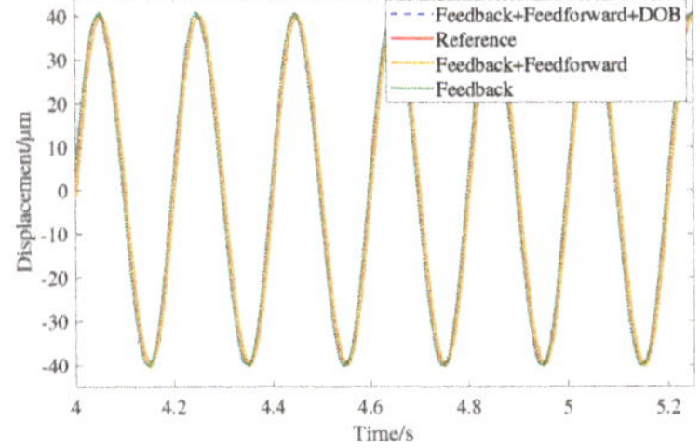

(a)

Figure 18. *Cont.*

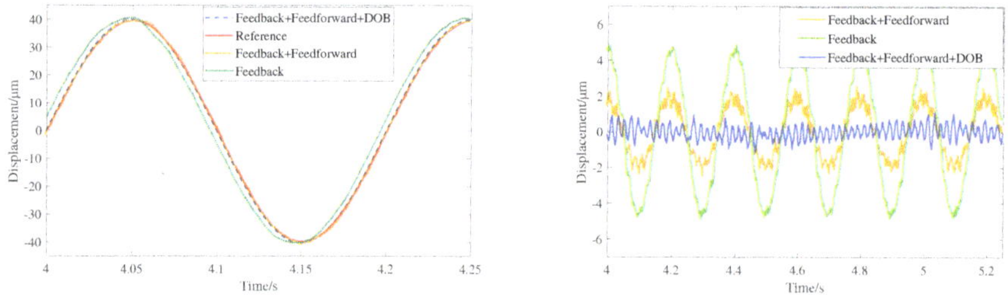

(b) (c)

Figure 18. Trajectory tracking performance (**a**) Trajectory tracking results. (**b**) Partial enlargement. (**c**) Tracking errors.

Table 4. Tracking performance of the mechanism with different control methods

Method	MTE (μm)	RMSTE (μm)
Feedback	±4.98	10.50
Feedback + Feedforward	±2.60	1.88
Feedback + Feedforward + DOB	±1.40	0.13

6. Conclusions

This paper presents a fully flexure displacement reduction mechanism, which is composed of three groups of compound bridge-type displacement reduction mechanisms to realize two-stage displacement reduction. Firstly, an analytical model for the predictions of the theoretical displacement reduction ratio, input stiffness, and natural frequency was derived based on the stiffness matrix method. Then, after determining the size of the mechanism based on the analytical model, the 3D model of the mechanism was established through SolidWorks and then imported into Ansys Workbench for FEA. The experimental tests were also carried out to validate the performances of the proposed displacement reduction mechanism. The motion range of 197.43 μm, the reduction ratio of 0.1369, the input stiffness of 28.50 N/mm, the motion resolution of 40 nm, and the natural frequency of 291 Hz were obtained from the experiment. The experimental test results were basically consistent with the analytical model and FEA model, and the maximum error was the error of the natural frequency, which was less than 10%. Finally, a two-DOF controller consisting of a feedback controller, a dynamic inverse feedforward controller combined with DOB anti-interference compensation was designed for the control of positioning stage. Excellent positioning and tracking performances were achieved, which verifies the effectiveness of the proposed displacement reduction mechanism and the designed controller. Our subsequent main work is to design a multi-DOF high-speed decoupled displacement reduction mechanism to improve the performance and application scope of the stage. In addition, taking this stage as a micro-stage to design a macro–micro dual-drive stage is also the focus of future research.

Author Contributions: Conceptualization, L.L. and Y.C.; methodology, L.L.; FEA and experiment tests, Y.C.; writing, Y.C. and L.L. All authors have read and agree to the published version of the manuscript.

Funding: This work was supported by the Natural Science Foundation of Shanghai (Grant No. 21ZR1426000), and the State Key Laboratory of Mechanical System and Vibration (Grant No. MSV202210).

Institutional Review Board Statement: Not applicable.

Informed Consent Statement: Not applicable.

Data Availability Statement: Not applicable.

Conflicts of Interest: The authors declare no conflict of interest.

References

1. Lee, C.; Lee, J.W.; Ryu, S.G.; Oh, J.H. Optimum design of a large area, flexure based XYθ mask alignment stage for a 12-inch wafer using grey relation analysis. *Robot. Comput. Integr. Manuf.* **2019**, *58*, 109–119. [CrossRef]
2. Zhu, W.L.; Yang, X.; Duan, F.; Zhu, Z.; Ju, B.F. Design and adaptive terminal sliding mode control of a fast tool servo system for diamond machining of freeform surfaces. *IEEE Trans. Ind. Electron.* **2017**, *66*, 4912–4922. [CrossRef]
3. Xiao, R.; Xu, M.; Shao, S.; Tian, Z. Design and wide-bandwidth control of large aperture fast steering mirror with integrated-sensing unit. *Mech. Syst. Signal Process.* **2019**, *126*, 211–226. [CrossRef]
4. Guo, D.; Nagel, W.S.; Clayton, G.M.; Leang, K.K. Spatial-temporal trajectory redesign for dual-stage nanopositioning systems with application in AFM. *IEEE/ASME Trans. Mechatron.* **2020**, *25*, 558–569. [CrossRef]
5. Yang, C.; Li, C.; Zhao, J. A nonlinear charge controller with tunable precision for highly linear operation of piezoelectric stack actuators. *IEEE Trans. Ind. Electron.* **2017**, *64*, 8618–8625. [CrossRef]
6. Tian, Y.; Cai, K.; Zhang, D.; Liu, X.; Wang, F.; Shirinzadeh, B. Development of a XYZ scanner for home-made atomic force microscope based on FPAA control. *Mech. Syst. Signal Process.* **2019**, *131*, 222–242. [CrossRef]
7. Csencsics, E.; Schitter, G. Exploring the pareto fronts of actuation technologies for high performance mechatronic systems. *IEEE/ASME Trans. Mechatron.* **2020**, *26*, 1053–1063. [CrossRef]
8. Rakuff, S.; Cuttino, J.F. Design and testing of a long-range, precision fast tool servo system for diamond turning. *Precis. Eng.* **2009**, *33*, 18–25. [CrossRef]
9. Ito, S.; Troppmair, S.; Lindner, B.; Cigarini, F.; Schitter, G. Long-range fast nanopositioner using nonlinearities of hybrid reluctance actuator for energy efficiency. *IEEE Trans. Ind. Electron.* **2018**, *66*, 3051–3059. [CrossRef]
10. Sun, L.; Chen, F.; Dong, W. A pzt actuated 6-dof positioning system for space optics alignment. *Flight Control Detect.* **2019**, *2*, 1–13.
11. Gutierrez, H.M.; Ro, P.I. Magnetic servo levitation by sliding-mode control of nonaffine systems with algebraic input invertibility. *IEEE Trans. Ind. Electron.* **2005**, *52*, 1449–1455. [CrossRef]
12. Awtar, S.; Parmar, G. Design of a large range XY nanopositioning system. *J. Mech. Robot.* **2013**, *5*, 021008. [CrossRef]
13. Hao, G.; Kong, X. A Novel Large-Range XY Compliant Parallel Manipulator With Enhanced Out-of-Plane Stiffness. *J. Mech. Des.* **2012**, *134*, 061009. [CrossRef]
14. Roy, N.K.; Cullinan, M.A. Design and characterization of a two-axis, flexure-based nanopositioning stage with 50 mm travel and reduced higher order modes. *Precis. Eng.* **2018**, *53*, 236–247. [CrossRef]
15. Chen, F.; Zhang, Q.; Gao, Y.; Dong, W. A review on the flexure-based displacement amplification mechanisms. *IEEE Access* **2020**, *8*, 205919–205937. [CrossRef]
16. Kim, J.H.; Kim, S.H.; Kwak, Y.K. Development and optimization of 3-D bridge-type hinge mechanisms. *Sens. Actuators A Phys.* **2004**, *116*, 530–538. [CrossRef]
17. Liang, C.; Wang, F.; Huo, Z.; Shi, B.; Tian, Y.; Zhao, X.; Zhang, D. A 2-DOF monolithic compliant rotation platform driven by piezoelectric actuators. *IEEE Trans. Ind. Electron.* **2019**, *67*, 6963–6974. [CrossRef]
18. Wu, H.; Lai, L.; Zhang, L.; Zhu, L. A novel compliant XY micropositioning stage using bridge-type displacement amplifier embedded with Scott-Russell mechanism. *Precis. Eng.* **2022**, *73*, 284–295. [CrossRef]
19. Choi, K.B.; Lee, J.J.; Hata, S. A piezo-driven compliant stage with double mechanical amplification mechanisms arranged in parallel. *Sens. Actuators A Phys.* **2010**, *161*, 173–181. [CrossRef]
20. Lai, L.J.; Zhu, Z.N. Design, modeling and testing of a novel flexure-based displacement amplification mechanism. *Sens. Actuators A Phys.* **2017**, *266*, 122–129. [CrossRef]
21. Yong, Y.K.; Aphale, S.S.; Moheimani, S.R. Design, identification, and control of a flexure-based XY stage for fast nanoscale positioning. *IEEE Trans. Nanotechnol.* **2008**, *8*, 46–54. [CrossRef]
22. Ling, M.; Cao, J.; Jiang, Z.; Lin, J. Theoretical modeling of attenuated displacement amplification for multistage compliant mechanism and its application. *Sens. Actuators A Phys.* **2016**, *249*, 15–22. [CrossRef]
23. Szafran, J.; Juszczyk, K.; Kamiński, M. Experiment-based reliability analysis of structural joints in a steel lattice tower. *J. Constr. Steel Res.* **2019**, *154*, 278–292. [CrossRef]
24. Valdebenito, M.; Jensen, H.; Schuëller, G.; Caro, F. Reliability sensitivity estimation of linear systems under stochastic excitation. *Comput. Struct.* **2012**, *92*, 257–268. [CrossRef]
25. Deng, W.; Yao, J. Extended-state-observer-based adaptive control of electrohydraulic servomechanisms without velocity measurement. *IEEE/ASME Trans. Mechatron.* **2019**, *25*, 1151–1161. [CrossRef]
26. Deng, W.; Yao, J.; Wang, Y.; Yang, X.; Chen, J. Output feedback backstepping control of hydraulic actuators with valve dynamics compensation. *Mech. Syst. Signal Process.* **2021**, *158*, 107769. [CrossRef]
27. Liu, Q.; Liu, M.; Jin, Q.; Liu, Y. Design of DOB-based control system in the presence of uncertain delays for low-order processes. *IEEE Trans. Control Syst. Technol.* **2018**, *28*, 558–565. [CrossRef]
28. Chen, M.; Shao, S.Y.; Jiang, B. Adaptive neural control of uncertain nonlinear systems using disturbance observer. *IEEE Trans. Cybern.* **2017**, *47*, 3110–3123. [CrossRef] [PubMed]
29. Gao, G.; Ye, M.; Zhang, M. Synchronous robust sliding mode control of a parallel robot for automobile electro-coating conveying. *IEEE Access* **2019**, *7*, 85838–85847. [CrossRef]

actuators

MDPI

Article

Design and Testing of a Hollow Continuum Magnetic Millirobot with Multimodal Motion

Yuanhe Chen †, Zichen Xu † and Qingsong Xu *

Department of Electromechanical Engineering, Faculty of Science and Technology, University of Macau, Avenida da Universidade, Taipa 999078, Macau, China
* Correspondence: qsxu@um.edu.mo
† These authors contributed equally to this work.

Abstract: Magnetic continuum millirobots have presented outstanding potential in ultrahigh-precision engineering including minimally invasive surgery, due to their flexible mechanical structures and dexterous manipulation. Traditional continuum millirobots exhibit limited cargo-loading capacity, which restricts their application. Herein, we propose a novel design scheme of a magnetically actuated untethered hollow continuum millirobot. The millirobot is composed of silicone as the mainframe structure and two tiny magnets for actuation. To improve the loading capacity, partial silicone is removed to create a flexible cavity, which enables cargo delivery and potential in vivo sampling functions under wireless magnetic actuation. Theoretical analysis and experimental testing are conducted to reveal the effectiveness of the proposed design. The soft structure brings a new strategy to achieve multimodal motion including rolling, tumbling, and swinging. Moreover, the magnet part can generate a powerful magnetic force output for dexterous manipulation. These functionalities lay a foundation for playing a greater role in next-generation biomedical applications.

Keywords: microrobotics; magnetic actuation; continuum millirobot; untethered soft robot; mechanism design

Citation: Chen, Y.; Xu, Z.; Xu, Q. Design and Testing of a Hollow Continuum Magnetic Millirobot with Multimodal Motion. *Actuators* **2022**, *11*, 269. https://doi.org/10.3390/act11100269

Academic Editors: Takeshi Mizuno and Steve Davis

Received: 3 July 2022
Accepted: 14 September 2022
Published: 20 September 2022

1. Introduction

Mobile microrobots have demonstrated attractive performance in various assignments at a small scale, especially in medical applications. The mini-level physical size of microrobots enables more precise operations in crowded and extreme environments. It facilitates reaching deeper tissues within the human body, where untethered actuation mechanisms [1,2] play a profound role. With the help of recent development and breakthrough in material [1,3], chemistry [4], and biotechnology [5], plenty of microrobots have been designed and put into concrete experimental applications, including actuation [6–8], navigation [9], control [10], and execution [1]. However, it remains a challenge to further promote microrobots' functionalities for more practical applications.

Among recent works, increasing attention has been focused on detailed clinical scenarios, such as cardiovascular diseases, which cause numerous death all over the world [11–13]. To overcome the current medical equipment's limitations in invasive surgeries, catheter-based continuum microrobots have been developed [1,14,15] to intrude into hard-to-reach corners, where the controlled magnetic fields contribute to high-efficiency untethered actuation. These works have revealed the attractive advantages of continuum structures, especially in endovascular surgeries. Nevertheless, from the perspective of functionalities and designs, relatively simple mechanical structures cannot enable more potential usage schemes. For example, polydimethylsiloxane (PDMS) was utilized to connect two permanent magnets, by which the continuum microrobot was produced for flexible magnetic actuation [16]. The solid continuum structure and complex structural design limits the diversity of applications [7,17–20], such as cargo loading, which is significant to achieve live sampling in deeper tissues, in view of the sampling procedure in the medical scenarios.

Proper mechanical design improvement and size optimization [2] is a potential solution for strengthening such functionalities.

Herein, we propose a novel design scheme for an untethered multifunctional hollow continuum magnetic millirobot (MHCMM). The MHCMM is constructed by integrating silicone as the main frame structure and several miniaturized magnets for actuation. The removal of some non-magnetic connecting parts in the millirobot creates a flexible cavity, which improves the cargo-loading capacity and enables cargo delivery during a potential bio-sampling role under a wireless magnetic drive. It is notable that the operating space of the MHCMM is limited to microscopic scales, such as multimodal motion range within 0.5 mm in each axis and a sampling capacity of 300–600 μL, which is beneficial for future exploration tasks in vivo. By controlling the external magnetic field, the MHCMM can offer multimodal motion and execute dexterous manipulation. In addition, the MHCMM is capable of producing an output force of 5–15 mN, which enables it to exert a sufficient force on the objects such as tumors in the human body. With the help of a magnetic navigation system powered by permanent magnets mounted on a robot arm, the wirelessly actuated millirobot is promising for minimally invasive surgery. It facilitates more flexible motion, where two-dimensional electromagnetic coils serve as the local actuation device. The reported multifunctional hollow continuum magnetic millirobot demonstrates great potential for minimally invasive surgery.

Compared with existing works [2,7,8], the novelty of the proposed MHCMM lies in the simplified sampling method, risk reduction of drug leakage, and more functions such as cargo transportation and output force test. The remaining parts of the paper are organized as follows. The design and modeling of the magnetic millirobot are presented in Section 2. The prototype fabrication and experimental investigations are outlined in Section 3, where the experimental results are given and the discussion is addressed. Section 4 concludes this paper.

2. Design and Modeling

2.1. Design Scheme

At small scales, it is important to support more functionalities by microdevices in a constrained space. For continuum millirobots, previous work utilized various soft materials to construct solid structures, which serve as the connecting parts or main structure of the millirobots [20,21]. Such a design achieved outstanding deformation and control results. However, the fully solid structures cause some wasted space, which is designed to integrate other functionalities. Here, we transform the soft structure into a hollow one. The redundant parts are supposed to be removed, which results in a novel hollow millirobot. As shown in Figure 1a, this design provides more usable space for cargo loading and sampling.

To form a cavity structure, we used rapid-forming silicone over the top of a thin glass tube (outer diameter of 0.5 mm) to allow it to settle naturally. In addition, we can control the magnetism at the end of the magnets (at both ends) to perform different functions. By tuning the distribution of the poles at both ends of the MHCMM, a circle structure and an arch structure can be obtained, respectively, as shown in Figure 1b. Under the steering of external magnetic fields, this concept design brought a new solution to the mechanical design of soft millirobots.

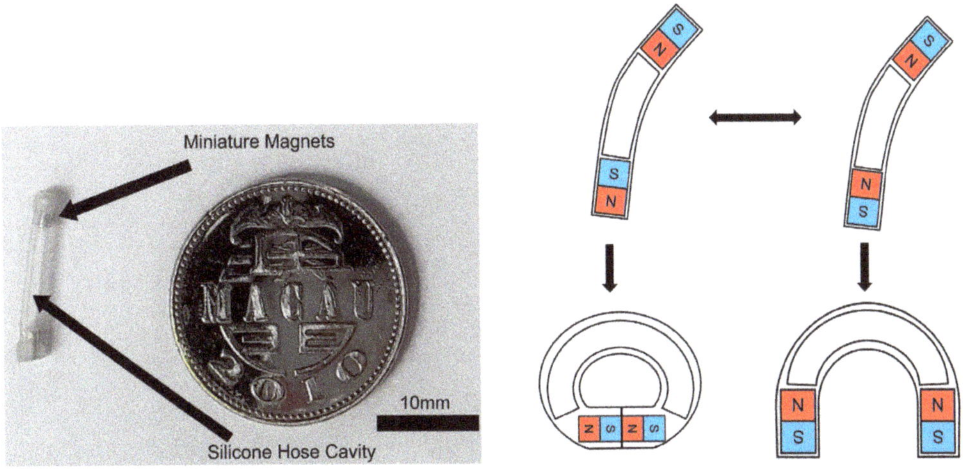

(**a**) (**b**)

Figure 1. Physical prototype and illustration of the working principle of the MHCMM. (**a**) The main structure of the MHCMM is composed of two miniature magnets at the two terminals and a silicone hose cavity in the middle. (**b**) The magnetism at the end of two magnets can be controlled to perform different functions. When the poles at both ends of the MHCMM are tuned to be the same polarity, an arch shape can be produced by applying an external gradient magnetic field. When they have different polarity magnetic poles, a circumferential shape can be generated.

2.2. *Physical Modeling*

2.2.1. Magnetic Actuation

The introduction of magnetic parts plays a significant role in actuating and controlling the millirobot. For realizing magnetic field actuation, we consider an axially magnetized cylindrical NdFeB (N52-grade) magnet with an axial-symmetric magnetic field distribution. The magnet has a radius of R and a length of 2R. For the permanent magnet with the same diameter and length, the distribution of the magnetic field remains constant when the characteristic dimensions of the magnetized body are normalized [15,22,23]. Thus, we can express the magnetic field at a determined spatial location in the form of a normalized vector function as follows:

$$\mathbf{B}(\mathbf{p}) = B_m \mathcal{F}(\mathbf{p}/R) \tag{1}$$

where $\mathbf{p}$ denotes the position vector with respect to the center of the magnet in cylindrical coordinates, B_m is the remanent magnetization of the magnet, and $\mathcal{F}$ represents the vector function (its implicit form is given in [24–26]).

Along the central axis of the magnet, we can explicitly express the magnitude of the magnetic field in the normalized form below:

$$B = \frac{B_m}{2}\left(\frac{d/R+1}{\sqrt{(d/R+1)^2+1}} - \frac{d/R-1}{\sqrt{(d/R+1)^2+1}}\right) \tag{2}$$

where d is the distance from the center of mass of the permanent magnet to the measurement point along the centerline.

From Equation (2), we can see that the magnetic field strength decreases as the normalized distance d/R increases. Thus, a large-enough external magnet is required to impose an actuation effect to guide the MHCMM within a reasonable working distance. Generally, the task of controlling the MHCMM requires a maximal magnetic field strength of 20–40 mT. Then, d/R = 0.66–1.34 can be derived according to Equation (2). Thus, we can calculate the required external magnet, i.e., a cylindrical permanent magnet with a diameter and length of 300 mm (with R = 150 mm) and B_m= 0.574 T. The normalized distance is translated into a distance, which is at least 284 mm from the center of the magnet (or 134 mm from the surface of the magnet).

In principle, a magnetized object in a magnetic field suffers from a magnetic force and a magnetic torque. The equations for the magnetic force and magnetic torque of an object subjected to magnetization in a magnetic field are both related to the gradient of the magnetic field, as follows:

$$\tau_m = \int_{V_m} \mathbf{M} \times \mathbf{B} dV_m \tag{3}$$

$$\mathbf{F}_m = \int_{V_m} (\mathbf{M} \cdot \nabla) \mathbf{B} dV_m \tag{4}$$

where **M** is the magnetization of the object and V_m is the volume of the magnetized object. All of these magnetic interactions determine the effects of magnetic actuation.

2.2.2. Robot Deformation

When manipulating the MHCMM with different polarities at both ends, the S-pole of the permanent magnet is initially placed in the positive Z-axis direction. This causes the MHCMM to remain standing at one end, along with an inclined angle for the other end. Ideally, the MHCMM should be purely vertical if the external permanent magnets are placed in the given position. In practice, the silicone material does not exhibit a uniform density, resulting in a slight bending deformation. This state is the result of several forces interacting together, such as the attractive force of opposite-pole magnets F_{M1}, repulsive force of same-pole magnets F_{M2}, and supporting force of materials F_{e1}, as shown in Figure 2A.

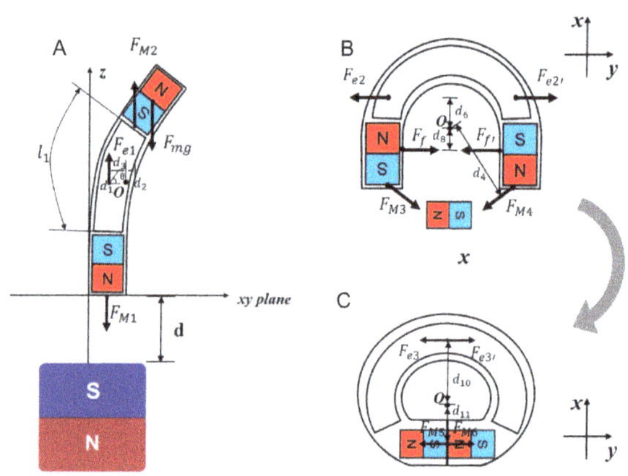

Figure 2. Physical analysis of MHCMM in three force states (lateral view). (**A**) Standing at one end; (**B**) leading the attraction process at both ends; (**C**) knotting into a circle.

Its static deformation is described by a linear elastic strain model, which is given by:

$$F_{e1} = \frac{E_1 dLS}{L} \tag{5}$$

where S is the maximum cross-sectional area of the MHCMM cavity, E_1 is the Young's modulus, and dL/L is the elongation of the cavity divided by the original length.

Through the above equation, the bearing force of the cavity F_{e1} is analyzed below.

$$0 = F_{M1} \cdot d_1 + F_{M2} \cdot d_2 + F_{e1} \cdot d_3 \tag{6}$$

where F_{M1} and F_{M2} are the forces applied to the magnets at both ends. d_1, d_2, and d_3 are the distances from the point of force to the center of mass of the cavity. Their sum is 0, which ensures a stable standing condition for the MHCMM.

Referring to Figure 2B, we use a small magnet to attract both ends of the MHCMM (with double-ended anisotropic poles) to contact each other in a plane. Due to the friction force and suction force of the anisotropic poles canceling each other, the elastic tension of silica gel is derived as follows:

$$0 = F_{M3} \cdot d_4 + F_{M4} \cdot d_5 + F_{e2} \cdot d_6 + F_{e2'} \cdot d_7 + F_f \cdot d_8 + F_f \cdot d_9 \tag{7}$$

where F_M is the shaped magnet attraction, F_f is the friction between the MHCMM and the plane, and F_e is the elastic tension of silica gel. In this way, MHCMM will form the shape as shown in Figure 2B. Subsequently, we remove the external small magnet, and the magnetic poles at both ends of MHCMM will attract each other to form a circle, as shown in Figure 2C.

3. Prototype Fabrication and Experimental Results

3.1. Prototype Fabrication and Experimental Setup

3.1.1. Material and Fabrication

For the potential clinical application of this millirobot, we selected several types of fabrication materials by considering biocompatibility, optimal working temperature, and other factors. In particular, five different soft materials (i.e., Dragon Skin 10 Medium (from Smooth-on Inc., East Texas, PA, USA), Dragon Skin 30 (from Smooth-on Inc.), Finished Silicone (from Runze Fluid Co. Ltd., Nanjing, China), E610 (from Hongye Technology Co. Ltd., Shenzhen, China), and E630 (from Hongye Technology Co. Ltd.)) are adopted as candidates to assemble the MHCMMs (e.g., making thin tubes out of silicone inverted molds). The cavities are created by heating and solidifying after natural dropping on a thin glass tube with a radius of 0.3 mm.

Due to the material property, only three of the five materials can be used to produce the prototype of the MHCMM. The adopted materials and their properties are shown in Table 1. The other two materials (E610 and E630) are not feasible because they are too viscous, which has a negative impact on forming cavities. For each of the three MHCMM samples, after fabricating the MHCMM sample with a length of 20 mm, it was stretched by increasing the distance between two magnets at the terminals. The original and maximally deformed lengths were measured for each stretch. The experiment was repeated six times, and the experimental results are shown in Figure 3. We can observe that the silicone material of Dragon Skin 10 Medium provides the best tensile property, which is stretched about 1.55 times on average.

We used rapid-forming silicone over the top of a thin glass tube (outer diameter: 0.5 mm) to allow it to settle naturally to form a cavity structure. Furthermore, its effective working temperature (−65 °F to 450 °F) also meets the requirements of our working environment (i.e., 99.5 °F, the internal ambient temperature of humans). Thus, the prototype with this material is chosen for the subsequent tests.

Table 1. Millirobot materials and properties.

Material Name	Material Properties	Stretching Rate
Dragon Skin 10 Medium	Shore hardness: 10A, Specific gravity 1.07 g/mL, Tensile strength: 475 psi	1.55
Dragon Skin 30	Shore hardness: 30A, Specific gravity 1.08 g/mL, Tensile strength: 500 psi	1.34
Finished Silicone Tube	Rockwell hardness: 70A	1.42

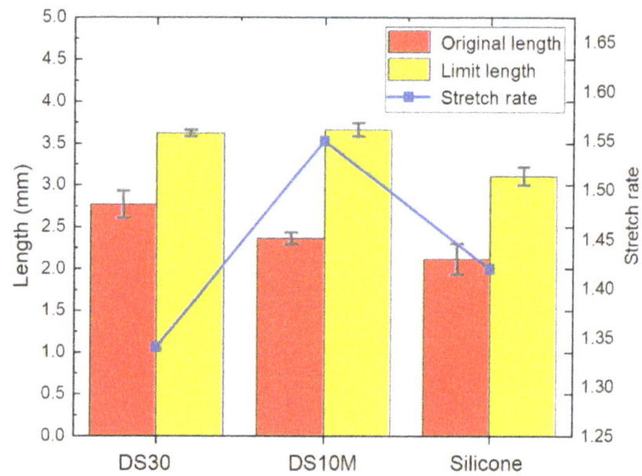

Figure 3. Test results of three types of materials. In the histogram, error bars compare the properties before and after stretching.

3.1.2. Experimental Setup

For the experimental study, we adopted a magnetic driving system that consists of a three-axis orthogonal electromagnetic coil device, a tube vessel (35 mm diameter) containing the millirobot, a top camera, and a side movable camera, as shown in Figure 4. A triaxial electromagnetic-coil actuation system generates the oscillating magnetic field, with each of the three orthogonally arranged coils driven by a servo amplifier (KXN-6020D DC, from Zhaoxin Inc., Shanghai, China). The amplifier was controlled via a digital I/O board by a developed LabVIEW program (National Instrument Inc., Pinehurst, NC, USA) running on a personal computer.

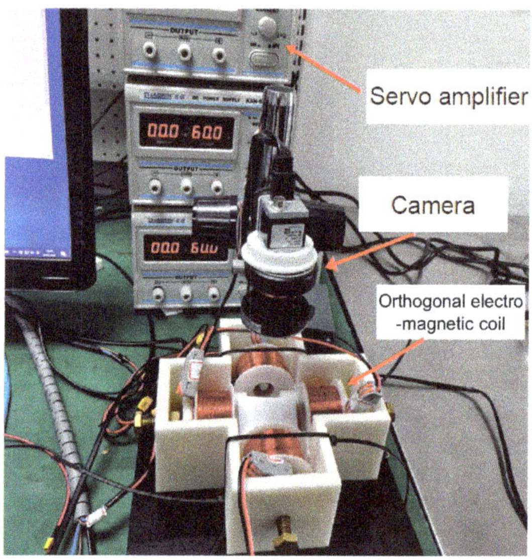

Figure 4. Experimental setup of a magnetic actuation system consisting of a three-axis orthogonal electromagnetic coil device, a Petri dishes (35 mm in diameter) with the robot inside, a top camera, and a side movable camera.

3.1.3. MHCMM Characterization

Since understanding the support capability of the MHCMM's inner cavity structure is the basis for the next step in the control of its unidirectional oscillation, it is necessary to measure the bending angle of the MHCMM under different magnetic field strengths. Here, we fixed one end of the MHCMM, and then applied different currents to the electromagnetic coil on the Z-axis of the MHCMM as shown in Figure 5A. The experimental test diagram and results are shown in Figure 5B.

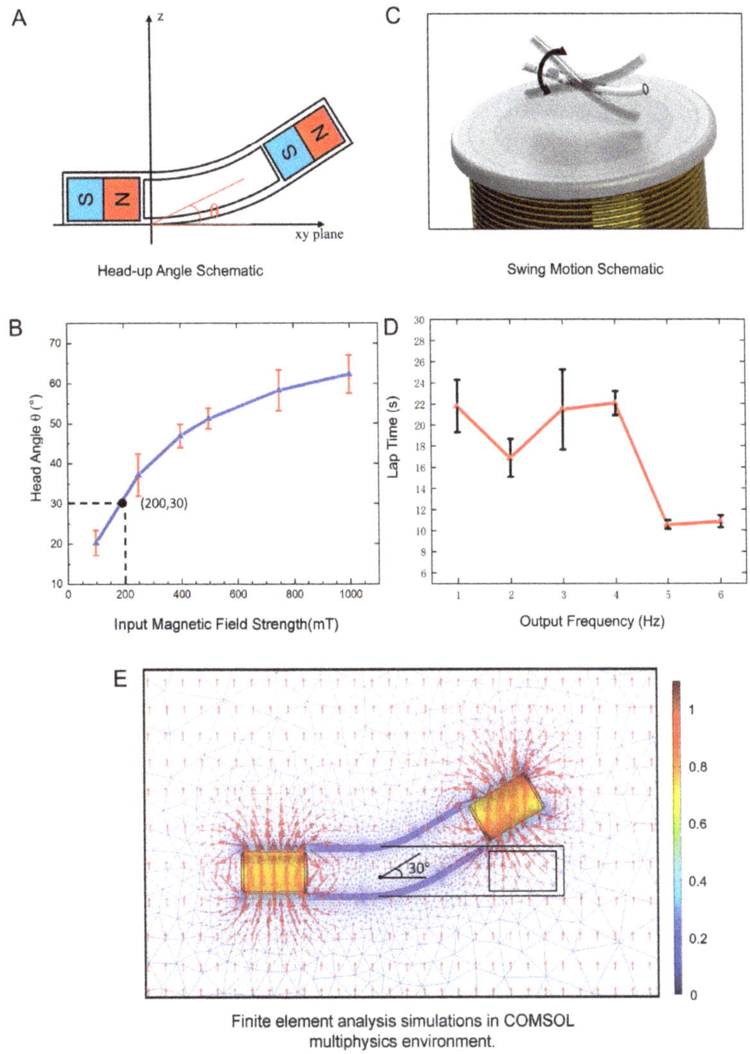

Figure 5. Head-up test and swing motion results. (**A**) The head-up angle of the MHCMM itself can be obtained by changing the input signal to the electromagnetic coil amplifier. (**B**) Experimental results of six sets of head-up angles versus input magnetic field strength. (**C**) A 3D schematic of oscillating forward motion with an electromagnetic coil underneath, which produces an oscillating magnetic field. (**D**) The electromagnetic coil, which has an output magnetic field strength of 300 mt, is controlled to produce an oscillating magnetic field of different frequencies, and the lap time is the time taken to make one revolution along a 35 mm diameter Petri dish. (**E**) The deformation of MHCMM under the action of an external magnetic field of strength 210 mT. The analysis result is consistent with the test result.

3.2. Experimental Results

3.2.1. Navigation Control Results

After fabricating the MHCMM, we installed permanent magnets at the end of a robot arm (R14000 dual-arm robot, from ABB Ltd., Zürich, Switzerland) with custom-built parts (THREE-M 3D printer, from Wiiboox.net Inc., Nanjing, China). A model structure of human blood vessels was built and manufactured using transparent resin by 3D printing. The main parts of the aorta, renal artery, common skeletal artery, and superior mesenteric artery are involved in the 3D vascular model. In consideration of the plasma density of the human body [27], an aqueous solution of 30% glycerol was chosen for filling the 3D vascular model in the experimental study.

The robot arm is programmed to guide the MHCMM to move along a predefined motion trajectory in the vascular pipeline for arriving at the target point. Due to the multidimensional motion of the robot arm, we not only can program the motion trajectory in the two-dimensional plane but also can set the motion program in three dimensions. Based on the previously calculated attraction force between the permanent magnet of the robot arm and the miniature magnetic robot, we found that the robot arm can carry the end-effector magnet well, and thus guide the MHCMM to deliver the catheter inside the 3D vascular model, as shown in Figure 6.

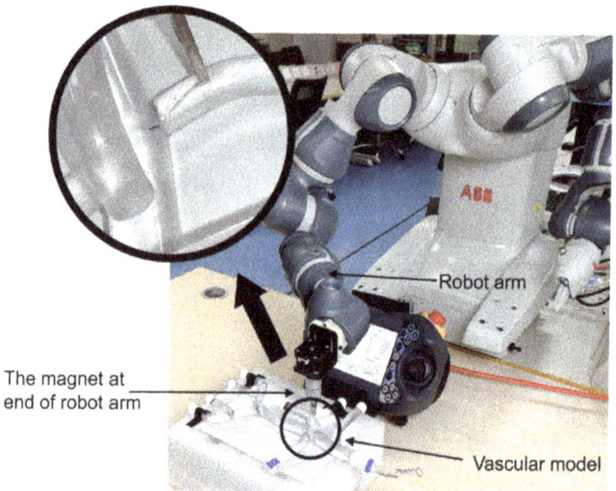

Figure 6. The MHCMM is moved to the perimeter of the target position by a permanent magnet fixed at the end of a robot arm.

Considering the limitation of precision, the robot arm cannot guarantee local precise navigation control. Hence, a coarse-fine strategy is adopted to improve the precision by the combination of a robot arm and an electromagnetic coil system. After a long-trip navigation by the robot arm, the electromagnetic coil system serves as a subsequent actuation device for more precise local navigation. We used an oscillating magnetic field to drive the MHCMM for swing motion, as shown in Figure 5C. The MHCMM produces turbulence at frequencies above 7 Hz. Thus, its performance is measured at frequencies from 1 to 6 Hz. The MHCMM has the best frequency response at 5 Hz, and it can go around the Petri dish in a shorter time with satisfactory stability.

To further demonstrate the advantages of the MHCMM in multimodal motion, we manipulate the external permanent magnet by hand to actuate the MHCMM. Both the circle structure and arch structure of MHCMM can tumble to the assigned target point, as shown in Figure 7A,B. Additionally, the tumbling motion of the MHCMM successfully crossed some obstacles, proving the ability to cross obstacles as shown in Figure 7C. Motion mechanics analysis of tumbling motion is shown in Figure 8. In addition, we explored a worm-like motion of the MHCMM, as shown in Figure 9. Due to this motion being mainly pulled by magnetic fields, it does not become a full-fledged motion, as shown in Figure 9.

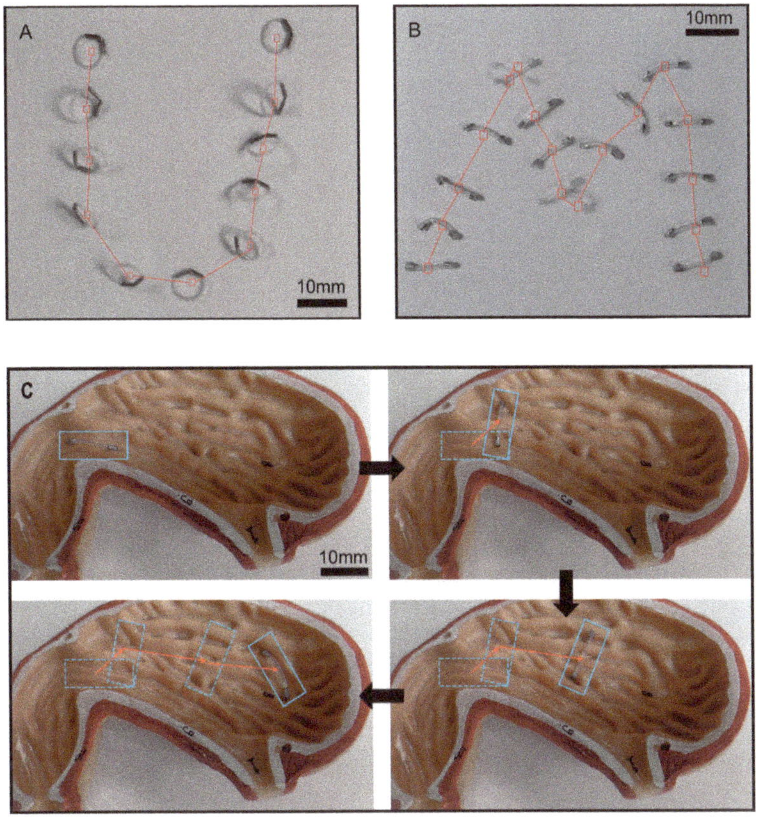

Figure 7. The multimodal motion of MHCMM. (**A**) The "U"-shaped trajectory by circular tumbling; (**B**) the "M"-shaped trajectory by the tumbling motion; (**C**) the MHCMM demonstrated the ability of the tumbling motion to cross obstacles by crawling by path in a human stomach model.

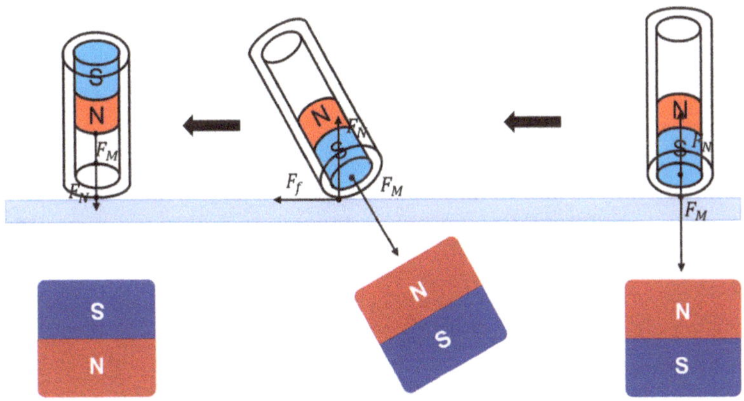

Figure 8. Motion mechanics analysis of tumbling motion. Different morphological forces are experienced when the MHCMM undergoes tumbling motion in the side view perspective.

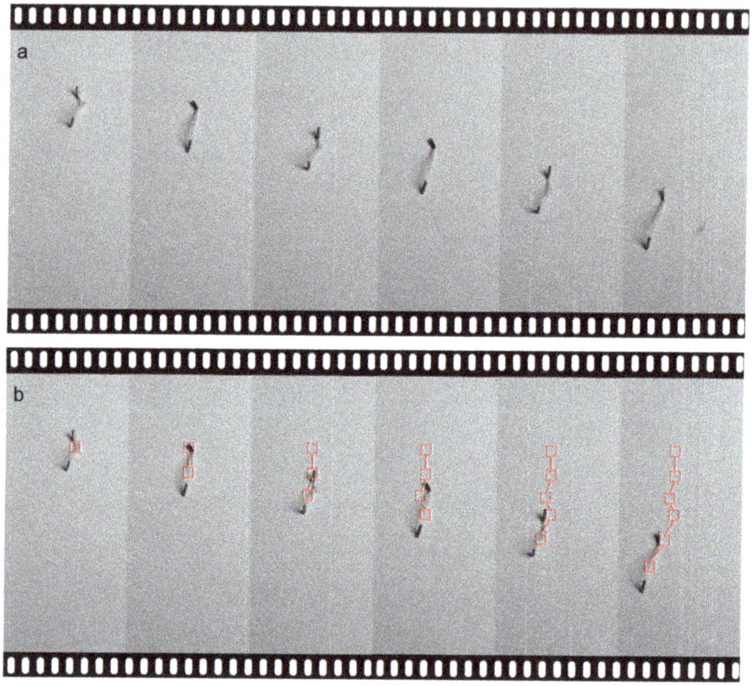

Figure 9. The worm-like motion and path of MHCMM. (**a**) By changing the distance of the external permanent magnet, the MHCMM with opposite magnet polarity at the end can perform a worm-like motion. (**b**) The path of worm-like motion of the MHCMM is marked.

3.2.2. MHCMM Function Test Results

The soft hollow structure design enables various potential medical applications. To test the functions of the MHCMM in drug delivery, material sampling, stirring, and manipulation, a series of experiments were conducted as follows.

The cavity of the MHCMM makes it accessible to load more cargo, which provides a tool for in vivo biological sampling and drug delivery. To demonstrate this functionality, stretch sampling was performed. In particular, the magnets at both ends of the MHCMM were tugged by the permanent magnets in hand for executing the outward stretching motion. In this way, the volume of the internal cavity of the MHCMM was increased, leading to filling in the cavity with external ambient fluid. After the external ambient fluid is mixed with the preservation fluid in the cavity, the external magnetic field is removed, and the MHCMM retracts. We repeated this operation several times to ensure the presence of external ambient solution in the MHCMM cavity. The final measurement can reach 300–600 uL of liquid volume per sample, which completely encompasses the requirements of biological sampling for the number of cells. The scheme of the biological sampling process is presented in Figure 10. What is more, the MHCMM uses a sample sampling port similar to the principle of a heart valve manufactured by needle hole injection, and the external pressure will effectively reduce its loss of internal cargo. In addition, if the MHCMM is loaded with drugs, we can also achieve targeted drug transport through the liquid exchange in the cavity.

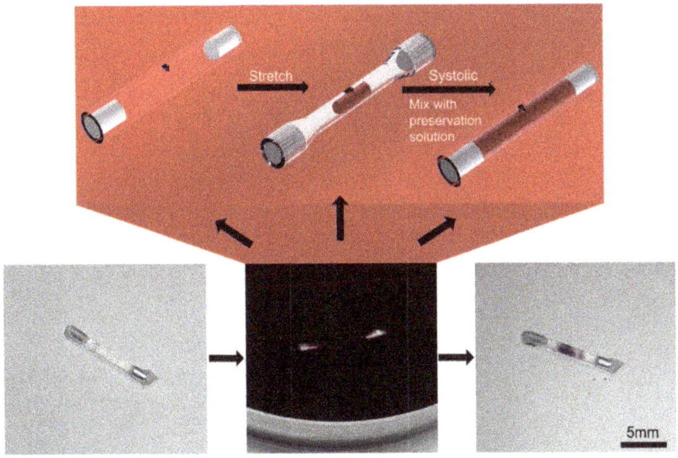

Figure 10. Illustration of material sampling by the millirobot. The MHCMM cavity was filled with water and then put into a colored solution for sampling testing.

For testing the stirring function of MHCMM, we regulated the external permanent magnet by hand to form a rotating permanent magnetic field. The experiment was conducted by using a double-ended anisotropic magnet MHCMM. According to a preliminary experiment, a double-ended magnet with a different polarity MHCMM easily forms an upright shape, which is suitable for stirring, as shown in Figure 11. We performed finite element analysis simulation by COMSOL software. The results demonstrate that with the same motion velocity, the string structure contributes more to the induced flow field, as shown in Figure 12. Whereas a double-ended magnet with the same polarity as MHCMM will exhibit an arch-like shape under the attraction of an external permanent magnet, which affects the stirring function. Driven by an external magnetic field, the MHCMM exhibits the function of stirring in water. According to this function, we can make the drug diffuse faster in the human body.

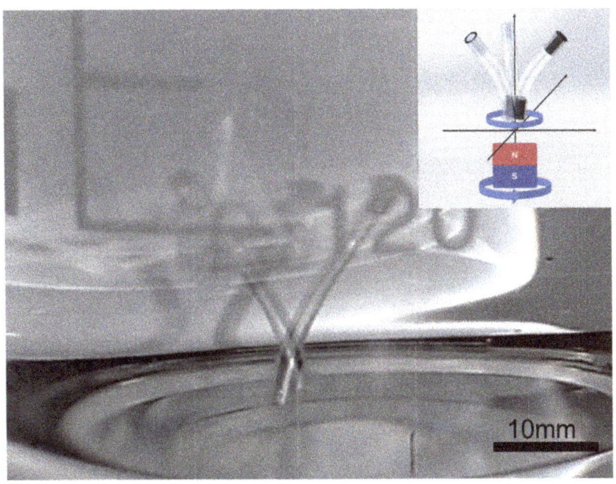

Figure 11. Stirring mode and physical demonstration. The MHCMM has the function of stirring in water, which is driven by the external magnetic field.

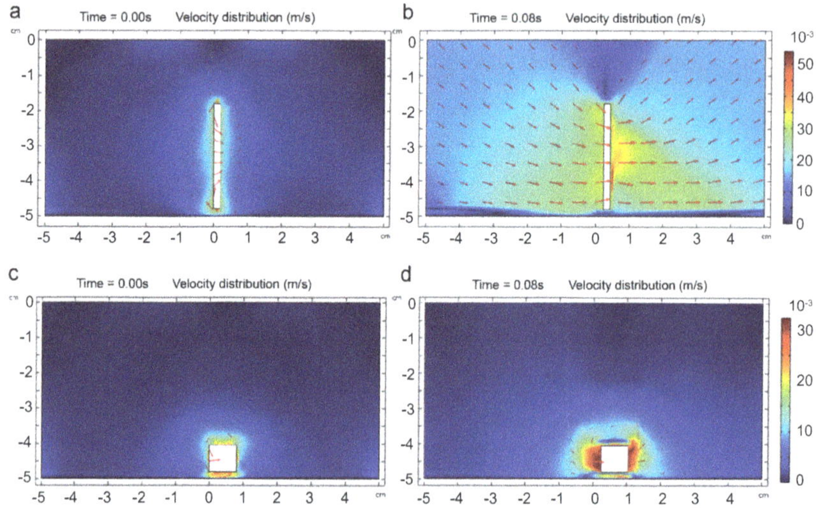

Figure 12. Simulation results of flow induction by the MHCMM. (**a**) The initial state (at 0 s) of the millirobot with the string structure is denoted by the slender square. (**b**) At 0.08 s, the slender square moves with a velocity of 3.0 cm/s. In the simulation, the total areas of the slender square and the flat square are identical. (**c**) The initial state (at 0 s) of the millirobot with normal structures, which are denoted by flat squares. (**d**) At 0.08 s, the squares move with a velocity of 3.0 cm/s.

Figure 13A shows the experimental setup for measuring the tensile force. In view of the potential of the circle-structure MHCMM for robotic manipulation, we accomplished target capture and measured its output forces, as shown in Figure 14. The experimental setup includes a microforce sensor (model 308445, from Zero Draft Inc.), which is an extended measurement range (extended force sensing range with a thin glass tube). One circle-structure MHCMM can manipulate the objects, as shown in Figure 13B. Its force signal is given in Figure 14a. This provides the experimental basis for the next step of MHCMM micromanipulation (e.g., pulling and snapping) of tumors in the vasculature.

From the experimental results, we found that two MHCMMs can form a bigger circle for executing the capture task of a larger target. Its output force signal is given in Figure 14b. Moreover, it is seen that the output force of the ring formed by the connection of two MHCMMs is greater than that of a single ring, which provides the indication for the next step of study in cluster control.

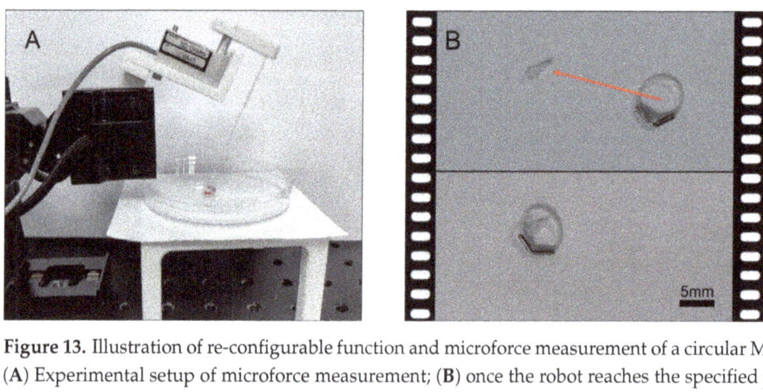

Figure 13. Illustration of re-configurable function and microforce measurement of a circular MHCMM. (**A**) Experimental setup of microforce measurement; (**B**) once the robot reaches the specified position, the hypothetical tumor (heteromorph) nesting is completed.

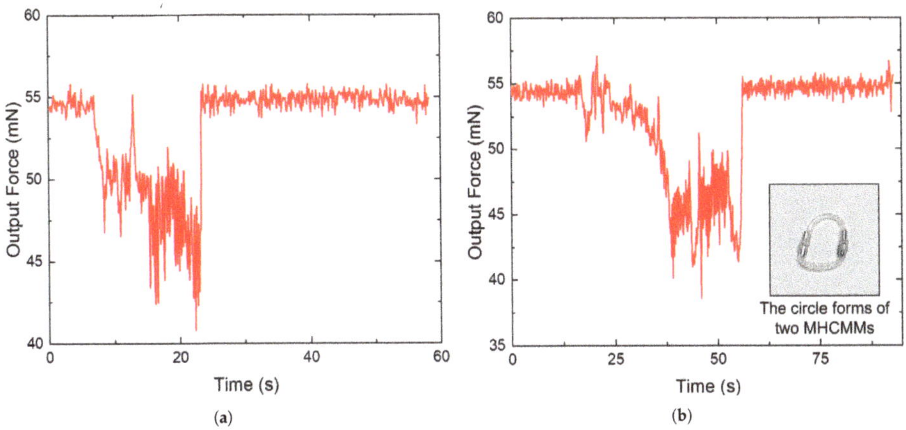

Figure 14. Microforce measurement result of circular MHCMM. (**a**) One ring is measured with the microforce measurement instrument for determining the tensile force, which can be released under constant magnetic field traction; (**b**) microforce measurement results with two MHCMMs, which provides an indication for the next step study of cluster control.

3.3. Discussion

We introduced a hollow mechanical structure to endow the ability to carry more cargo or potential functional components. Small permanent magnets can be easily actuated under various external magnetic fields, which brings diverse manipulation schemes for potential applications. Similar designs can be found in the literature [28], whose string-like mechanical structure is composed of a central magnetic beam with two identical buoyant components attached to its free ends. Its key actuation part is the magnetic beam between the two ends. With the help of the harmonic magnetization profile of the beam with a phase shift angle, six-degrees-of-freedom multimodal locomotion was obtained. However, the mechanical structure cannot provide available space for cargo loading. To solve this problem, in the literature [29], a magnetic soft capsule was introduced to overcome the disadvantage in cargo delivery. In addition, its robot footpad design brings more intriguing strategies in 3D locomotion. However, the introduced capsules increase the physical size, and environmental sampling remains a challenge. Compared with these works, our hollow mechanical design is more practical during long-trip navigation, guaranteeing the completeness of desired cargos.

In the literature, previous works have conducted fine attempts at hollow structure design for magnetic microrobots [2,7,8]. As compared with the endoscopic robot in the reference [2], the novelty of our proposed MHCMM lies in the optimized size of the robot and simplified sampling method for bio-liquids. As compared with the magnetic microrobot in [8], the novelty of the proposed MHCMM is the risk reduction of drug leakage. Unlike the sample output and input ports of the previous magnetic microrobot, our MHCMM introduces a simple sampling port, which is similar to the principle of a heart valve. The external pressure will effectively reduce the risk of internal cargo loss. In addition, as compared with the magnetic microrobot in [7], our proposed MHCMM has more practical functions, such as cargo transportation and output force test.

For many current continuum microrobots [15,16,21], there is still plenty of accessible improvement in their mechanical design, which should be paid more attention to remove unnecessary parts of the continuum millirobots to obtain more available space. In fact, the experimental results given in this paper indicate that the proper removal of solid structures has not severely reduced the functionalities of continuum millirobots. The increment of space is essential for further promoting the continuum millirobot, and the decrease in volume size reduces the potential dangers of blockages in vessels. The hollow design allows more space for integrating other necessary components, such as sensors and other components. For live sampling tasks, this design helps to reach deep tissues within the human internal body and carry more medical material for potential treatments. However, this promotion is limited because the long and hollow continuum part lacks sufficient rigidity, leading to poor performance in robotic manipulation tasks. More research will be conducted to reveal the optimal distribution of the hollow parts as well as the small permanent magnets in future work. It is also a big challenge to realize the precise control of sampling and releasing tasks, which is significant for safety and efficiency in medical scenarios. Moreover, free from the guidewire, the untethered millirobot can demonstrate more personalized operations, such as tumbling and deformation functions. Such untethered designs reveal potential research in multiple millirobot system control, which is a promising solution for further promoting the performance of millirobot systems in future work.

4. Conclusions

In this paper, we proposed a new soft continuum millirobot design promoting the functionalities and locomotion flexibility for potential intravascular applications. This millirobot is composed of two parts: (1) two small permanent magnets at the two ends, enabling powerful actuation for getting rid of the reliance on guidewire-based equipment; (2) soft materials as the combination part for the two magnets, which can be flexibly designed as a hollow mechanical structure. Force analysis reveals the effectiveness of magnetic actuation. Based on this, a movable magnetic field actuation system supplied by a robot arm is presented by fixing a permanent magnet as the end-effector for wirelessly navigating the millirobot in a long-trip 3D vascular model. Three-dimensional electromagnetic coils enable the achievement of more precise local motion control. In addition, with multimodal movements at a space of 0.5–0.7 mm, liquid sampling and cargo delivery of 300–600 μL volume were accomplished under the control of a magnet moved by hand. The output force of 5–20 mN was achieved by both single and multiple MHCMMs, indicating their outstanding potential for complex operations in a confined environment. In future work, we will try to integrate more powerful components (such as sensors) into the microrobots and provide reconfigurable control methods for multiple robots dedicated to various applications.

Author Contributions: Conceptualization, Y.C., Z.X. and Q.X.; methodology, Y.C., Z.X. and Q.X.; software, Y.C. and Z.X.; validation, Y.C. and Z.X.; formal analysis, Y.C. and Z.X.; investigation, Y.C. and Z.X.; data curation, Y.C. and Z.X.; writing—original draft preparation, Y.C., Z.X. and Q.X.; resources, Q.X.; writing—review and editing, Y.C., Z.X. and Q.X.; visualization, Y.C. and Z.X.; supervision, Q.X.; project administration, Q.X.; funding acquisition, Q.X. All authors have read and agreed to the published version of the manuscript.

Funding: This work was funded in part by the National Natural Science Foundation of China (File no. 52175556), The Science and Technology Development Fund, Macau SAR (File no. 0022/2019/AKP and 0153/2019/A3), and University of Macau (File no. MYRG2018-00034-FST and MYRG2019-00133-FST).

Data Availability Statement: Not applicable.

Conflicts of Interest: The authors declare no conflict of interest. The founding sponsors had no role in the design of the study; in the collection, analyses, or interpretation of data; in the writing of the manuscript; or in the decision to publish the results.

References

1. Li, M.; Pal, A.; Aghakhani, A.; Pena-Francesch, A.; Sitti, M. Soft actuators for real-world applications. *Nat. Rev. Mater.* **2022**, *7*, 235–249. [CrossRef]
2. Son, D.; Gilbert, H.; Sitti, M. Magnetically actuated soft capsule endoscope for fine-needle biopsy. *Soft Robot.* **2020**, *7*, 10–21. [CrossRef]
3. Nitta, T.; Wang, Y.; Du, Z.; Morishima, K.; Hiratsuka, Y. A printable active network actuator built from an engineered biomolecular motor. *Nat. Mater.* **2021**, *20*, 1149–1155. [CrossRef]

4. Jin, D.; Yuan, K.; Du, X.; Wang, Q.; Wang, S.; Zhang, L. Domino reaction encoded heterogeneous colloidal microswarm with on-demand morphological adaptability. *Adv. Mater.* **2021**, *33*, 2100070. [CrossRef]
5. Akolpoglu, M.B.; Alapan, Y.; Dogan, N.O.; Baltaci, S.F.; Yasa, O.; Aybar Tural, G.; Sitti, M. Magnetically steerable bacterial microrobots moving in 3D biological matrices for stimuli-responsive cargo delivery. *Sci. Adv.* **2022**, *8*, eabo6163. [CrossRef]
6. Cheng, Y.; Li, S.; Liu, J. Abnormal deformation and negative pressure of a hard magnetic disc under the action of a magnet. *Sens. Actuators Phys.* **2021**, *332*, 113065. [CrossRef]
7. Kim, Y.; Yuk, H.; Zhao, R.; Chester, S.A.; Zhao, X. Printing ferromagnetic domains for untethered fast-transforming soft materials. *Nature* **2018**, *558*, 274–279. [CrossRef]
8. Zhang, J.; Ren, Z.; Hu, W.; Soon, R.H.; Yasa, I.C.; Liu, Z.; Sitti, M. Voxelated three-dimensional miniature magnetic soft machines via multimaterial heterogeneous assembly. *Sci. Robot.* **2021**, *6*, eabf0112. [CrossRef]
9. Wu, Z.; Zhang, Y.; Ai, N.; Chen, H.; Ge, W.; Xu, Q. Magnetic Mobile Microrobots for Upstream and Downstream Navigation in Biofluids with Variable Flow Rate. *Adv. Intell. Syst.* **2022**, *4*, 2100266. [CrossRef]
10. Wu, Z.; Zhang, Y.; Chi, Z.; Xu, Q. Design and Development of a New Rotating Electromagnetic Field Generation System for Driving Microrobots. *IEEE Trans. Magn.* **2021**, *58*, 1–8. [CrossRef]
11. Kearney, K.; Hira, R.S.; Riley, R.F.; Kalyanasundaram, A.; Lombardi, W.L. Update on the management of chronic total occlusions in coronary artery disease. *Curr. Atheroscler. Rep.* **2017**, *19*, 1–10. [CrossRef]
12. Song, P.; Rudan, D.; Zhu, Y.; Fowkes, F.J.; Rahimi, K.; Fowkes, F.G.R.; Rudan, I. Global, regional, and national prevalence and risk factors for peripheral artery disease in 2015: An updated systematic review and analysis. *Lancet Glob. Health* **2019**, *7*, e1020–e1030. [CrossRef]
13. Scales, C.D., Jr.; Smith, A.C.; Hanley, J.M.; Saigal, C.S.; Urologic Diseases in America Project. Prevalence of kidney stones in the United States. *Eur. Urol.* **2012**, *62*, 160–165. [CrossRef]
14. Azizi, A.; Tremblay, C.C.; Gagné, K.; Martel, S. Using the fringe field of a clinical MRI scanner enables robotic navigation of tethered instruments in deeper vascular regions. *Sci. Robot.* **2019**, *4*, eaax7342. [CrossRef]
15. Kim, Y.; Parada, G.A.; Liu, S.; Zhao, X. Ferromagnetic soft continuum robots. *Sci. Robot.* **2019**, *4*, eaax7329. [CrossRef]
16. Jeon, S.; Hoshiar, A.K.; Kim, K.; Lee, S.; Kim, E.; Lee, S.; Kim, J.Y.; Nelson, B.J.; Cha, H.J.; Yi, B.J.; et al. A magnetically controlled soft microrobot steering a guidewire in a three-dimensional phantom vascular network. *Soft Robot.* **2019**, *6*, 54–68. [CrossRef]
17. Khalil, I.S.; Tabak, A.F.; Hosney, A.; Mohamed, A.; Klingner, A.; Ghoneima, M.; Sitti, M. Sperm-shaped magnetic microrobots: Fabrication using electrospinning, modeling, and characterization. In Proceedings of the 2016 IEEE International Conference on Robotics and Automation (ICRA), Stockholm, Sweden, 16–21 May 2016; pp. 1939–1944.
18. Park, M.; Le, T.A.; Yoon, J. Offline Programming Guidance for Swarm Steering of Micro-/Nano Magnetic Particles in a Dynamic Multichannel Vascular Model. *IEEE Robot. Autom. Lett.* **2022**, *7*, 3977–3984. [CrossRef]
19. Dai, Y.; Jia, L.; Wang, L.; Sun, H.; Ji, Y.; Wang, C.; Song, L.; Liang, S.; Chen, D.; Feng, Y.; et al. Magnetically Actuated Cell-Robot System: Precise Control, Manipulation, and Multimode Conversion. *Small* **2022**, *18*, 2105414. [CrossRef]
20. Lin, D.; Wang, J.; Jiao, N.; Wang, Z.; Liu, L. A Flexible Magnetically Controlled Continuum Robot Steering in the Enlarged Effective Workspace with Constraints for Retrograde Intrarenal Surgery. *Adv. Intell. Syst.* **2021**, *3*, 2000211. [CrossRef]
21. Kafash Hoshiar, A.; Jeon, S.; Kim, K.; Lee, S.; Kim, J.y.; Choi, H. Steering algorithm for a flexible microrobot to enhance guidewire control in a coronary angioplasty application. *Micromachines* **2018**, *9*, 617. [CrossRef]
22. Kim, Y.; Genevriere, E.; Harker, P.; Choe, J.; Balicki, M.; Regenhardt, R.W.; Vranic, J.E.; Dmytriw, A.A.; Patel, A.B.; Zhao, X. Telerobotic neurovascular interventions with magnetic manipulation. *Sci. Robot.* **2022**, *7*, eabg9907. [CrossRef] [PubMed]
23. Wang, L.; Kim, Y.; Guo, C.F.; Zhao, X. Hard-magnetic elastica. *J. Mech. Phys. Solids* **2020**, *142*, 104045. [CrossRef]
24. Craik, D.J. *Magnetism: Principles and Applications*; John Wiley & Sons, Inc.: Hoboken, NJ, USA, 2003.
25. Agashe, J.S.; Arnold, D.P. A study of scaling and geometry effects on the forces between cuboidal and cylindrical magnets using analytical force solutions. *J. Phys. Appl. Phys.* **2008**, *41*, 105001. [CrossRef]
26. Di Natali, C.; Beccani, M.; Valdastri, P. Real-time pose detection for magnetic medical devices. *IEEE Trans. Magn.* **2013**, *49*, 3524–3527. [CrossRef]
27. Geiger, M. *Fundamentals of Vascular Biology*; Springer: Cham, Switzerland, 2019.
28. Xu, C.; Yang, Z.; Tan, S.W.K.; Li, J.; Lum, G.Z. Magnetic Miniature Actuators with Six-Degrees-of-Freedom Multimodal Soft-Bodied Locomotion. *Adv. Intell. Syst.* **2022**, *4*, 2100259. [CrossRef]
29. Wu, Y.; Dong, X.; Kim, J.K.; Wang, C.; Sitti, M. Wireless soft millirobots for climbing three-dimensional surfaces in confined spaces. *Sci. Adv.* **2022**, *8*, eabn3431. [CrossRef]

actuators

Article

Development of a Sinusoidal Corrugated Dual-Axial Flexure Mechanism for Planar Nanopositioning

Yuhan Niu, Xingyou Chen, Li Chen, Zhiwei Zhu and Peng Huang *

School of Mechanical Engineering, Nanjing University of Science and Technology, Nanjing 210094, China
* Correspondence: hp@njust.edu.cn

Abstract: Taking advantage of the concurrent stretching and bending property of corrugated flexure hinges, a sinusoidal corrugated flexure linkage was proposed and applied for the construction of a corrugated dual-axial mechanism with structural symmetry and decoupled planar motion guidance. Castigliano's second theorem was employed to derive the complete compliance for a basic sinusoidal corrugated flexure unit, and matrix-based compliance modeling was then applied to find the stiffness of the sinusoidal corrugated flexure linkage and the corrugated dual-axial mechanism. Using established analytical models, the influence of structural parameters on the stiffness of both the corrugated flexure linkage and the dual-axial mechanism were investigated, with further verification by finite element analysis, with errors less than 20% compared to the analytical results for all cases. In addition, the stiffness of the corrugated flexure mechanism was practically tested, and its deviation between practical and analytical was around 7.4%. Further, the feasibility of the mechanism was demonstrated by successfully applying it for a magnetic planar nanopositioning stage, for which both open-loop and closed-loop performances were systematically examined. The stage has a stroke around 130 μm for the two axes and a maximum cross-talk less than 2.5%, and the natural frequency is around 590 Hz.

Keywords: dual-axial nanopositioning; corrugated flexure hinge; system modeling; trajectory tracking

Citation: Niu, Y.; Chen, X.; Chen, L.; Zhu, Z.; Huang, P. Development of a Sinusoidal Corrugated Dual-Axial Flexure Mechanism for Planar Nanopositioning. *Actuators* **2022**, *11*, 276. https://doi.org/10.3390/act11100276

Academic Editors: Jinchuan Zheng and Ioan Ursu

Received: 18 August 2022
Accepted: 27 September 2022
Published: 29 September 2022

1. Introduction

Flexure hinge-based planar XY nanopositioning stages in parallel are widely employed in the fields of micro-machining, micro-manipulation, and scanning-based surface metrology, to mention a few of the applications [1–3]. For state-of-the-art designs, dual-axial compliant mechanisms are symmetrically constructed by combining several flexure hinges having a single degree-of-freedom (DOF) of planar bending [4]. For example, four sets of "T"-shape mechanisms were employed in [4], for a total of 12 leaf-spring flexure hinges, and a combination of the bending motions of all the hinges jointly contributed to the decoupled dual-axial motions for the end-effector. Taking advantage of the right circular flexure hinges, a compound mechanism combining a separated prismatic joint and a parallelogram was developed to construct a decoupled XY nanopositioning stage in [5].

Although a structural configuration combining flexure hinges with the bending DOF dominates the current design of dual-axial nanopositioning stages, this increases structural complexity and the equivalent moving inertia caused by the employment of multiple flexure hinges [4,5]. To reduce structural complexity, simplified non-symmetric mechanisms having only two orthogonal parallelogram mechanisms were developed to guide the dual-axial motions, as reported in [6,7]. However, this simplified structure may lead to an unconstrained parasitic motion for the end-effector, which may greatly deteriorate the positioning accuracy of the planar stages. To overcome this defect, an alternative solution might be the adoption of flexure hinges to directly connect the end-effector and the base in a symmetric manner, which may simultaneously simplify the structure and guarantee accurate motion. In this condition, concurrent transition of planar bending and axial stretching deformations is essentially required for the flexure hinges to enable planar motions. As a candidate, "L"-shape flexure linkages are promising, and four sets of "L"-shape linkages may provide symmetric guidance for the motion along the two directions [8–10]. However, considering the corner structure of the linkage, it is difficult to arrange a parallel configuration for the multiple linkages to improve the resistance capability.

Recently, the concept of corrugated flexure hinges was developed and applied to construct dual-axial XY nanopositioning stages [11–13]. Along with a circular curve segment [14], a cone-shaped [15] segment was also developed to enrich the grouping of the corrugated flexural hinges. Following the conventional "T"-shape structure, corrugated hinges have also been applied to construct dual-axial XY nanopositioning stages [11–13]. As for those designs, the corrugated hinges were adopted to reduce the stress and extend the deformation range of conventional flexure hinges within a limited space [14], and the unique stretchable feature has not been explored for motion delivery. Since the "T"-shape structure was adopted for those designs, the resulting corrugated hinge-based planar stages may inevitably have similar defects as aforementioned.

To gain high-performance dual-axial guidance, we develop a dual-axial corrugated flexure mechanism with a simple structure and low moving mass. The main contributions of this study are: (a) A sinusoidal corrugated flexure hinge with a simple mathematical description is developed and comprehensively characterized by both analytical and finite element simulation methods. (b) The unique stretchable property is explored to realize concurrent stretching and bending of the corrugated flexure hinge for dual-axial motion guidance. (c) A corrugated dual-axial flexure mechanism is developed and demonstrated by applying it to a dual-axial electromagnetic stage.

The remainder of this paper is organized as follows: Section 2 introduces the sinusoidal corrugated hinge and derives its static compliance, Section 3 presents the developed corrugated dual-axial mechanism, and Section 4 demonstrates the parameter selection for constructing the dual-axial nanopositioning stage. The experimental testing of mechanism stiffness and the basic performance of the constructed dual-axial stage is detailed in Section 5, and the main conclusions are drawn in Section 6.

2. The Sinusoidal Corrugated Flexure Linkage

Although a sinusoidal corrugated beam was developed in [12] to clamp the movers to eliminate under-constraining, the complex stretching and bending property was not investigated for the corrugated beam. Herein, we introduce a monolithic sinusoidal corrugated flexure linkage, and the three-dimensional (3D) structure is illustrated in Figure 1. Along with the bending property, the linkage is intrinsically stretchable due to the structure corrugation, and multiple cycles may further enhance the flexibility for tuning the stretching stiffness.

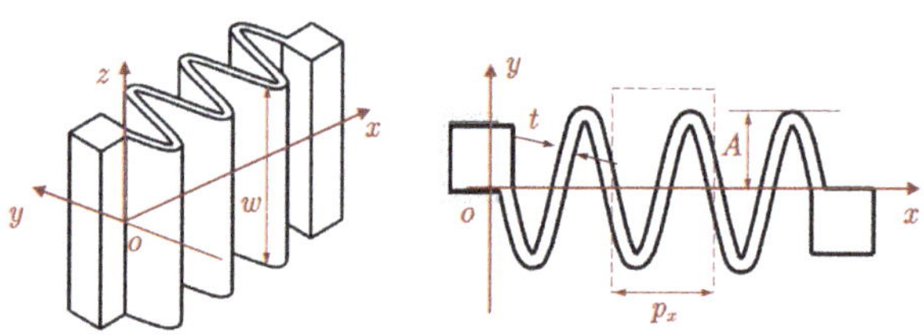

Figure 1. Structure of the sinusoidal corrugated flexure linkage with $n = 3$ for illustration.

Mathematically, the sinusoidal corrugated flexure linkage can be described by a simple and continuous equation as

$$z(x) = -A\sin(2\pi p_x^{-1} x),\ x \in [0, np_x] \tag{1}$$

where A, p_x, and n denote the amplitude, spatial periodicity, and cycle number of the linkage, respectively.

2.1. Stiffness Modeling of the Sinusoidal Corrugated Flexure Linkage

The sinusoidal corrugated linkage shown in Figure 1 can be decomposed into several (n) serially connected flexure units ($x \in [0, p_x]$). Assuming the load and corresponding deformation at the free end of the unit are, respectively, $\mathbf{F} = [f_x, f_y, f_z, m_x, m_y, m_z]^{\mathrm{T}}$ and $\mathbf{u} = [u_x, u_y, u_z, \theta_x, \theta_y, \theta_z]^{\mathrm{T}}$, the bending torques at position x can be expressed as

$$M_b(x) := \begin{cases} M_{xy}(x) = m_z + f_y(p_x - x) + f_x z(x) \\ M_{xz}(x) = m_y + f_z\sqrt{(p_x - x)^2 + z^2(x)} \\ M_{yz}(x) = m_x \end{cases} \quad (2)$$

and the planar normal and shear force are

$$\begin{cases} N(x) = f_y \sin\varphi + f_x \cos\varphi; \\ S(x) = -f_y \cos\varphi + f_x \sin\varphi \end{cases} \quad (3)$$

with

$$\varphi = \begin{cases} \arctan|\dot{z}|, \; x \in [\frac{1}{4}p_x, \frac{3}{4}p_x] \\ \pi - \arctan|\dot{z}|, \; \text{otherwise} \end{cases} \quad (4)$$

Therefore, following Castigliano's second theorem, the elastic strain energy can be expressed as [16]

$$\begin{aligned} U = & \int_0^{\mathcal{L}} \left(\frac{M_b^2(x)}{2EI} + \frac{N^2(x)}{2EA_c} + \frac{\alpha_s S^2(x)}{2GA_c} \right) \mathrm{d}s \\ = & \int_0^{p_x} \left(\frac{M_b^2(x)}{2EI} + \frac{N^2(x)}{2EA_c} + \frac{\alpha_s S^2(x)}{2GA_c} \right) \sqrt{1 + \dot{z}^2(x)} \mathrm{d}x \end{aligned} \quad (5)$$

where I is the second moment of the rectangular section of the corrugated beam, and $A_c = wt$ is the cross-sectional area. In addition, E represents the Young's modulus of the material, $G = \frac{E}{2(1+\mu)}$ is the shear modulus, with μ denoting the Passion's ratio, and $\alpha_s = \frac{12 + 11\mu}{10(1+\mu)}$ is the shear coefficient.

Accordingly, the spatial six-DOF compliance for the sinusoidal corrugated flexure unit can be obtained as [16,17]

$$\begin{cases} C_{x,f_x} = \dfrac{\partial^2 U}{\partial f_x^2}, \; C_{y,f_y} = \dfrac{\partial^2 U}{\partial f_y^2}, \; C_{z,f_z} = \dfrac{\partial^2 U}{\partial f_z^2}, \\ C_{\theta_x,m_x} = \dfrac{\partial^2 U}{\partial m_x^2}, \; C_{\theta_y,m_y} = \dfrac{\partial^2 U}{\partial m_y^2}, \; C_{\theta_z,m_z} = \dfrac{\partial^2 U}{\partial m_z^2}, \\ C_{\theta_z,f_y} = C_{y,m_z} = \dfrac{\partial^2 U}{\partial m_z \partial f_y}, \\ C_{\theta_y,f_z} = C_{z,m_y} = \dfrac{\partial^2 U}{\partial m_y \partial f_z}. \end{cases} \quad (6)$$

By substituting Equations (1)–(5) into Equation (6), the compliance items can be derived, which are presented in detail in Appendix A. Accordingly, the relationship between the deformations and loads for the flexure unit yields

$$\mathbf{u} = \mathbf{C}_s \mathbf{F} \quad (7)$$

where the compliance matrix $\mathbf{C}_s$ is defined as

$$\mathbf{C}_s = \begin{bmatrix} C_{x,f_x} & 0 & 0 & 0 & 0 & 0 \\ 0 & C_{y,f_y} & 0 & 0 & 0 & C_{y,m_z} \\ 0 & 0 & C_{z,f_z} & 0 & C_{z,m_y} & 0 \\ 0 & 0 & 0 & C_{\theta_x,m_x} & 0 & 0 \\ 0 & 0 & C_{\theta_y,f_z} & 0 & C_{\theta_y,m_y} & 0 \\ 0 & C_{\theta_z,f_y} & 0 & 0 & 0 & C_{\theta_z,m_z} \end{bmatrix} \quad (8)$$

Considering the serially connected flexure units (n) as shown in Figure 1, the complete compliance of the sinusoidal corrugated linkage can be derived following matrix-based compliance modeling (MCM) [18,19] as

$$\mathbf{C}_u = \sum_{i=1}^{n} \mathbf{T}_i \mathbf{C}_s \mathbf{T}_i^{\mathrm{T}} \quad (9)$$

where $\mathbf{T}_i$ is the compliance transformation matrix (CTM) to transfer the local coordinate system of the i-th flexure unit to the coordinate system of the linkage [18,19]. Accordingly, the stiffness matrix of the corrugated linkage can be derived as $\mathbf{K}_s = \mathbf{C}_s^{-1}$.

2.2. FEA-Based Stiffness Verification

FEA is conducted via commercial software ANSYS/Workbench to characterize the performance and to verify the stiffness model of the corrugated flexure linkage. Aluminum alloy (AL7075-T651) is selected as the material for the mechanism, with Young's modulus of $E = 72$ GPa and Poisson's ratio of $\mu = 0.33$. In addition, the adaptive meshing method provided by ANSYS/Workbench modulus is employed for element meshing. To guarantee simulation accuracy, the simulation was conducted using different scales of the element size, with the final scale chosen when further refinement did not lead to variation of the simulated results.

By setting the overall height $w = 8$ mm, there are four main parameters (A, t, p_x, and n) that can flexibly determine the deformation behavior of the flexure linkage. By fixing three of them, the stretchable (x-axis) and bending (y-axis) stiffness related to the one other parameter are obtained through both the analytical and FEA model, which are then comparatively illustrated in Figures 2 and 3, respectively. In general, the stretching stiffness is much higher than the bending stiffness, which may be attributed to the much larger equivalent bending length.

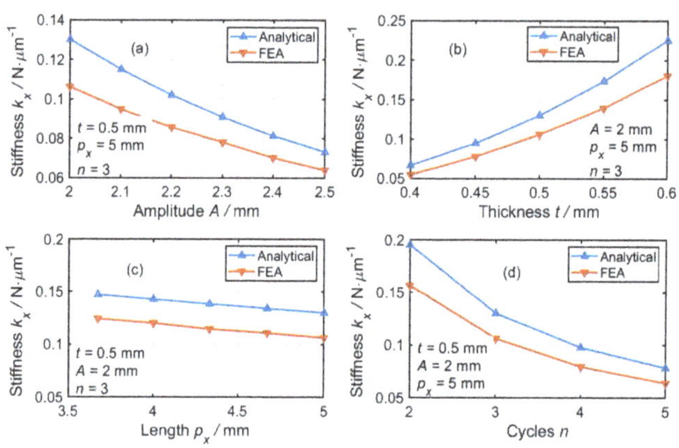

Figure 2. The x-axis stiffness k_x of the sinusoidal corrugated flexure hinge related to (**a**) amplitude A, (**b**) thickness t, (**c**) length p_x, and (**d**) cycle number n.

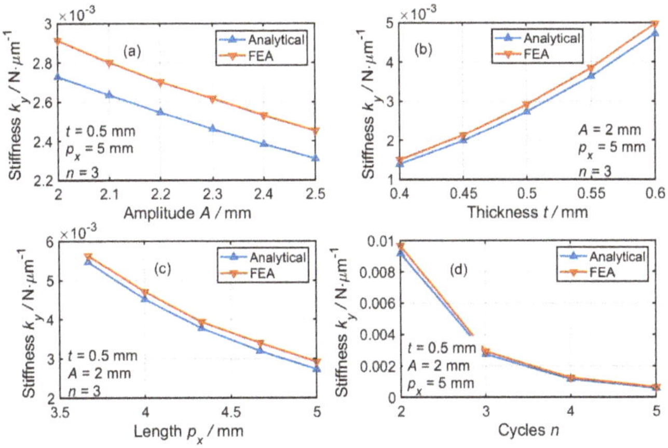

Figure 3. The y-axis stiffness k_y of the sinusoidal corrugated flexure hinge related to (**a**) amplitude A, (**b**) thickness t, (**c**) length p_x, and (**d**) cycle number n.

Although the stiffness values for the two directions are different for the linkage, their variation in relation to structural parameter changes are similar. As shown in Figures 2 and 3, an increase in amplitude A, unit length p_x, and cycle number n may decrease both the stretching and bending stiffness. By contrast, an exponential increase in stiffness may occur in terms of a linear increase in the thickness t. Overall, good agreement is observed between the analytical and FEA result, and all deviations between the analytical and FEA results for all cases are within 20%, verifying the effectiveness of the developed stiffness model for sinusoidal corrugated flexure hinges.

3. The Sinusoidal Corrugated Dual-Axial Mechanism

Taking advantage of the linkage, a sinusoidal corrugated dual-axial mechanism is constructed, as shown in Figure 4a. It mainly consists of four pairs of parallelograms arranged symmetrically, and each parallelogram has two parallel sinusoidal corrugated flexure linkages. For force balancing during axial elongation/compression, the two parallel linkages for each parallelogram are specially designed to have mirror-symmetry.

As illustrated in Figure 4b, when an actuation force (F_x for example) is applied, the double parallelograms along the actuation direction (x-axis) will stretch, and the other two double parallelograms (y-axis) will mainly bend to generate the x-axial motion with suppressed parasitic motions. Accordingly, dual-axial actuation forces on the end-effector may simultaneously generate axial stretching and planar bending for planar dual-axial motions.

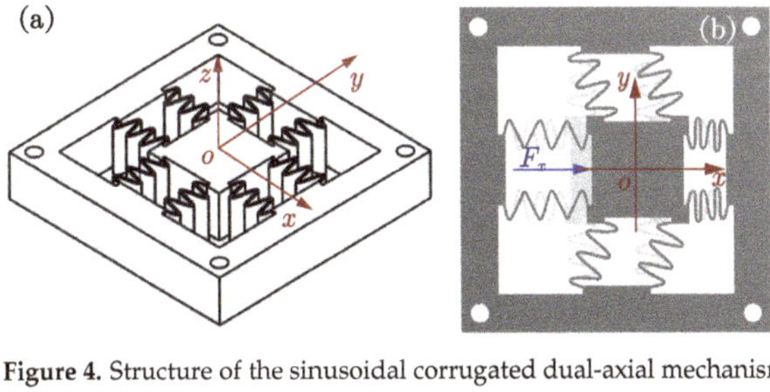

Figure 4. Structure of the sinusoidal corrugated dual-axial mechanism: (**a**) 3-D structure and (**b**) deformation principle.

3.1. *Modeling of the Dual-Axial Mechanism*

3.1.1. Stiffness Modeling

With respect to the k-th corrugated linkage, its compliance in the global coordinate system $o - xyz$ as shown in Figure 4 can be derived through MCM as

$$\mathbf{C}_L^{(k)} = \mathbf{T}_k \mathbf{C}_u \mathbf{T}_k^{\mathrm{T}} \tag{10}$$

where $\mathbf{T}_k$ is the CTM transferring the coordinate system of the k-th linkage to the global coordinate system of the mechanism.

Since the dual-axial mechanism is constructed by eight linkages in parallel, the stiffness for the mechanism can be derived as

$$\mathbf{K} = \sum_{k=1}^{8} \left(\mathbf{C}_L^{(k)}\right)^{-1} \tag{11}$$

Accordingly, x- and y-directional stiffness for the mechanism at point o is

$$k_x = k_y = \mathbf{K}(1,1) = \mathbf{K}(2,2) \tag{12}$$

3.1.2. Dynamics Modeling

Assume the generalized coordinate for the end-effector is $\mathbf{u} = [u_x, u_y]$. Following Lagrange's equation, by ignoring the damping effect, the dynamics equation for the mechanism can be expressed as

$$\mathbf{M}\ddot{\mathbf{u}} + \mathbf{K}\mathbf{u} = \mathbf{F} \tag{13}$$

where $\mathbf{M}$ represents the equivalent mass matrix, and $\mathbb{F}$ is the generalized force. In addition, the stiffness matrix for the mechanism is $\mathbf{K} = diag(k_x, k_y)$.

Considering the structural symmetry, the kinetic energy for the mechanism is

$$T = \frac{1}{2}\left(m\dot{u}_x^2 + m\dot{u}_y^2 + 4J\frac{\dot{u}_x^2}{l^2} + 4J\frac{\dot{u}_y^2}{l^2}\right) \tag{14}$$

where m is the equivalent moving mass, and J and $l = np_x$ are the rotational inertia and length of the corrugated hinge, respectively.

Accordingly, the mass matrix $\mathbf{M}$ yields

$$\mathbf{M} = \frac{\partial T}{\partial \dot{\mathbf{u}}}\dot{\mathbf{u}}^{-1} \tag{15}$$

and the resonant frequencies can be obtained through solving

$$\left|(2\pi f_i)^2\mathbf{M} - \mathbf{K}\right| = 0,\ i = x, y \tag{16}$$

3.2. *FEA Verification of the Mechanism*

Similarly, FEA-based simulation is further conducted using the same material and software for stiffness verification of the corrugated dual-axial flexure mechanism. As performed in Section 2.2, by fixing three of the four parameters, the directional stiffness k_x or k_y related to the selected parameter is presented in Figure 5. An increase in the amplitude A, unit length p_x, and cycle number n may lead to a decrease in the stiffness, whereas increasing the thickness t results in an increase in the stiffness. In addition, the maximum deviation between the analytical and FEA results is also smaller than 20% for all cases.

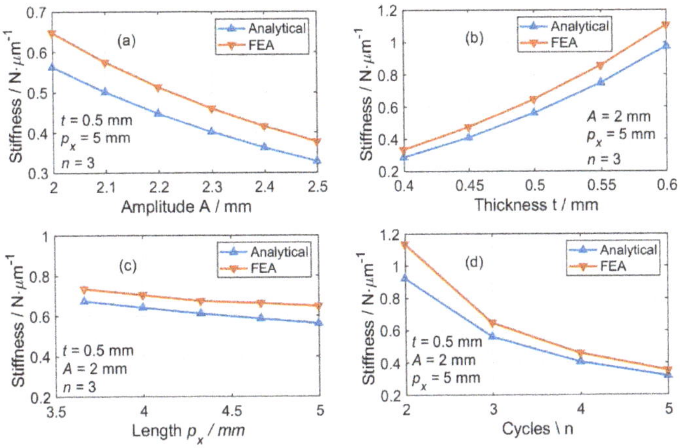

Figure 5. The stage's stiffness related to (**a**) amplitude A, (**b**) thickness t, (**c**) length p_x, and (**d**) cycle number n.

4. Structure Parameter Determination for a Planar Nanopositioning Stage

Since the mechanical structure is simple, the structure of the corrugated hinge may have a very limited influence on the equivalent moving mass. The working performance, including both the stroke and natural frequency, may highly depend on the stiffness of the mechanism. Therefore, only the axial stiffness is adopted as the design target for the construction of the planar nanopositioning stage, and it is set as $k_x = k_y = 0.4$ N/μm, taking into full consideration the actuation force.

Through trial-and-error, the dimensional parameters for the flexure linkages are determined as $A = 2.3$ mm, $t = 0.44$ mm, w = 9.4 mm, p_x = 3.33 mm, and $n = 3$, which lead to an analytical stiffness of 0.408 N/μm. By adopting the same FEA model as in Section 3.2, the directional deformation when subjected to an x-axial force of 100 N on the platform is illustrated in Figure 6a. Through dividing the driving force (100 N) by the deformation (199.76 μm), the stiffness is derived to be about 0.496 N/μm. Taking the FEA result as the benchmark, the analytical stiffness has an acceptable deviation around 20%. In addition, in-plane rotation is subjected to a torque (1 N·m) around the z-axis of the platform, which is illustrated in Figure 6b. The rotation angle of the platform is estimated to be about 0.0227 rad, which suggests an in-plane rotational stiffness of about 4.405×10^7 N·μm/rad. The analytical result is then found to have a deviation of about 19.4%.

As this paper mainly studies the flexible mechanism, only the moving part of the stage, including the flexible mechanism and the part directly connected to it, is considered in the dynamic simulation. By assembling all the necessary accessories for the planar stage demonstrated in Section 5.2, FEA simulation is employed to characterize the first four resonant modes of the dual-axial flexure mechanism, which are illustrated in Figure 7. As expected, the first two resonances have a nearly identical resonant frequency around 679 Hz, and the mode shapes are consistent with the desired dual-axial motions. Considering the structural symmetry, the analytical resonant frequencies for the first two resonant frequencies are calculated to be identical as 599.341 Hz, which deviates about 11.7% compared with the FEA result. In addition, out-of-plane translation and rotation are observed for the third and fourth mode, and the corresponding resonant frequencies are about 976.54 Hz and 1007.5 Hz, as shown in Figure 7c,d.

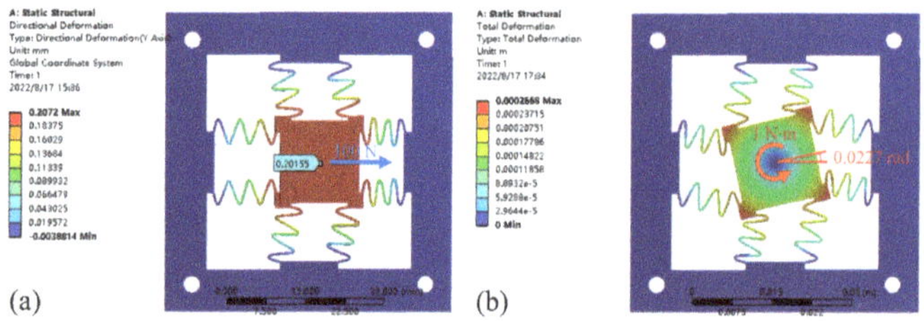

Figure 6. Simulated deformation of the platform: (**a**) axial deformation when subjected to an axial force of 100 N, and (**b**) rotation when subjected to a torque around the *z*-axis of 1 N·m.

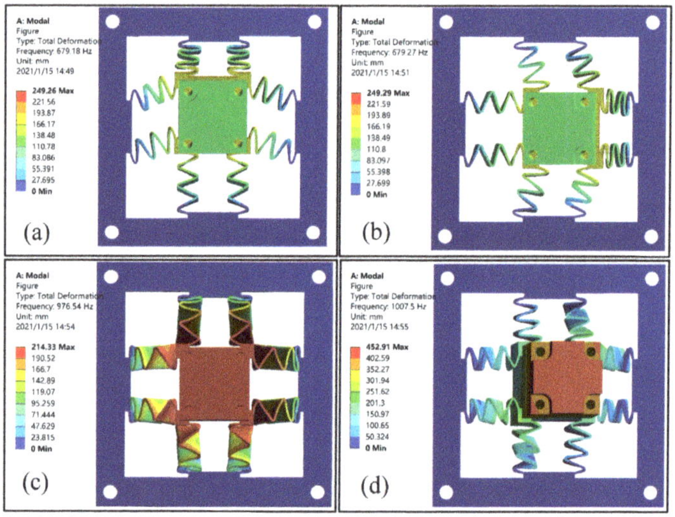

Figure 7. Simulated mode shapes: (**a**) first, (**b**) second, (**c**) third, and (**d**) fourth modes.

Note 1 The compliance and stiffness matrices of the designed stage are presented in Appendix A.3. From the calculated stiffness matrix **K**, the *z*-axial stiffness (10.8 N/μm) is about two orders of magnitude larger than the *x*- and *y*-axial stiffness (0.408 N/μm), suggesting that the out-of-plane DOF is well-constrained.

Note 2 Compared with the in-plane rotation, the much larger tilting stiffness around the *x*- and *y*-axes suggests that the two tilting DOFs are also constrained. Although the induced rotation is relatively small (1 N disturbance on the sidewall of the platform may only lead to a slight rotation of 0.17 mrad), undesired overly large in-plane torques must be carefully avoided to eliminate in-plane rotation errors for practical applications.

5. Experimental Results and Discussion

5.1. Stiffness-Testing the Mechanism

Using the selected structural parameters, the produced prototype of the sinusoidal corrugated dual-axial mechanism is shown in Figure 8a. The experimental setup in Figure 8b is developed for testing the stiffness of the mechanism. A force gauge with a digital display is employed to apply a directional force on the end-effector, and a dial indicator is used to record the resulting displacement.

Figure 8. Photograph of (**a**) the mechanism prototype and (**b**) the experimental setup for stiffness measurement.

The obtained displacement related to the applied force for each direction is presented in Figure 9. From the best-fitted linear lines, the stiffness along the two directions is almost identical as $k_x \approx k_y = 0.38$ N/µm. Taking the practically tested stiffness as the benchmark, the analytical modeling error is about 7.4%.

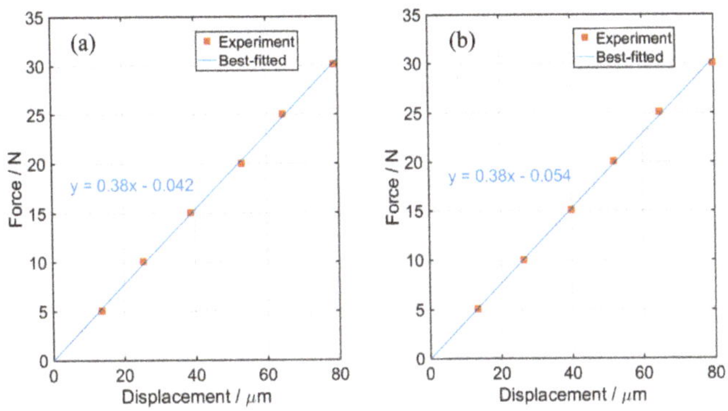

Figure 9. Relationship between applied force and resulting displacement for (**a**) the x-axis and (**b**) the y-axis.

5.2. Performance Testing for the Planar Stage

Practically, the feasibility of the sinusoidal corrugated dual-axial mechanism is demonstrated by applying it to a planar nanopositioning stage. With the stage, dual-axial normal-stressed electromagnetic forces are applied on the side surfaces of the armature, which is directly attached to the end-effector of the mechanism. Therefore, the actuation force is imposed on the end-effector and guides the corrugated mechanism during work. A detailed description of the newly developed electromagnetic actuator is presented in [20].

The photograph of the assembled planar nanopositioning stage is illustrated in Figure 10a, and the main components, including the electromagnetic and mechanical parts, are shown in Figure 10b. Two linear servo amplifiers (SMA5005-1, Glenteck Corporation, El Segundo, CA, USA) were adopted to amplify the command for driving the stage, and two ultra-precise capacitive sensors (Micro-sense-5810, Micro-sense Corporation, Lowell, MA, USA) were used to measure the end-effector motion. All the signals were collected and sent out through a data acquisition board (PCI-6259, NI Corporation, Austin, TX, USA) with a sampling frequency of 20 kHz.

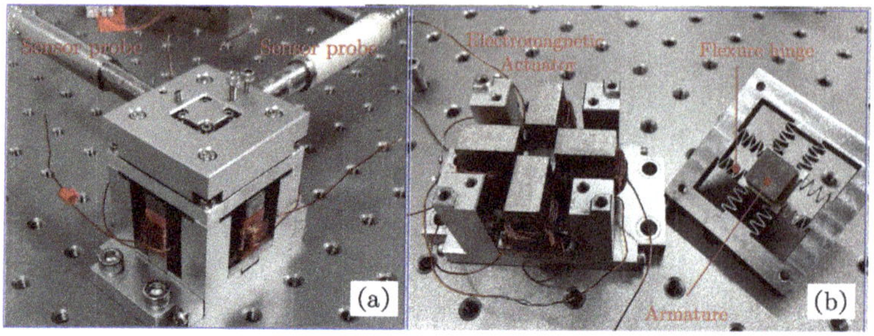

Figure 10. Photograph of the magnetic stage: (**a**) the assembled stage and (**b**) the key components.

5.2.1. Open-Loop Performance

A maximum allowable current of 3A was independently applied to the excitation coil winding for each direction, and the resulting primary motion and its cross-talk are shown in Figure 11. Overall, nearly identical strokes are observed: 133.5 μm and 132.6 μm for the *x*- and *y*-axis, respectively. The practical cross-talks are around 3.3 μm (2.47%) and 1.26 μm (0.95%), which might be caused by manufacturing errors in both the compliant mechanism and the electromagnetic actuator.

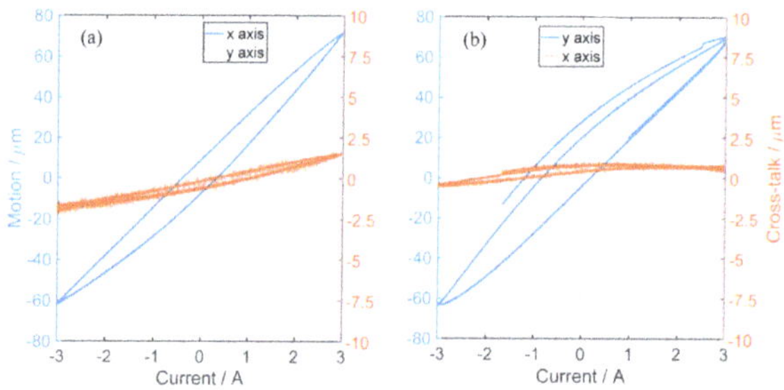

Figure 11. Motion and resulting cross-talk with maximum actuation along (**a**) the *x*-axis and (**b**) the *y*-axis.

In addition, taking advantage of the independent sweep excitation for each axis, the resulting frequency response functions for both axes are obtained and illustrated in Figure 12. As shown in Figure 12, the primary resonant frequencies for the two axes are identified as about $f_1 = 586$ Hz and $f_2 = 596$ Hz, and the smaller resonant peaks at $f_3 = 816$ Hz and $f_4 = 882$ Hz may correspond to the high-order resonances. Overall, the resonances exhibit good agreement with the simulation results obtained by FEA.

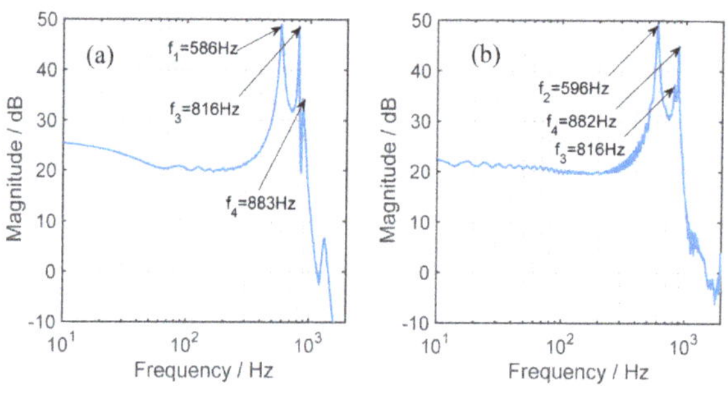

Figure 12. The amplitude response function of (**a**) the *x*-axis and (**b**) the *y*-axis.

5.2.2. Closed-Loop Performance

To perform trajectory tracking, a closed-loop system is constructed for the dual-axial nanopositioning stage. With the control system, each axis is treated as a single-input–single-output system (SISO), and the practical cross-talks and system hysteresis are lumped as the general disturbance to be compensated for by the feedback control [21].

The developed control system for each axis is schematically illustrated in Figure 13. The control system employs a typical proportional–integral–differential (PID) controller combining a system dynamics inversion-based feedforward compensator as the main controller for trajectory tracking [22,23]. Considering the lumped disturbance, a system model-based disturbance observer (DOB) is employed for disturbance compensation, as detailed in Figure 13. To simplify the controller design, the system nominal models for both axes are approximated by the system gains relating to the static input voltage and output motion, namely $P_m(s) \approx g_m$, and the PID parameters are tuned manually through trail-and-error.

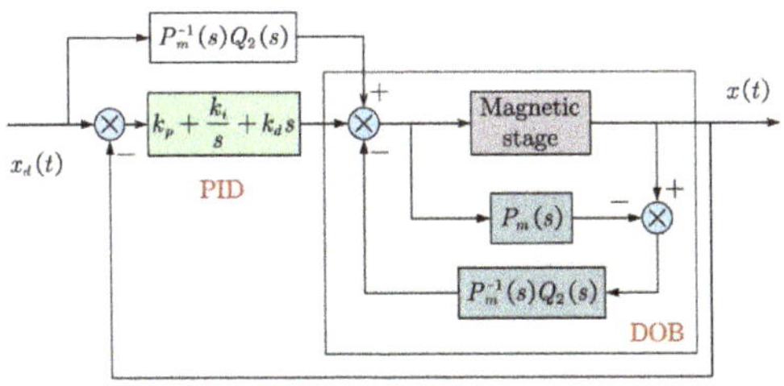

Figure 13. Schematic of the control loop, where k_p, k_i, and k_d are, respectively, the proportional, integral, and differential gains of the PID controller, and $Q_1(s)$ and $Q_2(s)$ are the low-pass filters.

The closed-loop performance of the stage is demonstrated by tracking a spiral trajectory. By setting $f = 30$ Hz as the motion frequency, the spiral trajectory can be mathematically expressed as

$$\begin{cases} x = 125t\sin(2\pi ft)\ \mu m \\ y = 125t\cos(2\pi ft)\ \mu m \end{cases} \tag{17}$$

The tracking result shown in Figure 14a suggests good accordance between the desired and practical motion. In addition, from the tracking error shown in Figure 14b,c, the maximum tracking error for both axes is observed to be less than ±0.5 µm, which is about ±1% of the full motion span.

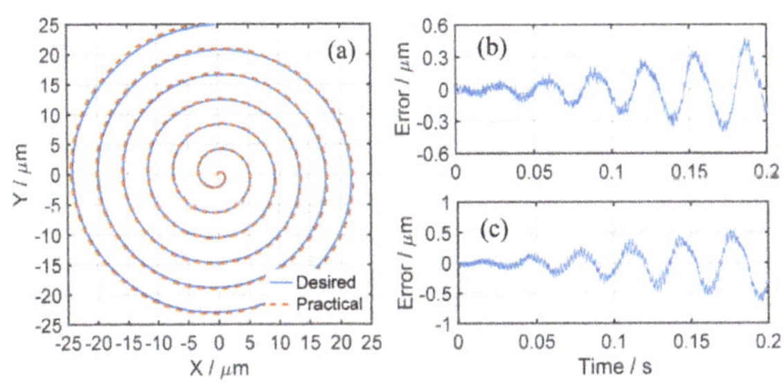

Figure 14. Trajectory tracking performance: (**a**) spiral trajectory tracking, and the error along (**b**) the x-axis and (**c**) the y-axis.

5.3. Results Comparison and Discussion

For comparison, the theoretical and practical performances including the stiffness and resonant frequency are summarized in Table 1. Considering the structure symmetry, only the stiffness k_x and the first resonant frequency f_x are presented, and the error between the analytical and practical results is also included.

Table 1. Performance comparison of the mechanism.

	k_x (N/μm)	f_x (Hz)
Analytical	0.408	599
FEA	0.496	679
Experiment	0.38	586
Error	7.4%	2.2%

As shown in Table 1, both the stiffness and natural frequency obtained by FEA are slightly larger than the analytical and experimental results, and better agreement is obtained between the theoretical and practical results. As for stiffness testing, the inevitable deformation of the clamped structure of the mechanism leads to overestimated motion when subjected to an external force. In addition, since the force is not consistent with the central axis of the end-effector, an extra torque is also imposed on the end-effector. The two factors may jointly lead to smaller practical stiffness.

In addition, the theoretical result only considers the mechanical structure, and the influence of the electromagnetic system on the final natural frequency of the stage is ignored. With the magnetic actuator, the actuation force is linearly related to the position of the armature, which may lead to a "negative" stiffness phenomenon for the nanopositioning stage [20,24]. Therefore, considering the "negative" stiffness effect, the natural frequency of the flexure mechanism might be slightly larger than the practically tested one presented in Table 1.

To illustrate the performance of the designed platform, comparisons with some typical dual-axial stages are presented in Table 2. As shown in Table 2, a much smaller size, a larger workspace, and a relatively high bandwidth are achieved in this work. Different from traditional designs, the proposed sinusoidal flexible hinge has two-degree-of-freedom planar bending, so it can realize a large range of two-axis motion in a limited volume. In addition, the designed mechanism does not have to consider complex input decoupling and driver protection issues due to the contactless electromagnetic actuation at the input ends, so high frequencies and large strokes can be achieved in a compact dimension. The coupling ratio of the designed mechanism is higher than that of the compared platforms. This may be due to the complex structure of the sinusoidal flexure hinge, which results in higher manufacturing errors than with the traditional design, and the assembly induced two-axis crosstalk at the input end.

Table 2. Performance comparison with state-of-the-art dual-axial stages.

Reference	Dimension (mm^2)	Workspace ($μm^2$)	Bandwidth (Hz)	Coupling Ratio (%)
[4]	165 × 145	31.5 × 31.5	570	0.7/0.9
[5]	142 × 142	40.2 × 42.9	483	0.58/0.56
[11]	-	1800 × 1820	72	-
[25]	160 × 160	55.4 × 53.2	253	0.42/0.45
[26]	190 × 190	19.5 × 19	2k	0.62/0.99
This work	54.4 × 54.4	133.5 × 132.6	586	2.47/0.95

6. Conclusions

A sinusoidal corrugated flexure linkage is proposed to concurrently stretch and bend, which is crucial to directly deliver dual-axial motions. Taking into full consideration this unique property, a corrugated flexure mechanism using four pairs of corrugated linkage-based parallelograms is developed for the planar nanopositioning stages to support and guide the dual-axial motions. The stiffness is analytically modeled for both the sinusoidal corrugated flexure linkage and dual-axial mechanism, and their deformation behavior related to the structural parameters are further investigated with finite element simulation-based verification. The deviation between the analytical and finite element result is less than 20% for all cases.

As for the dual-axial mechanism, the stiffness of the produced prototype is practically tested, and the deviation between the practical and analytical stiffness is around 7.4%. By applying the mechanism for the construction of a planar magnetic nanopositioning stage, a nearly identical stroke and resonant frequency are observed for the two axes, which are, respectively, around 130 μm and 590 Hz, and the maximum cross-talk is observed to be less than 2.5%. Finally, by adopting a PID-based main controller with the disturbance observer, the closed-loop performance of the dual-axis stage is also demonstrated.

Author Contributions: Conceptualization, Z.Z. and P.H.; methodology, Y.N.; software, Y.N. and X.C.; validation, Y.N., X.C. and L.C.; formal analysis, Y.N. and X.C.; investigation, Y.N. and X.C.; resources, Z.Z. and P.H.; data curation,Y.N. and X.C.; writing—original draft preparation, Y.N.; writing—review and editing, P.H. and Z.Z.; visualization, Y.N., X.C. and L.C.; supervision, Z.Z. and P.H.; project administration, Z.Z. and P.H.; funding acquisition, Z.Z. and P.H. All authors have read and agreed to the published version of the manuscript.

Funding: This research was funded by the National Natural Science Foundation of China(U2013211, 52275437, and 52105454), the Fundamental Research Funds for the Central Universities (30921013102), and the Guangdong Basic and Applied Basic Research Foundation (2020A1515111011).

Data Availability Statement: The data that support the findings of this study are available from the corresponding author upon reasonable request.

Conflicts of Interest: The authors declare no conflict of interest.

Appendix A

Appendix A.1. The Compliance Equations for the Sinusoidal Corrugated Flexure Unit

Let $\omega_x = \frac{2\pi}{p_x}$, $D_x = \sqrt{p_x^2 + 4A^2\pi^2}$, $E_1 = \text{EllipticF}\left(\omega_x x, \frac{4A^2\pi^2}{p_x^2 + 4A^2\pi^2}\right)$, and $E_2 = \text{EllipticE}\left(\omega_x x, \frac{4A^2\pi^2}{p_x^2 + 4A^2\pi^2}\right)$, for which EllipticF and EllipticE stand for the incomplete elliptic integral of the first and second kind, respectively. Accordingly, the detailed equation for the compliance term can be expressed as follows:

$$C_{x,f_x} = \frac{(16A^4\pi^4 - p_x^4)E_2 + p_x^2 D_x^2 E_1}{2w\pi^3 t^3 E D_x} - \frac{A^2\sqrt{p_x^2 + 2A^2\pi^2 + 2A^2\pi^2\cos(2\omega_x x)}\sin(2\omega_x x)}{\pi w t^3 E} + \frac{(12+11\mu)D_x^2 E_2 - (7+11\mu)p_x^2 E_1}{10\pi w t E D_x} \tag{A1}$$

$$C_{y,f_y} = \frac{12\int_0^{p_x}(p_x - x)^2\sqrt{1 + A^2(\omega_x)^2\cos^2(\omega_x x)}\,dx}{wt^3E} + \frac{5D_x^2E_2 - p_x^2(17+11\mu)E_1}{10\pi w t E D_x} \tag{A2}$$

$$C_{\theta_z,m_z} = \frac{6D_xE_2}{\pi w t^3 E} \tag{A3}$$

$$C_{\theta_z,f_y} = C_{y,m_z} = \frac{12\int_0^{p_x}(p_x - x)\sqrt{1 + A^2\omega_x^2\cos^2(\omega_x x)}\,dx}{wt^3E} \tag{A4}$$

$$C_{z,f_z} = \frac{12\int_0^{p_x}\sqrt{1 + A^2\omega_x^2\cos^2(\omega_x x)}\left[(p_x - x)^2 + A^2\sin^2(\omega_x x)\right]dx}{tw^3E} \tag{A5}$$

$$C_{\theta_y,m_y} = \frac{6D_xE_2}{\pi tw^3E} \tag{A6}$$

$$C_{\theta_y,f_z} = C_{z,m_y} = \frac{12\int_0^{p_x}\sqrt{[1+A^2\omega_x^2\cos^2(\omega_x x)]\left[(p_x-x)^2+A^2\sin^2(\omega_x x)\right]}\,dx}{tw^3E} \tag{A7}$$

$$C_{\theta_x,m_x} = \frac{12(1+\mu)D_xE_2}{\pi(tw^3+wt^3)E} \tag{A8}$$

Appendix A.2. The Compliance Transformation Matrix

The compliance transformation matrix (CTM) is defined by [18,19]

$$\mathbf{T} = \begin{bmatrix} \mathbf{R}(\theta) & \mathbf{S}(\mathbf{r})\mathbf{R}(\theta) \\ \mathbf{O} & \mathbf{R}(\theta) \end{bmatrix} \tag{A9}$$

where $\mathbf{R}(\theta)$ is the required rotation operation for the local coordinate system to rotate to the global coordinate system in terms of a rotation angle of θ. For this study, the rotation is around the z-axis, which has the form of

$$\mathbf{R}(\theta) = \begin{bmatrix} \cos\phi & \sin\phi & 0 \\ -\sin\phi & \cos\phi & 0 \\ 0 & 0 & 1 \end{bmatrix} \tag{A10}$$

In Equation (A9), $\mathbf{S}(\mathbf{r})$ represents the position transformation, which is defined as

$$\mathbf{S} = \begin{bmatrix} 0 & -z_r & y_r \\ z_r & 0 & -x_r \\ -y_r & x_r & 0 \end{bmatrix} \tag{A11}$$

where $\mathbf{r} = [x_r, y_r, z_r]$ is the relative position of the local coordinate system in the global coordinate system.

Appendix A.3. The Characteristic Matrix of the Designed Stage

The compliance matrix for the sinusoidal corrugated flexure hinge designed for the mechanism is

$$\mathbf{C}_s = \begin{bmatrix} 4.02 & 0 & 0 & 0 & 0 & 0 \\ 0 & 7.88 & 0 & 0 & 0 & 0.0035 \\ 0 & 0 & 0.236 & 0 & 1.01\times10^{-5} & 0 \\ 0 & 0 & 0 & 1.21\times10^{-8} & 0 & 0 \\ 0 & 0 & 1.01\times10^{-5} & 0 & 4.57\times10^{-9} & 0 \\ 0 & 0.0035 & 0 & 0 & 0 & 2.09\times10^{-6} \end{bmatrix} \tag{A12}$$

The stiffness matrix for the sinusoidal corrugated dual-axial mechanism is

$$\mathbf{K} = \begin{bmatrix} 0.408 & 0 & 0 & 0 & 0 & 0 \\ 0 & 0.408 & 0 & 0 & 0 & 0 \\ 0 & 0 & 10.8 & 0 & 0 & 0 \\ 0 & 0 & 0 & 1.07\times10^{9} & 0 & 0 \\ 0 & 0 & 0 & 0 & 1.07\times10^{9} & 0 \\ 0 & 0 & 0 & 0 & 0 & 3.55\times10^{7} \end{bmatrix} \tag{A13}$$

References

1. Yong, Y.K.; Moheimani, S.O.R.; Kenton, B.J.; Leang, K.K. Invited Review Article: High-speed flexure-guided nanopositioning: Mechanical design and control issues. *Rev. Sci. Instruments* **2012**, *83*, 121101. [CrossRef] [PubMed]
2. Ortlepp, I.; Fröhlich, T.; Füßl, R.; Reger, J.; Schäffel, C.; Sinzinger, S.; Strehle, S.; Theska, R.; Zentner, L.; Zöllner, J.P.; et al. Tip-and Laser-based 3D Nanofabrication in Extended Macroscopic Working Areas. *Nanomanuf. Metrol.* **2021**, *4*, 132–148. [CrossRef]
3. Matsukuma, H.; Adachi, K.; Sugawara, T.; Shimizu, Y.; Gao, W.; Niwa, E.; Sasaki, Y. Closed-Loop Control of an XYZ Micro-Stage and Designing of Mechanical Structure for Reduction in Motion Errors. *Nanomanuf. Metrol.* **2021**, *4*, 53–66. [CrossRef]
4. Zhu, W.L.; Zhu, Z.; Guo, P.; Ju, B.F. A novel hybrid actuation mechanism based XY nanopositioning stage with totally decoupled kinematics. *Mech. Syst. Signal Process.* **2018**, *99*, 747–759. [CrossRef]
5. Wang, F.; Zhao, X.; Huo, Z.; Shi, B.; Liang, C.; Tian, Y.; Zhang, D. A 2-DOF nano-positioning scanner with novel compound decoupling-guiding mechanism. *Mech. Mach. Theory* **2021**, *155*, 104066. [CrossRef]
6. Zhu, Z.; Tong, Z.; To, S.; Jiang, X. Tuned diamond turning of micro-structured surfaces on brittle materials for the improvement of machining efficiency. *CIRP Ann.* **2019**, *68*, 559–562. [CrossRef]
7. Polit, S.; Dong, J. Development of a high-bandwidth XY nanopositioning stage for high-rate micro-/nanomanufacturing. *IEEE/ASME Trans. Mechatronics* **2010**, *16*, 724–733. [CrossRef]
8. Tian, Y.; Ma, Y.; Wang, F.; Lu, K.; Zhang, D. A novel XYZ micro/nano positioner with an amplifier based on L-shape levers and half-bridge structure. *Sens. Actuators A Phys.* **2020**, *302*, 111777. [CrossRef]
9. Zhu, Z.; To, S.; Ehmann, K.F.; Zhou, X. Design, analysis, and realization of a novel piezoelectrically actuated rotary spatial vibration system for micro-/nanomachining. *IEEE/ASME Trans. Mechatronics* **2017**, *22*, 1227–1237. [CrossRef]
10. Howell, L.L. Compliant mechanisms. In *21st Century Kinematics*; Springer: Berlin/Heidelberg, Germany, 2013; pp. 189–216.
11. Wang, N.; Zhang, Z.; Zhang, X.; Cui, C. Optimization of a 2-DOF micro-positioning stage using corrugated flexure units. *Mech. Mach. Theory* **2018**, *121*, 683–696. [CrossRef]
12. Xi, S.; Lai, L.J. Paired double parallelogram flexure mechanism clamped by corrugated beam for underconstraint elimination. *Rev. Sci. Instruments* **2020**, *91*, 086102. [CrossRef]
13. Li, C.; Wang, N.; Yue, F.; Zhang, X. Optimization of Translational Flexure Joints Using Corrugated Units Under Stress Constraints. *J. Mech. Robot.* **2021**, *13*, 061006. [CrossRef]
14. Wang, N.; Liang, X.; Zhang, X. Stiffness Analysis of Corrugated Flexure Beam Used in Compliant Mechanisms. *Chin. J. Mech. Eng.* **2015**, *28*, 776–784. [CrossRef]
15. Wang, N.; Zhang, Z.; Yue, F.; Zhang, X. Design and analysis of translational joints using corrugated flexural beams with conic curve segments. *Mech. Mach. Theory* **2019**, *132*, 223–235. [CrossRef]
16. Wu, J.; Zhang, Y.; Cai, S.; Cui, J. Modeling and analysis of conical-shaped notch flexure hinges based on NURBS. *Mech. Mach. Theory* **2018**, *128*, 560–568. [CrossRef]
17. Lobontiu, N.; Cullin, M. In-plane elastic response of two-segment circular-axis symmetric notch flexure hinges: The right circular design. *Precis. Eng.* **2013**, *37*, 542–555. [CrossRef]
18. Li, Y.; Xu, Q. Design and analysis of a totally decoupled flexure-based XY parallel micromanipulator. *IEEE Trans. Robot.* **2009**, *25*, 645–657.
19. Zhu, Z.; Zhou, X.; Liu, Z.; Wang, R.; Zhu, L. Development of a piezoelectrically actuated two-degree-of-freedom fast tool servo with decoupled motions for micro-/nanomachining. *Precis. Eng.* **2014**, *38*, 809–820. [CrossRef]
20. Zhu, Z.; Chen, L.; To, S. A novel direct drive electromagnetic XY nanopositioning stage. *CIRP Ann.* **2021**, *70*, 415–418. [CrossRef]
21. Khawwaf, J.; Zheng, J.; Chai, R.; Lu, R.; Man, Z. Adaptive Microtracking Control for an Underwater IPMC Actuator Using New Hyperplane-Based Sliding Mode. *IEEE/ASME Trans. Mechatronics* **2019**, *24*, 2108–2117. [CrossRef]
22. Yi, J.; Chang, S.; Shen, Y. Disturbance-observer-based hysteresis compensation for piezoelectric actuators. *IEEE/ASME Trans. Mechatronics* **2009**, *14*, 456–464.
23. Zhu, Z.; Chen, L.; Huang, P.; Schönemann, L.; Riemer, O.; Yao, J.; To, S.; Zhu, W.L. Design and control of a piezoelectrically actuated fast tool servo for diamond turning of microstructured surfaces. *IEEE Trans. Ind. Electron.* **2020**, *67*, 6688–6697. [CrossRef]
24. Ito, S.; Troppmair, S.; Lindner, B.; Cigarini, F.; Schitter, G. Long-range fast nanopositioner using nonlinearities of hybrid reluctance actuator for energy efficiency. *IEEE Trans. Ind. Electron.* **2018**, *66*, 3051–3059. [CrossRef]
25. Wang, F.; Huo, Z.; Liang, C.; Shi, B.; Tian, Y.; Zhao, X.; Zhang, D. A Novel Actuator-Internal Micro/Nano Positioning Stage with an Arch-Shape Bridge-Type Amplifier. *IEEE Trans. Ind. Electron.* **2019**, *66*, 9161–9172. [CrossRef]
26. Du, Y.; Li, T.; Jiang, Y.; Wang, H. Design and analysis of a 2-degree-of-freedom flexure-based micro-motion stage. *Adv. Mech. Eng.* **2016**, *8*, 168781401663830. [CrossRef]

actuators

MDPI

Article

Design and Analysis of a Compliant End-Effector for Robotic Polishing Using Flexible Beams

Yanding Qin [1,2], Haitao Wu [1,2], Zhiyuan Li [1,2], Ning Sun [1,2] and Lei Sun [1,2,*]

1 Tianjin Key Laboratory of Intelligent Robotics, College of Artificial Intelligence, Nankai University, Tianjin 300350, China
2 Institute of Intelligence Technology and Robotic Systems, Shenzhen Research Institute of Nankai University, Shenzhen 518083, China
* Correspondence: sunl@nankai.edu.cn

Abstract: The contact force between the polishing tool and the workpiece is crucial in determining the surface quality in robotic polishing. Different from rigid end-effectors, this paper presents a novel compliant end-effector (CEE) for robotic polishing using flexible beams. The flexibility of the CEE helps to suppress the excessive displacement caused by the inertia of the polishing robot and avoids damaging the polishing tool and workpiece surface. In addition, the contact force can also be precisely estimated via the measurement of the CEE's displacement using a capacitive position sensor. The design, modeling and experimental validation of the CEE are presented. Firstly, the analytical model of the CEE is established using the stiffness matrix method. Subsequently, the analytical model is verified by finite element analysis. Further, a prototype is manufactured, and its characteristics and performance are experimentally tested. The equivalent stiffness is measured to be 0.335 N/μm, and the first natural frequency along its working direction is 42.1 Hz. Finally, the contact force measurement using the CEE is compared with a force sensor. Under open-loop condition, the resolution of the contact force measurement is found to be 0.025 N, which makes the fine tuning of the contact force possible in robotic polishing.

Keywords: robotic polishing; compliant end-effector; finite element analysis; equivalent stiffness

Citation: Qin, Y.; Wu, H.; Li, Z.; Sun, N.; Sun, L. Design and Analysis of a Compliant End-Effector for Robotic Polishing Using Flexible Beams. *Actuators* **2022**, *11*, 284. https://doi.org/10.3390/act11100284

Academic Editors: Limin Zhu, Yuen Kuan Yong, Yanling Tian and Yingxiang Liu

Received: 9 September 2022
Accepted: 3 October 2022
Published: 5 October 2022

1. Introduction

Polishing is an extremely important processing technology in improving the surface quality of the workpiece [1]. The control of the contact force between the polishing tool and the workpiece is crucial in polishing. Currently, manual polishing is still very common, especially in the fine polishing of complex surfaces [2]. However, the quality of manual polishing is highly dependent on the skills and experience of the practitioner, resulting in low production efficiency and poor consistency [3]. On the contrary, in various precision industries, robots have been more and more frequently used for deburring and polishing the workpiece surface [4]. Compared with manual polishing, in robotic polishing, the contact force can be precisely regulated using force sensing techniques. Therefore, how to precisely sense and maintain the contact force has become an important issue in the robotic polishing process [5,6].

In order to realize contact force control, there are two kinds of control methods: passive compliance and active compliance. The most straightforward way to control the contact force between the polishing tool and the workpiece surface is to use a linear spring or a compliance mechanism [7], i.e., the passive compliance control. The active compliance control is to realize the closed-loop control of the contact force through the force sensor. In active compliance system, the most widely used drivers are pneumatic cylinder [8] and voice coil motor [9]. In order to avoid excessive contact between the polishing tool and the workpiece surface, the polishing end effector should exhibit certain compliance characteristics. Currently, the end-effectors can be briefly divided into mechanical [10], optimization program, so it is set manually herein. Finally, the main parameters of the flexible beam used are shown in Table 1.

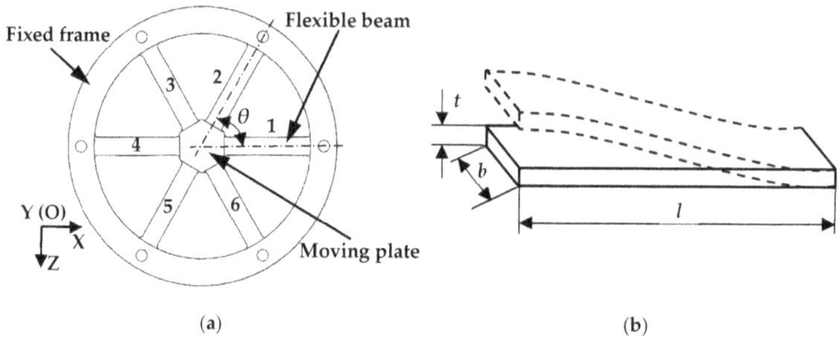

Figure 2. Design parameters of the CEE: (**a**) overall structure; (**b**) the parameters of the flexible beam.

Table 1. Key parameters of the flexible beam.

Parameter	l (mm)	b (mm)	t (mm)	θ (rad)
Value	50	10	2	$\pi/3$

2.2. Analytical Modeling of the Compliant End-Effector

When the polishing tool touches the workpiece surface, the CEE can generate elastic deformations. The deformation of the CEE is dependent on its stiffness and the allowable stress. In order to obtain the displacement of the moving plate, a capacitive displacement sensor is placed behind the moving plate. The stiffness matrix method [28] shown below is used to obtain stiffness of the CEE.

$$\mathbf{M}\ddot{\mathbf{q}} + \mathbf{K}\mathbf{q} = \mathbf{F}, \tag{1}$$

where the mass matrix $\mathbf{M} = \text{diag}[m_x\ m_y\ J_z]$ corresponds to the inertia mass and moment of inertia of the moving plate in its generalized coordinates $\mathbf{q}$, m_x and m_y represent the mass of the moving plate, and J_z indicates the moment of inertia of the moving plate on $\mathbf{Z}$ axis. The stiffness matrix $\mathbf{K}$ is

$$\mathbf{K} = \sum_{i=1}^{6} \mathbf{T}_i \widetilde{\mathbf{K}} \mathbf{T}_i^T, \tag{2}$$

where $\widetilde{\mathbf{K}}$ is the stiffness matrix of a single flexible beam, and

$$\mathbf{T}_i = \begin{bmatrix} 1 & 0 & 0 \\ d_{zi} & 1 & 0 \\ 0 & d_{xi} & 1 \end{bmatrix} \begin{bmatrix} \cos\theta_i & 0 & -\sin\theta_i \\ 0 & 1 & 0 \\ \sin\theta_i & 0 & \cos\theta_i \end{bmatrix}, \tag{3}$$

As presented in Equation (3), $[d_{xj}\ 0\ d_{zi}]^{\mathrm{T}}$ represents the distance from O-XYZ to O_i-$X_iY_iZ_i$. θ_i is the rotating angle of O_i-$X_iY_iZ_i$ with respect to O-XYZ, as displayed in Figure 3. The values of all the above parameters are shown in Table 2.

Table 2. Detailed parameters of matrix $\mathbf{T}_i$.

i	d_{xi} (mm)	d_{zi} (mm)	θ_i (rad)
1	12.99	0	0
2	6.50	−11.25	$\pi/3$
3	−6.50	−11.25	$2\pi/3$
4	−12.99	0	π
5	−6.50	11.25	$4\pi/3$
6	6.50	11.25	$5\pi/3$

actuators MDPI

Article

Design and Analysis of a Compliant End-Effector for Robotic Polishing Using Flexible Beams

Yanding Qin [1,2], Haitao Wu [1,2], Zhiyuan Li [1,2], Ning Sun [1,2] and Lei Sun [1,2,*]

1 Tianjin Key Laboratory of Intelligent Robotics, College of Artificial Intelligence, Nankai University, Tianjin 300350, China
2 Institute of Intelligence Technology and Robotic Systems, Shenzhen Research Institute of Nankai University, Shenzhen 518083, China
* Correspondence: sunl@nankai.edu.cn

Abstract: The contact force between the polishing tool and the workpiece is crucial in determining the surface quality in robotic polishing. Different from rigid end-effectors, this paper presents a novel compliant end-effector (CEE) for robotic polishing using flexible beams. The flexibility of the CEE helps to suppress the excessive displacement caused by the inertia of the polishing robot and avoids damaging the polishing tool and workpiece surface. In addition, the contact force can also be precisely estimated via the measurement of the CEE's displacement using a capacitive position sensor. The design, modeling and experimental validation of the CEE are presented. Firstly, the analytical model of the CEE is established using the stiffness matrix method. Subsequently, the analytical model is verified by finite element analysis. Further, a prototype is manufactured, and its characteristics and performance are experimentally tested. The equivalent stiffness is measured to be 0.335 N/μm, and the first natural frequency along its working direction is 42.1 Hz. Finally, the contact force measurement using the CEE is compared with a force sensor. Under open-loop condition, the resolution of the contact force measurement is found to be 0.025 N, which makes the fine tuning of the contact force possible in robotic polishing.

Keywords: robotic polishing; compliant end-effector; finite element analysis; equivalent stiffness

Citation: Qin, Y.; Wu, H.; Li, Z.; Sun, N.; Sun, L. Design and Analysis of a Compliant End-Effector for Robotic Polishing Using Flexible Beams. *Actuators* **2022**, *11*, 284. https://doi.org/10.3390/act11100284

Academic Editors: Limin Zhu, Yuen Kuan Yong, Yanling Tian and Yingxiang Liu

Received: 9 September 2022
Accepted: 3 October 2022
Published: 5 October 2022

1. Introduction

Polishing is an extremely important processing technology in improving the surface quality of the workpiece [1]. The control of the contact force between the polishing tool and the workpiece is crucial in polishing. Currently, manual polishing is still very common, especially in the fine polishing of complex surfaces [2]. However, the quality of manual polishing is highly dependent on the skills and experience of the practitioner, resulting in low production efficiency and poor consistency [3]. On the contrary, in various precision industries, robots have been more and more frequently used for deburring and polishing the workpiece surface [4]. Compared with manual polishing, in robotic polishing, the contact force can be precisely regulated using force sensing techniques. Therefore, how to precisely sense and maintain the contact force has become an important issue in the robotic polishing process [5,6].

In order to realize contact force control, there are two kinds of control methods: passive compliance and active compliance. The most straightforward way to control the contact force between the polishing tool and the workpiece surface is to use a linear spring or a compliance mechanism [7], i.e., the passive compliance control. The active compliance control is to realize the closed-loop control of the contact force through the force sensor. In active compliance system, the most widely used drivers are pneumatic cylinder [8] and voice coil motor [9]. In order to avoid excessive contact between the polishing tool and the workpiece surface, the polishing end effector should exhibit certain compliance characteristics. Currently, the end-effectors can be briefly divided into mechanical [10], pneumatic [11], electrical [12] and electromagnetic [13]. With the force feedback, the active compliance control can obtain a nearly constant contact force via adjusting the displacement between the polishing tool and the workpiece surface [14,15].

Li et al. proposed a novel macro-mini robot with active force control for robotic polishing [9]. In this design, a macro robot provides the posture control during polishing operations, and a high-bandwidth end-effector, the mini robot, realizes constant contact force control. Tian et al. set up an active and passive compliance control polishing model by explicit force control based on position, using a tilting polishing tool with an elastic sponge disk to achieve a relatively constant force control effect [16]. Fan et al. presented a novel smart end-effector for active contact force control by using a gravity compensated force controller and two novel eddy current dampers for vibration suppression in the robotic polishing of thin-walled blisks [17].

Although the above researchers realized the constant force control in polishing, they did not consider the impact of the excessive displacement caused by the inertia of the robot on the workpiece and the polishing tool. Du et al. designed a compliant end-effector, where the deformation of the rubber support was used to improve the compliance of the polishing system. At the same time, an adaptive anti-saturation integral separated fuzzy proportional–integral (PI) controller was designed to control the contact force by using a force sensor [18]. Wu et al. proposed a novel force-controlled spherical polishing tool combined with self-rotation and co-rotation motion, providing compliance and polishing force in the polishing process [19]. Mohammad et al. presented a novel design of a force-controlled end-effector for automated polishing. In this design, the polishing tool can be extended and retracted by a linear hollow voice coil actuator to provide compliance [1].

The compliant mechanism has the advantage of no gap and friction, no lubrication and assembly, and integrated design and processing [20–22]. It has been utilized in a wide range of applications, such as micro-electromechanical systems, scanning probe microscopes, ultra-precision machining, and biological cell operations [23–25]. Wei et al. proposed a novel end-effector based on constant force mechanism (CFM), a specially designed compliant mechanism featuring almost zero stiffness within the effective range. This is the first time that CFM is used in robotic polishing. The constant-force motion range acts as a damper to counteract the excessive displacement caused by the inertia, and thus the end-effector regulates the contact force passively [26]. Ding et al. proposed a novel CFM based on the combination of positive and negative stiffness mechanism by using folding beam and bi-stable beam mechanisms. Without using any additional sensors and control algorithms, the proposed CFM can produce a travel range in constant force manner [27]. However, in the above CFMs, the value of the constant force is predefined cannot be adjusted. Therefore, when the contact force needs to be adjusted to another value, the current CFM has to be replaced to another CFM with a new set of dimensions.

In this paper, a flexible beam based compliant end-effector (CEE) for robotic polishing is proposed, which is beneficial to solve the problem that the polishing tool and workpiece surface are damaged due to excessive displacement caused by the inertia when the polishing robot approaches the workpiece surface quickly. When the polishing tool contacts the workpiece surface, the elastic deformation of the CEE can act as a damper. Further, the contact force can also be calculated from the displacement and the stiffness coefficient of the CEE without the use of the force sensor. As a capacitive position sensor is used to measure the displacement, the accuracy of the contact force sensing can be guaranteed, which is important to improve the force control accuracy of robotic polishing.

The rest of this paper is arranged as follows: Section 2 gives the mechanical structure design and analytical stiffness model of the CEE for robotic polishing. Sections 3 and 4 give the finite element analysis (FEA) results and the experimental verification, respectively. Finally, the conclusion is provided in Section 5.

2. Design and Analysis of the Compliant End-Effector

2.1. Mechanical Design of the Compliant End-Effector

The schematic diagram of robotic polishing is illustrated in Figure 1, where a robot arm is used to hold and moves the polishing tool across the workpiece surface. A pneumatic cylinder is used to push the polishing tool against the workpiece with controllable contact force. Different from the conventional active compliance design, an additional CEE is inserted between the slider of the pneumatic cylinder and the polishing tool. During polishing, the pneumatic cylinder moves the polishing tool toward the workpiece. When the polishing tool touches the workpiece surface, the moving plate of the CEE can move back and forth, following the variation of the contact force, as shown in the inset of Figure 1. The elastic deformation of the CEE can act as a damper, which is beneficial to stabilize the contact force.

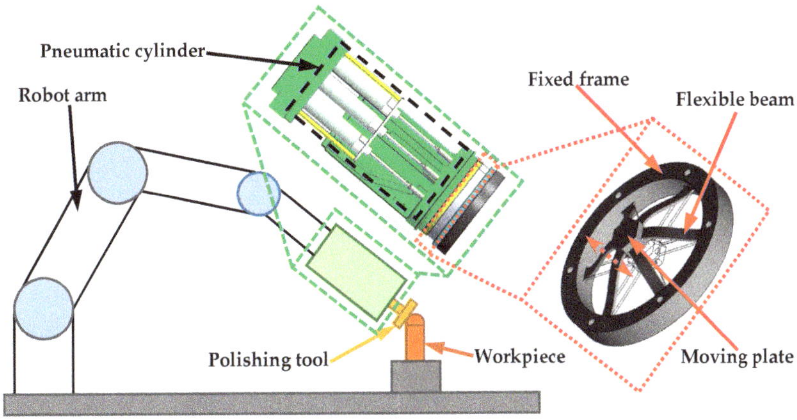

Figure 1. Schematic diagram of the proposed CEE installed on an industrial robot arm.

Considering the possibility of the CEE's failure and making it compatible for different robotic polishing tasks, a modular design is adopted herein such that the CEE can be easily replaced, as shown in Figure 1. In addition, due to the harsh polishing environment and the long working time, it is necessary to improve the adaptability of the CEE. Therefore, in the mechanical design of the CEE, very complex and tiny structures are not pursued. Considering the eccentric force generated by the rotation of the motor and the transverse force generated during polishing, six groups of flexible beams with completely symmetrical distribution are adopted in the design of the CEE, as shown in the inset of Figure 1. A flexible beam is adopted in construct the CEE because of its lower stress concentration and ease of manufacture when compared with the other types of flexure hinges. In order to further reduce the stress concentration, fillets are adopted at both ends of the flexible beams.

The proposed CEE also features precise contact force sensing capability. Within the elastic deformation range, the CEE can be treated as a linear spring with constant stiffness. In applications, the stiffness of the CEE can be calibrated in advance, and thus the contact force can be obtained via the displacement measurement of the CEE without the use of force sensor. In the prototype, high-precision capacitive displacement sensor is used as displacement feedback. Therefore, precise contact force sensing is also available.

The key design parameters of the CEE are the length l, the width b, the thickness t, and the rotating angle θ, as shown in Figure 2. As the designed CEE is used to obtain the contact force during robotic polishing, the design target value of the resolution of contact force is set to be below 0.1 N. Moreover, the deformation range should be as large as possible, because the CEE also acts as a damper when the polishing tool contacts the workpiece surface. In addition, the structure of this design is simple, and it does not need a complicated optimization program, so it is set manually herein. Finally, the main parameters of the flexible beam used are shown in Table 1.

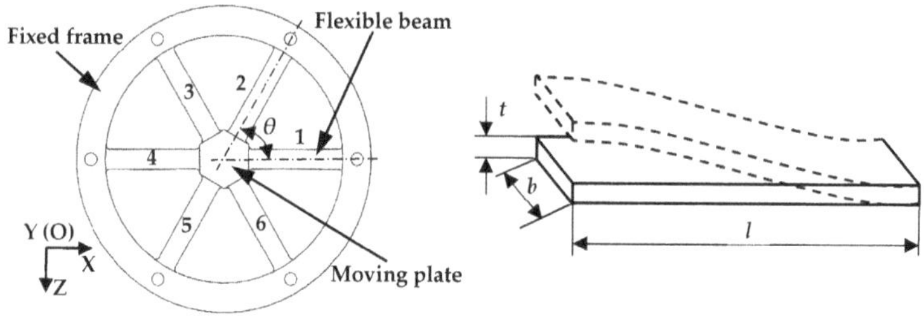

(a) (b)

Figure 2. Design parameters of the CEE: (**a**) overall structure; (**b**) the parameters of the flexible beam.

Table 1. Key parameters of the flexible beam.

Parameter	l (mm)	b (mm)	t (mm)	θ (rad)
Value	50	10	2	$\pi/3$

2.2. Analytical Modeling of the Compliant End-Effector

When the polishing tool touches the workpiece surface, the CEE can generate elastic deformations. The deformation of the CEE is dependent on its stiffness and the allowable stress. In order to obtain the displacement of the moving plate, a capacitive displacement sensor is placed behind the moving plate. The stiffness matrix method [28] shown below is used to obtain stiffness of the CEE.

$$\mathbf{M}\,\ddot{\mathbf{q}} + \mathbf{K}\mathbf{q} = \mathbf{F}, \tag{1}$$

where the mass matrix $\mathbf{M} = \text{diag}[m_x\ m_y\ J_z]$ corresponds to the inertia mass and moment of inertia of the moving plate in its generalized coordinates $\mathbf{q}$, m_x and m_y represent the mass of the moving plate, and J_z indicates the moment of inertia of the moving plate on $\mathbf{Z}$ axis. The stiffness matrix $\mathbf{K}$ is

$$\mathbf{K} = \sum_{i=1}^{6} \mathbf{T}_i\,\widetilde{\mathbf{K}}\,\mathbf{T}_i^T, \tag{2}$$

where $\widetilde{\mathbf{K}}$ is the stiffness matrix of a single flexible beam, and

$$\mathbf{T}_i = \begin{bmatrix} 1 & 0 & 0 \\ d_{zi} & 1 & 0 \\ 0 & d_{xi} & 1 \end{bmatrix} \begin{bmatrix} \cos\theta_i & 0 & -\sin\theta_i \\ 0 & 1 & 0 \\ \sin\theta_i & 0 & \cos\theta_i \end{bmatrix}, \tag{3}$$

As presented in Equation (3), $[d_{xj}\ 0\ d_{zi}]^{\mathrm{T}}$ represents the distance from O-XYZ to O_i-$X_iY_iZ_i$. θ_i is the rotating angle of O_i-$X_iY_iZ_i$ with respect to O-XYZ, as displayed in Figure 3. The values of all the above parameters are shown in Table 2.

Table 2. Detailed parameters of matrix $\mathbf{T}_i$.

i	d_{xi} (mm)	d_{zi} (mm)	θ_i (rad)
1	12.99	0	0
2	6.50	−11.25	$\pi/3$
3	−6.50	−11.25	$2\pi/3$
4	−12.99	0	π
5	−6.50	11.25	$4\pi/3$
6	6.50	11.25	$5\pi/3$

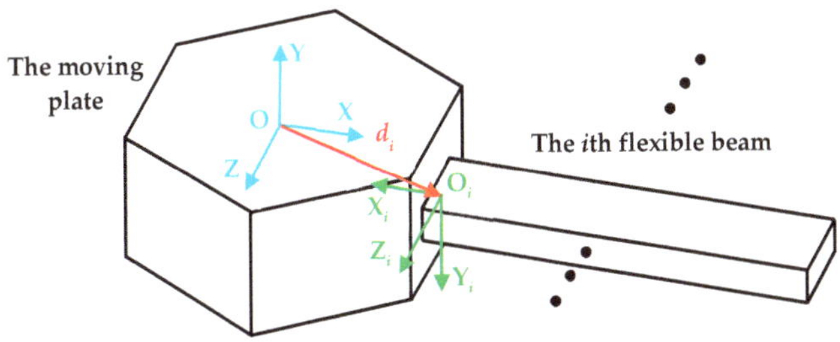

Figure 3. The moving plate connected with the *i*th flexible beam.

The flexible beam used in this CEE is a straight beam, and its flexibility matrix $\widetilde{\mathbf{C}}$ expressed in its local coordinate is

$$\widetilde{\mathbf{C}} = \begin{bmatrix} \frac{1}{Ebt} & 0 & 0 \\ 0 & \frac{4l^3}{Ebt^3} + \frac{3l}{2Gbt} & \frac{6l^2}{Ebt^3} \\ 0 & \frac{6l^2}{Ebt^3} & \frac{12l}{Ebt^3} \end{bmatrix}, \tag{4}$$

where E and G are the elastic modulus and shear modulus of the material, respectively, and the stiffness matrix $\widetilde{\mathbf{K}}$ of the flexible beam is

$$\widetilde{\mathbf{K}} = \widetilde{\mathbf{C}}^{-1}, \tag{5}$$

The generalized force, $\mathbf{F} = \text{diag}[F_x \; F_y \; M_z]$, is applied on the moving plate. In terms of the static analysis, the relationship between the static force and displacement can be expressed as follows:

$$\mathbf{q} = \mathbf{K}^{-1}\mathbf{F}, \tag{6}$$

According to the vibration theory, the natural frequency f of the CEE can be obtained by solving the characteristic equation of $|\lambda\mathbf{I} - \mathbf{M}^{-1}\mathbf{K}| = 0$,

$$f = \frac{\sqrt{\lambda}}{2\pi}, \tag{7}$$

When the contact force F_y is generated on the polishing tool, the output motion of the CEE can be calculated by Equation (6). Based on the above-mentioned method, the theoretical displacement u and stiffness k of the CEE along its working direction can be given by the following formulas:

$$k = \frac{1}{\mathbf{C}_{2,2}}, \tag{8}$$

$$u = \frac{F_y}{k}, \tag{9}$$

3. Finite Element Analysis and Verification

In order to verify the static and dynamic performance of the CEE, the 3D model is constructed and imported into ANSYS Workbench for the FEA. The selected material is aluminum alloy 7075, and the elastic modulus E and Poisson's ratio v are 70 GPa and 0.3, respectively.

3.1. Static Analysis Validation

The maximum force output of the pneumatic cylinder is 350 N. Therefore, in order to detect the maximum deformation of the proposed CEE, a force of 350 N is applied along its working direction. In this case, the deformation along the CEE working direction can reach

1.14 mm, as shown in Figure 4a. Through calculation, the stiffness k of the proposed CEE is 0.307 N/μm.

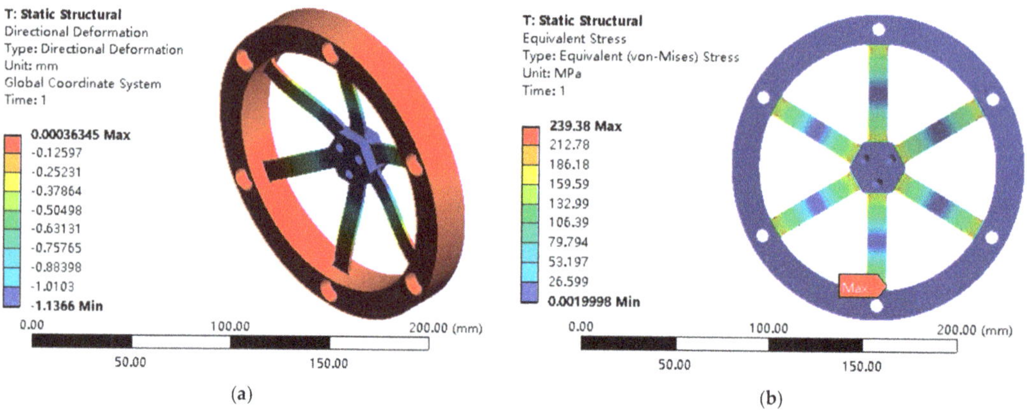

Figure 4. FEA results of the CEE for robotic polishing: (**a**) deformation; (**b**) stress distribution.

Moreover, the static stress analysis is conducted to verify whether the safety factor is qualified, as shown in Figure 4b. The yield strength of Al 7075 is 503 MPa, and the maximum stress is 239.38 MPa. Hence, the safety factor is calculated as 2.10 (>1).

According to the different polishing requirements, the contact force generally varies within 10~40 N [1,16]. In this force range, the deformation and stress of the CEE are also obtained, which are 32.48~129.90 μm and 6.84~27.36 MPa, respectively, which are within the allowable elastic deformation range of the material.

3.2. Dynamic Analysis Validation

When the load of the CEE is 0 kg, the first natural frequency along its working direction is 501.93 Hz, as shown in Figure 5a. It is consistent with the analytic result (498.84 Hz). The second and third mode shapes are the rotations in the vertical direction and the horizontal direction, respectively, as shown in Figure 5b,c.

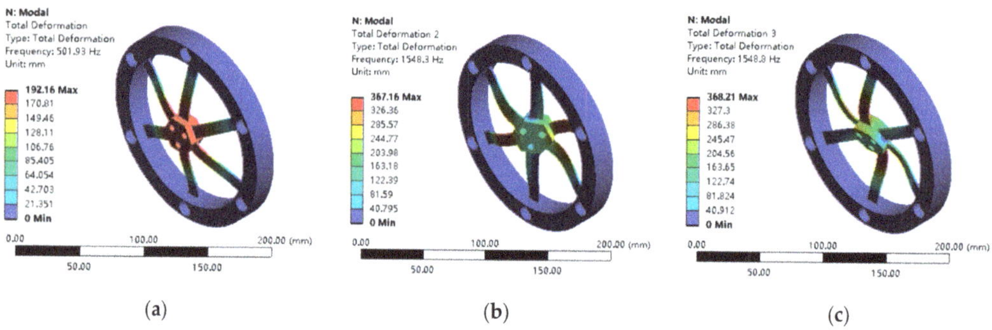

Figure 5. The first three mode shapes of the CEE without load: (**a**) the first mode; (**b**) the second mode; and (**c**) the third mode.

In applications, the polishing tool and the connectors need to be installed on the moving plate of the CEE, i.e., the external loads to the CEE. In this case, the dynamic performance of the CEE will be influenced. Therefore, the corresponding theoretical calculation and FEA are also carried out to evaluate the influence of the external loads. For example, the load of the connecting flange installed on the CEE's moving plate is 13.82 N, and the loads of the plastic and aluminum polishing tools (model: YQ060501, Shenzhen Han's Robot Co., Ltd., Shenzhen, China) are 6.86 N and 9.02 N, respectively. When these loads are installed on CEE's moving plate, the first natural frequency will be decreased. The variation of the first natural frequency against the external load is listed in Table 3.

Table 3. Analytical model and FEA results of first natural frequency under different loads.

Load	No Load (0 N)	Flange Only (13.82 N)	Flange + Plastic Tool (20.68 N)	Flange + Aluminum Tool (22.83 N)
Anal. (Hz)	498.84	77.32	58.48	55.65
FEA (Hz)	501.93	68.87	56.48	53.78

4. Prototype Fabrication and Experimental Test

4.1. Experimental Setup

In order to further verify the performance of the proposed CEE for robotic polishing, a prototype was fabricated using Al 7075 in a monolithic piece. In the fabricated prototype, the width b and the thickness t of the flexible beam slightly vary from the nominal values, and an increment of 0.1 mm is found. This magnitude of manufacturing error cannot be ignored. As a result, the measured dimensions of the CEE are adopted in the subsequent calculations. A force sensor (model: ZNLBS-V1-30 kg, Bengbu chino sensor Co., Ltd., Bengbu, China) is used to calibrate the contact force. A capacitive displacement sensor (model: NS-CDCS10L-400 with a resolution of 35 nm, Sanying Motion Control Instruments, Ltd., Tianjin, China) is used to measure the displacement. All force and displacement signals are acquired by a real-time target machine (model: Performance with a data acquisition card of IO133, Speedgoat). The overall experimental setup is shown in Figure 6.

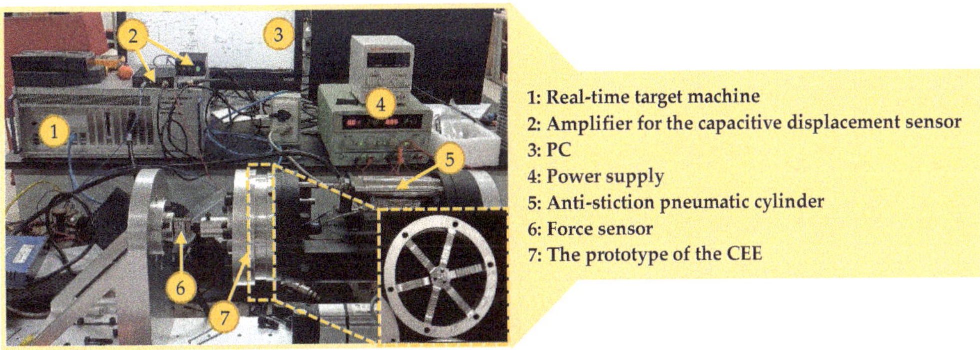

Figure 6. Experimental setup of the proposed CEE for robotic polishing.

4.2. Test of Equivalent Stiffness

Firstly, the equivalent stiffness of the CEE is calibrated. A real-time control system is built on the real-time target machine. A stair signal with a step of 180 μm is set to the pneumatic cylinder under the open-loop condition. Under the excitation of this signal, the pneumatic cylinder pushes the CEE against the force sensor. The reaction force is measured by the force sensor, and the displacement of the moving plate of the CEE is measured by the capacitive displacement sensor. The measured force–displacement relationship is provided in Figure 7. Based on the measurements, the equivalent stiffness k of the CEE along its working direction can be calculated as follows:

$$\hat{F}_y = ku + b, \tag{10}$$

where

$$k = \frac{\sum_{i=1}^{n} u_i F_{yi} - \sum_{i=1}^{n} u_i \sum_{i=1}^{n} \frac{F_{yi}}{n}}{\sum_{i=1}^{n} u_i^2 - \left(\sum_{i=1}^{n} u_i\right)^2 / n}, \tag{11}$$

$$b = \sum_{i=1}^{n} \frac{F_{yi}}{n} - k \sum_{i=1}^{n} \frac{u_i}{n}, \tag{12}$$

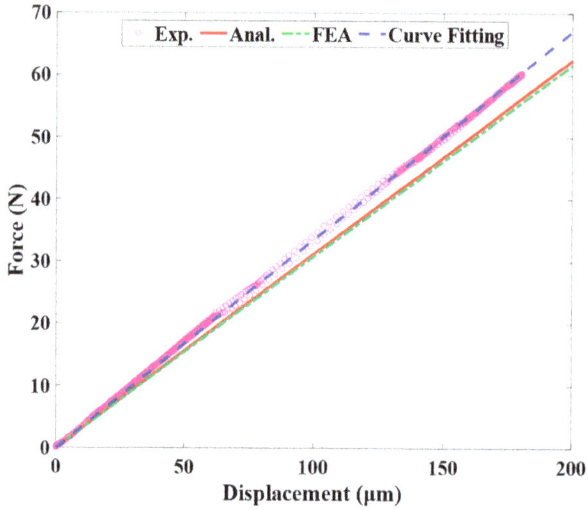

Figure 7. Comparative analysis of the stiffness k of the CEE.

After the above treatment, the equivalent stiffness k of the CEE is calculated to be 0.335 N/μm. The equivalent stiffness k obtained from the analytical model and FEA results are 0.312 N/μm and 0.307 N/μm, respectively. The analytical and FEA results are also shown in Figure 7 as a comparison. Both the analytical model and FEA results slightly underestimate the equivalent stiffness, and the errors are calculated to be 6.87% and 8.36%, respectively.

4.3. *Test of Contact Force Sensing*

In order to verify the consistency between the contact force estimated by the CEE and that measured by the force sensor, a triangular wave is used to drive the cylinder to reciprocate and record the signal of force sensor and deformation displacement of the CEE in real time. The comparison results are shown in Figure 8a. It can be seen that the contact force estimated by the CEE is consistent with the force sensor. It further verifies that the contact force can also be calculated from the displacement and the stiffness coefficient of the CEE without the use of force sensor. Within the contact force range shown in Figure 8a, the maximum deformation and maximum stress of the CEE are 65.67 μm and 15.05 MPa, respectively, within the allowable elastic deformation range of the material.

In the meantime, a stair signal with 0.025 N height is used to drive the cylinder for force resolution test. As shown in Figure 8b, it can be found that the noise level of the force sensor is in the magnitude of 0.2 N, whereas the noise level of the CEE measurement is in the magnitude of 0.04 N. The contact force resolution obtained by using CEE instead of force sensor is 0.025 N, which is important to improve the force control of the robotic polishing process. All the above experimental tests are carried out under the open-loop condition.

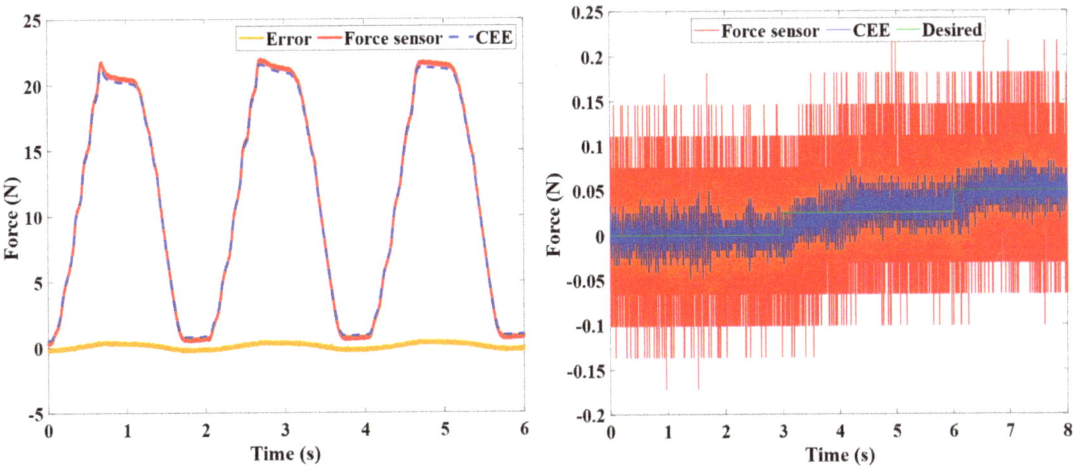

(**a**) (**b**)

Figure 8. Comparison of contact force obtained by force sensor and CEE: (**a**) consistency of measurement results; (**b**) resolution of contact force.

4.4. Test of Natural Frequency

In order to obtain the natural frequency of CEE for robotic polishing, a modal hammer is used to apply an impact load along its working direction to excite the CEE. At the same time, the capacitive displacement sensor is used to measure its displacement. The time domain signal is recorded and shown in Figure 9a. The natural frequency of the signal is 42.1 Hz after fast Fourier transform processing, as shown in Figure 9b. The measured first natural frequency is lower than the FEA results. This might result from the compliance of the pneumatic cylinder and the additional loads and accessories installed on the CEE during the test, such as the end cap.

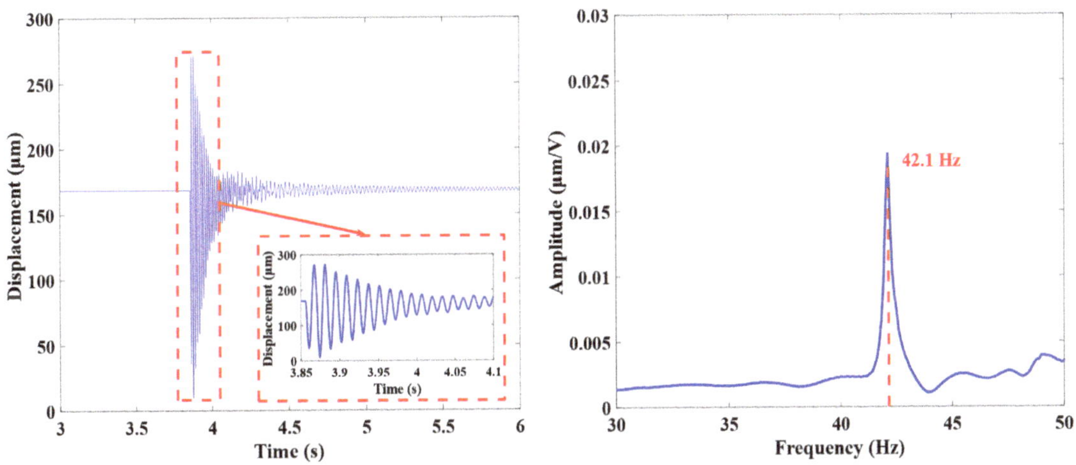

(**a**) (**b**)

Figure 9. Natural frequency test for the CEE: (**a**) time domain signal; (**b**) frequency domain signal.

5. Conclusions

This paper presents a novel design of the CEE for robot polishing using flexible beams. Compared with the conventional rigid end-effector, the elastic deformation of the CEE can act as a damper when the polishing robot installed polishing tool approaches the workpiece surface quickly. This design can help to solve the problem of excessive displacement caused by the inertia of the polishing robot and avoid damaging the polishing tool and workpiece surface during the polishing process. Further, the contact force can also be estimated from the displacement of the CEE without the use of a force sensor. Firstly, the equivalent stiffness and dynamic characteristics of the proposed CEE are carried out by means of analytical modeling and FEA. Then, a series of experimental tests are performed to verify its comprehensive performances. The experimental results demonstrate that the stiffness of this CEE is 0.335 N/μm. The resolution of the contact force measurement under open loop is found to be 0.025 N. Finally, the first natural frequency along its working direction is 42.1 Hz. All of the above work makes the fine tuning of the contact force possible in robotic polishing.

Author Contributions: Y.Q.: Conceptualization, methodology, resources, review and editing, supervision, project administration, funding acquisition; H.W.: conceptualization, methodology, validation, formal analysis, data curation, writing—original draft preparation; Z.L.: validation, formal analysis, data curation; N.S.: conceptualization, review and editing; L.S.: conceptualization, review and editing, resources, project administration, funding acquisition. All authors have read and agreed to the published version of the manuscript.

Funding: This work was supported in part by the National Natural Science Foundation of China under Grant 61873133, Grant 62173192, and Grant 52005270; and in part by the Natural Science Foundation of Tianjin under Grant 21JCZDJC00090.

Institutional Review Board Statement: Not applicable.

Informed Consent Statement: Not applicable.

Data Availability Statement: The data that support the findings of this study are available from the corresponding author upon reasonable request.

Conflicts of Interest: The authors declare no conflict of interest.

References

1. Mohammad, A.E.K.; Hong, J.; Wang, D. Design of a force-controlled end-effector with low-inertia effect for robotic polishing using macro-mini robot approach. *Robot. Comput.-Integr. Manuf.* **2018**, *49*, 54–65. [CrossRef]
2. Mohammad, A.E.K.; Wang, D. Electrochemical mechanical polishing technology: Recent developments and future research and industrial needs. *Int. J. Adv. Manuf. Technol.* **2016**, *86*, 1909–1924. [CrossRef]
3. Tao, B.; Zhao, X.; Ding, H. Mobile-robotic machining for large complex components: A review study. *Sci. China Technol. Sci.* **2019**, *62*, 1388–1400. [CrossRef]
4. Wan, S.; Zhang, X.; Wang, W.; Xu, M. Effect of pad wear on tool influence function in robotic polishing of large optics. *Int. J. Adv. Manuf. Technol.* **2019**, *102*, 2521–2530. [CrossRef]
5. Tsai, M.J.; Huang, J.F.; Kao, W.L. Robotic polishing of precision molds with uniform material removal control. *Int. J. Mach. Tools Manuf.* **2009**, *49*, 885–895. [CrossRef]
6. Liao, L.; Xi, F.; Liu, K. Modeling and control of automated polishing/deburring process using a dual-purpose compliant toolhead. *Int. J. Mach. Tools Manuf.* **2008**, *48*, 1454–1463. [CrossRef]
7. Liu, C.; Chen, C.C.A.; Huang, J. The polishing of molds and dies using a compliance tool holder mechanism. *J. Mater. Process. Technol.* **2005**, *166*, 230–236. [CrossRef]
8. Cheng, F.; Qi, Z.; Lei, Z.; Hong, Z. Development of the Polishing Tool System Based on the Pneumatic Force Servo. In Proceedings of the 2017 2nd International Conference on Advanced Robotics and Mechatronics (ICARM), Hefei and Tai'an, China, 27–31 August 2017; pp. 126–131.
9. Li, J.; Guan, Y.; Chen, H.; Wang, B.; Zhang, T.; Liu, X.; Hong, J.; Wang, D.; Zhang, H. A High-Bandwidth End-Effector With Active Force Control for Robotic Polishing. *IEEE Access* **2020**, *8*, 169122–169135. [CrossRef]
10. Tang, J.; Wang, T.; Yan, Z.Q.; Wang, L.W. Design and Analysis of the End-Effector of the Flexible Polishing Robot. *Key Eng. Mater.* **2016**, *693*, 58–63. [CrossRef]
11. Tian-Soon, S.; Ang, M.H., Jr.; Kah-Bin, L. A compliant end-effector coupling for vertical assembly: Design and evaluation. *Robot. Comput.-Integr. Manuf.* **1997**, *13*, 21–27, 29–30. [CrossRef]

12. Lopes, A.; Almeida, F. A force–impedance controlled industrial robot using an active robotic auxiliary device. *Robot. Comput.-Integr. Manuf.* **2008**, *24*, 299–309. [CrossRef]
13. Kim, J.-D.; Noh, I.-H. Magnetic polishing of three dimensional die and mold surfaces. *Int. J. Adv. Manuf. Technol.* **2007**, *33*, 18–23. [CrossRef]
14. Wahrburg, A.; Bos, J.; Listmann, K.D.; Dai, F.; Matthias, B.; Ding, H. Motor-Current-Based Estimation of Cartesian Contact Forces and Torques for Robotic Manipulators and Its Application to Force Control. *IEEE Trans. Autom. Sci. Eng.* **2018**, *15*, 879–886. [CrossRef]
15. Ott, C.; Albu-Schaffer, A.; Kugi, A.; Hirzinger, G. On the Passivity-Based Impedance Control of Flexible Joint Robots. *IEEE Trans. Robot.* **2008**, *24*, 416–429. [CrossRef]
16. Tian, F.; Lv, C.; Li, Z.; Liu, G. Modeling and control of robotic automatic polishing for curved surfaces. *CIRP J. Manuf. Sci. Technol.* **2016**, *14*, 55–64. [CrossRef]
17. Chen, F.; Zhao, H.; Li, D.; Chen, L.; Tan, C.; Ding, H. Contact force control and vibration suppression in robotic polishing with a smart end effector. *Robot. Comput.-Integr. Manuf.* **2019**, *57*, 391–403. [CrossRef]
18. Du, H.; Sun, Y.; Feng, D.; Xu, J. Automatic robotic polishing on titanium alloy parts with compliant force/position control. *Proc. Inst. Mech. Eng. Part B J. Eng. Manuf.* **2015**, *229*, 1180–1192. [CrossRef]
19. Wu, X.; Huang, Z.; Wan, Y.; Liu, H.; Chen, X. A Novel Force-Controlled Spherical Polishing Tool Combined With Self-Rotation and Co-Rotation Motion. *IEEE Access* **2020**, *8*, 108191–108200. [CrossRef]
20. Qin, Y.; Zhao, X.; Shirinzadeh, B.; Tian, Y.; Zhang, D. Closed-Form Modeling and Analysis of an XY Flexure-Based Nano-Manipulator. *Chin. J. Mech. Eng.* **2018**, *31*, 7. [CrossRef]
21. He, S.; Tang, H.; Zhu, Z.; Zhang, P.; Xu, Y.; Chen, X. A Novel Flexure Piezomotor With Minimized Backward and Nonlinear Motion Effect. *IEEE Trans. Ind. Electron.* **2022**, *69*, 652–662. [CrossRef]
22. Ma, X.; Liu, Y.; Deng, J.; Gao, X.; Cheng, J. A compact inchworm piezoelectric actuator with high speed: Design, modeling, and experimental evaluation. *Mech. Syst. Signal Process.* **2023**, *184*, 109704. [CrossRef]
23. Wang, F.; Huo, Z.; Liang, C.; Shi, B.; Tian, Y.; Zhao, X.; Zhang, D. A Novel Actuator-Internal Micro/Nano Positioning Stage With an Arch-Shape Bridge-Type Amplifier. *IEEE Trans. Ind. Electron.* **2019**, *66*, 9161–9172. [CrossRef]
24. Meng, Y.; Wang, X.; Huang, W.-W.; Li, L.; Hu, C.; Zhang, X.; Zhu, L. Intelligent Tracking Error Prediction and Feedforward Compensation for Nanopositioning Stages with High-bandwidth Control. *IEEE Trans. Ind. Inform.* **2022**, 1–10. [CrossRef]
25. Ding, B.; Li, X.; Li, Y. Configuration design and experimental verification of a variable constant-force compliant mechanism. *Robotica* **2022**, *40*, 3463–3475. [CrossRef]
26. Wei, Y.; Xu, Q. Design of a new passive end-effector based on constant-force mechanism for robotic polishing. *Robot. Comput. -Integr. Manuf.* **2022**, *74*, 102278. [CrossRef]
27. Ding, B.; Zhao, J.; Li, Y. Design of a spatial constant-force end-effector for polishing/deburring operations. *Int. J. Adv. Manuf. Technol.* **2021**, *116*, 3507–3515. [CrossRef]
28. Lai, L.; Zhu, Z. Design, modeling and testing of a novel flexure-based displacement amplification mechanism. *Sens. Actuators A Phys.* **2017**, *266*, 122–129. [CrossRef]

actuators

Communication

Analytical Modeling of Density and Young's Modulus Identification of Adsorbate with Microcantilever Resonator

Yue Yang [1], Yanling Tian [1,*], Xianping Liu [1], Yumeng Song [1] and Hui Tang [2]

[1] School of Engineering, University of Warwick, Coventry CV4 7AL, UK
[2] State Key Laboratory of Precision Electronic Manufacturing Technology and Equipment, Guangdong University of Technology, Guangzhou 510006, China
* Correspondence: y.tian.1@warwick.ac.uk

Abstract: Density and Young's modulus are critical parameters in biological research, which can be used to characterize molecules, cells, or tissues in the diagnosis of severe diseases. Microcantilever resonators are ideal tools to measure the physical parameters of small objects at the micro/nanoscale. In this study, a mathematical model was built based on the Rayleigh–Ritz method with the consideration of the first five-order bending natural frequencies. The mathematical model can be used to detect the density and Young's modulus of an adsorbate on a cantilever resonator with a single measurement. The influence of different order natural frequencies and the adsorbate position on the measurement accuracy and reliability was analyzed. This study revealed that the frequency pairs and the relative position of the adsorbate on the cantilever are two important factors that affect the accuracy and reliability of the measurement. Choosing appropriate frequency pairs can help to improve the accuracy and reliability of measurement. Finally, the results of finite element analysis verified the proposed method.

Keywords: microcantilever resonator; Young's modulus measurement; density measurement; bending vibration mode

Citation: Yang, Y.; Tian, Y.; Liu, X.; Song, Y.; Tang, H. Analytical Modeling of Density and Young's Modulus Identification of Adsorbate with Microcantilever Resonator. *Actuators* **2022**, *11*, 335. https://doi.org/10.3390/act11110335

Academic Editor: Nicola Pio Belfiore

Received: 22 October 2022
Accepted: 16 November 2022
Published: 18 November 2022

1. Introduction

The microcantilever resonator has shown high versatility, sensitivity, and flexibility in the characterization of micro/nano-scale matters in chemistry and biology research [1]. It shows potential applications in particle detection [2–8], biochemical reactions [9], magnetics [10], force research [11], humidity sensing [12], liquid characterization [13], and biomolecular detection [14,15]. It is proved that the Young's modulus of cancer cells is dramatically different from that of normal ones [16]. It could be possible to diagnose cancer at an early stage by measuring the Young's modulus of the cells.

Young's modulus and density can be deciphered from the dynamic process of the measurement [17]. The adsorbate on the cantilever resonator increases the natural frequency of the vibration system, especially when it is located at the fixed end. At the free end, the mass of the adsorbate takes a significant role and decreases the natural frequency [18,19]. In previous research, without considering the stiffness effect, the mass of the added sample could be obtained with high accuracy [5,20–22]. Gil-Santos took nanowires as the carrier and measured the mass and stiffness of an adsorbate with high sensitivity by investigating the frequencies of the first two vibration directions caused by the imperfect axisymmetry of the nanowire [23]. In Belardinelli's study, the combination of the first and fifth-order natural frequencies improved the measurement accuracy of the density and Young's modulus of the polymer located at the fixed end of the cantilever [24]. However, it is still unclear whether higher frequency pairs can bring better results. For adsorbates with different sizes located at different positions on the resonator, the best frequency pairs to measure Young's modulus and density may vary, which is another question that needs to be investigated. Moreover, there are few studies about the impact of unexpected frequency shifts on the accuracy and stability of mechanical properties measurement. Hence, challenges remain to measure the mechanical properties of adsorbates with any size and at any position with high accuracy and reliability.

In this study, we built a novel mathematical model to measure the Young's modulus and density of a single adsorbate on a resonator based on the Rayleigh-Ritz method by using multiple order vertical bending mode natural frequencies. We considered adsorbates with different sizes and located at different positions along the cantilever. We further discussed the frequency error effects on measurement results and provided an elaborate method to obtain more accurate and reliable results by selecting specific frequency pairs. The study revealed that the best frequency pairs may vary when the adsorbate is located at different relative positions on the cantilever, but they are hardly affected by the geometry or mechanical parameters of the adsorbate or the resonator.

2. Mathematical Modeling

To calculate the Young's modulus and density of an adsorbate, the first step is to obtain the mode shapes of the resonator. Based on the Euler equation for a beam, the vertical bending vibration of the cantilever resonator can be obtained. The natural frequencies in torsion, lateral bending, and longitudinal vibration modes of the cantilevers are higher than in vertical bending mode, and these mode shapes are more challenging to be measured in the real world than the vertical bending mode shape. To simplify the measurement of vibration frequency and the mathematical calculation, this study focused on the vertical bending vibration modes of the cantilever resonator, thus the vibration modes in torsion, lateral bending, and longitudinal directions are ignored. For simplification and without loss of generality, mode shape changes of the resonator are ignored when the mass and stiffness of the adsorbate are significantly less than that of the resonator [24]. The air-damping effect is complex and unclear [25]. In an atmospheric environment, air damping causes an insignificant effect on the resonator vibration [26]. Furthermore, resonators are commonly applied in vacuum or low atmosphere environments to achieve high-quality factors, which realize high sensitivity and accuracy measurement [27]. So, the damping effect of air is ignored in this study. Materials of both the resonator and the adsorbate are assumed to be linearly elastic, one-layer, and isotropic. Damping, residual stress, and temperature are assumed not to affect the Young's modulus or density of the resonator and the adsorbate. The minimum dimensions of the mathematical model are larger than the nanoscale. In other words, the mathematical model is based on the assumption that the Young's modulus and the density of the resonator and the adsorbate are constant. The vertical mode shapes of the cantilever resonator are given as

$$\psi_n(x) = A_n(\cos \kappa_n x - \cosh \kappa_n x) + B_n(\sin \kappa_n x - \sinh \kappa_n x) \tag{1}$$

where $\psi_n(x)$ is the time-independent vibration mode shape of the nth order natural frequency, x is the position along the length of the cantilever from the fixed end. The modal wavenumbers κ_n are the solutions to $\cos(\kappa_n L)\cosh(\kappa_n L) = -1$, and $\kappa_n L$ = 1.875, 4.694, 7.855, 10.996, 14.137, respectively. L is the length of the cantilever. The mode coefficients fulfill $A_n/B_n = (\cos \kappa_n L + \cosh \kappa_n L)/(\sin \kappa_n L - \sinh \kappa_n L)$ and $A_n/B_n = -1.362$, -0.982, -1.001, -1.000, -1.000, respectively [5].

In this study, the adsorbate is assumed to be a cuboid. When it is coated on the resonator, it caused both mass and stiffness to increase in the system. The natural frequencies of the vibration system would increase as the stiffness increases and would decrease as the mass increases.

The schematic diagram of the resonator and the adsorbate is shown in Figure 1, where W_R, T_R, E_R, and ρ_R are the width, thickness, Young's modulus, and density parameters of the resonator, respectively. L_C, W_C, T_C, E_C and ρ_C are the length, width, thickness, Young's modulus and density of the adsorbate, respectively. In this study, E_C and ρ_C are the parameters that need to be measured. L_{Cini} is the start position of the adsorbate on the cantilever, and L_{Cend} is the end position of the adsorbate on the resonator. h_0 is the position of the neutral surface. y is the axis along the thickness direction and the zero point is set on the neutral surface of the section.

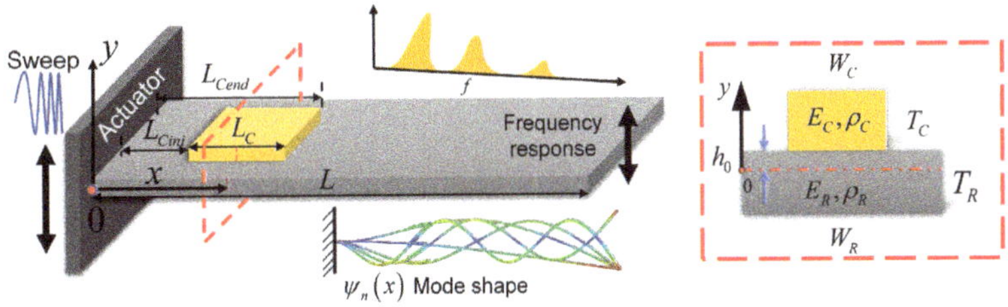

(**a**) (**b**)

Figure 1. (**a**) The schematic diagram of the cantilever resonator with an adsorbate. (**b**) The cross-section of the adsorbate and the resonator.

Next, the Rayleigh–Ritz method is used to calculate the natural frequencies of the resonator with an adsorbate. The numerator in Equation (2) denotes the total effective stiffness of the resonator and the adsorbate, while the denominator denotes the total effective mass of the resonator and the adsorbate from the fixed end to the free end of the cantilever [18,24].

$$\omega_n^2 = \frac{\int_0^{L_R} D(x)\left(\frac{\partial^2 \psi_n(x)}{\partial x^2}\right)^2 dx}{\int_0^{L_R} \rho(x)\psi_n^2(x)dx} \tag{2}$$

where ω_n is the nth-order angular natural frequency of the vibration system. $D(x)$ means the bending rigidity of the section. According to the mechanics of materials, the bending rigidity of the resonator section without the adsorbate equals the Young's modulus of the resonator multiplied by the moment of inertia of the resonator section.

$$D(x) = E_R \frac{W_R T_R^3}{12}(0 \leq x < L_{Cini}, L_{Cend} < x \leq L) \tag{3}$$

Figure 1b shows the section of the resonator where the adsorbate is located. The position of the neutral surface h_0 should be determined first. The position of the neutral surface is found from the condition that the resultant axial force acting on the cross-section is zero [28]. Therefore,

$$E_C \int_C ydA + E_R \int_R ydA = 0 \tag{4}$$

In this study, the widths of the adsorbate and the resonator are constant along the length direction of the resonator. So, Equation (4) can be expressed below

$$E_C W_C \int_{h_0}^{h_0+T_C} ydy + E_R W_R \int_{h_0-T_R}^{h_0} ydy = 0 \tag{5}$$

Then h_0 can be solved as,

$$h_0 = \frac{E_R W_R T_R^2 - E_C W_C T_C^2}{2(E_C W_C T_C + E_R W_R T_R)} \tag{6}$$

The flexural rigidity of the section is expressed as [28]

$$D(x) = E_C I_{Ch_0} + E_R I_{Rh_0}\ (L_{Cini} \leq x \leq L_{Cend}) \tag{7}$$

where I_{Ch_0} and I_{Rh_0} are the moments of inertia of the adsorbate and resonator. Then $D(x)$ is expressed as [18]

$$D(x) = \frac{E_C^2 W_C^2 T_C^4 + E_R^2 W_R^2 T_R^4 + E_C W_C T_C E_R W_R T_R (4T_C^2 + 4T_R^2 + 6T_C T_R)}{12(E_C W_C T_C + E_R W_R T_R)} \; (L_{Cini} \le x \le L_{Cend}) \tag{8}$$

$\rho(x)$ means the linear density of the section, presented as

$$\rho(x) = \begin{cases} \rho_R W_R T_R \; (0 \le x < L_{Cini}, L_{Cend} < x \le L) \\ \rho_R W_R T_R + \rho_C W_C T_C \; (L_{Cini} \le x \le L_{Cend}) \end{cases} \tag{9}$$

Commonly in experiments, the geometry parameters of the resonator and the adsorbate can be measured easily with a scanning electron microscope with nanoscale precision [24]. Resonators are built at nano precision with rigorously selected material, such as high-purity silicon. To calculate the density and Young's modulus of the adsorbate, it is necessary to obtain the relationship between the Young's modulus and the density of the adsorbate. Equation (8) contains the relationship between $D(x)$ and E_C. Hence, the next step is to find the relationship between $D(x)$ and ρ_C for different order natural frequencies. Here, Equation (2) is transformed into the following formula.

$$D(x) = \frac{\omega_n^2 \left[(\rho_R W_R T_R) \int_0^L \psi_n^2(x) dx + \rho_C W_C T_C \int_{L_{Cini}}^{L_{Cend}} \psi_n^2(x) dx \right] - E_R \frac{W_R T_R^3}{12} \left[\int_0^{L_{Cini}} \left(\frac{\partial^2 \psi_n(x)}{\partial x^2} \right)^2 dx + \int_{L_{Cend}}^{L} \left(\frac{\partial^2 \psi_n(x)}{\partial x^2} \right)^2 dx \right]}{\int_{L_{Cini}}^{L_{Cend}} \left(\frac{\partial^2 \psi_n(x)}{\partial x^2} \right)^2 dx} \tag{10}$$

By combining Equations (8) and (10), the relationship between the density and Young's modulus of the adsorbate for different order natural frequency is obtained and expressed as

$$E_C = f(\rho_C, \omega_n) \tag{11}$$

As E_C and ρ_C are the material properties of the adsorbate, they should fulfill Equation (11) for different order natural frequencies. The results can be calculated by solving equation sets generated from Equation (11) of different order natural frequency curves. Thus, the determined values of E_C and ρ_C can be read out from the intersection of the curves drawn from Equation (11).

3. Results and Discussion

3.1. Numerical Simulation with No Frequency Error

Based on Equation (2) and the parameters in Table 1, a simulation model was built to describe the effect of Young's modulus and density of the adsorbate on the natural frequencies of the resonator. The geometry and physical parameters of the resonator referred to the size and characteristics of an atomic force microscope silicon cantilever in common usage. The material of the adsorbate referred to the polymer mentioned in Belardinelli's experiment [24].

Table 1. The parameters of the silicon resonator and polymer adsorbate.

Resonator	Material	L	W_R	T_R	E_R	ρ_R
	Silicon	400 μm	40 μm	2 μm	168 GPa	2329 kg/m^3
Adsorbate	Material	L_C	W_C	T_C	E_C	ρ_C
	Polymer	30 μm	30 μm	2 μm	1.36 GPa	746 kg/m^3

Table 2 shows the calculated natural frequencies of the resonator with an adsorbate located at various positions from the fixed end toward the free end of the cantilever. Then the obtained natural frequencies in Table 2 and the parameters in Table 1 except E_C and ρ_C were substituted into Equation (11). The Young's modulus-density curves are shown in Figure 2.

Table 2. The natural frequency values when adsorbate located at various positions on the resonator.

L_{Cini}	0 μm	100 μm	200 μm	370 μm
ω_1 (rad/s)	108,892	108,160	107,296	104,421
ω_2 (rad/s)	680,706	669,885	668,214	659,332
ω_3 (rad/s)	1,901,794	1,863,729	1,889,331	1,847,176
ω_4 (rad/s)	3,722,077	3,679,178	3,669,068	3,697,473
ω_5 (rad/s)	6,141,323	6,117,132	6,102,113	6,114,159

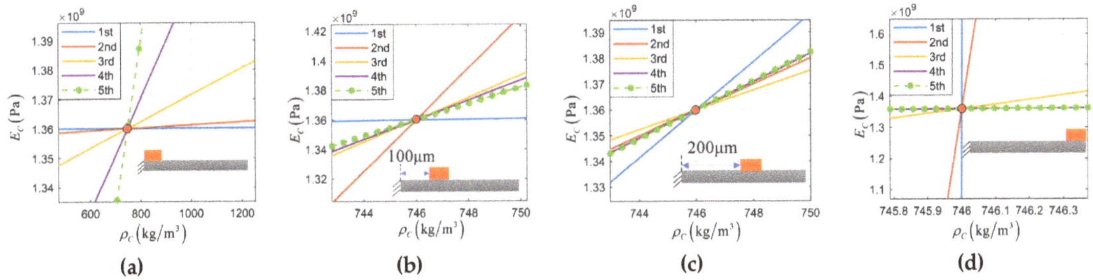

Figure 2. The first five order natural frequency E_C-ρ_C curves. The red points denote the ideal Young's modulus and density values of the adsorbate. The left side of the adsorbate L_{Cini} on the resonator was: (**a**) the fixed end. (**b**) 100 μm from the fixed end. (**c**) 200 μm from the fixed end. (**d**) the free end.

It is noted in Figure 2 that the E_C-ρ_C curves of the first five order natural frequencies intersect at the same point in these figures, and thus the density and Young's modulus values of the adsorbate can be obtained as $E_C = 1.36$ GPa and $\rho_C = 746\ \text{kg/m}^3$. The results are consistent with the pre-set ρ_C and E_C values in Table 1.

3.2. Analysis of Numerical Simulation Results with Frequency Errors

Errors in the size and material properties are inevitable in the measurement and manufacturing process, which will cause frequency errors in the resonator. So, it is necessary to evaluate how frequency errors affect the calculated E_C and ρ_C values. Due to frequency error, the intersections of different order natural frequency E_C-ρ_C curves will not coincide at the same point.

The errors of Young's modulus values can be presented as $\Delta E_C = \left| E_{C(\omega_i,\omega_j)} - E_C \right|$, in which $E_C = 1.36$ GPa is the ideal value and $E_{C(\omega_i,\omega_j)}$ is the Young's modulus value obtained from the *i*th and *j*th order natural frequencies. A relative frequency error of 0.001% was put into each of the first five order natural frequencies. Figure 3a illustrates the relative Young's modulus errors of the adsorbate calculated from different natural frequency pairs at different positions on the resonator. Similarly, the density error is $\Delta\rho_C = \left| \rho_{C(\omega_i,\omega_j)} - \rho_C \right|$, in which $\rho_C = 746\ \text{kg/m}^3$ is the ideal value and $\rho_{C(\omega_i,\omega_j)}$ is the density value obtained from the *i*th and *j*th order natural frequencies. The relative density errors are shown in Figure 3b. Secondly, the Young's modulus and density errors caused by 0.01% frequency error are shown in Figure 4a,b.

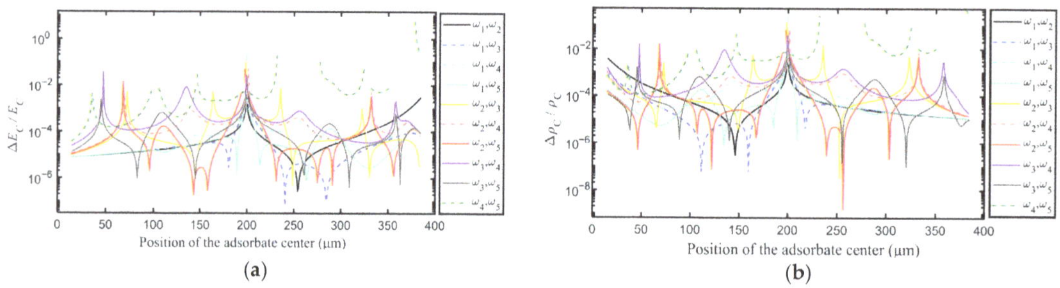

Figure 3. The relative physical parameter errors of the polymer adsorbate when 0.001% frequency error were added to the first five order natural frequencies. (**a**) Young's modulus errors, (**b**) Density errors.

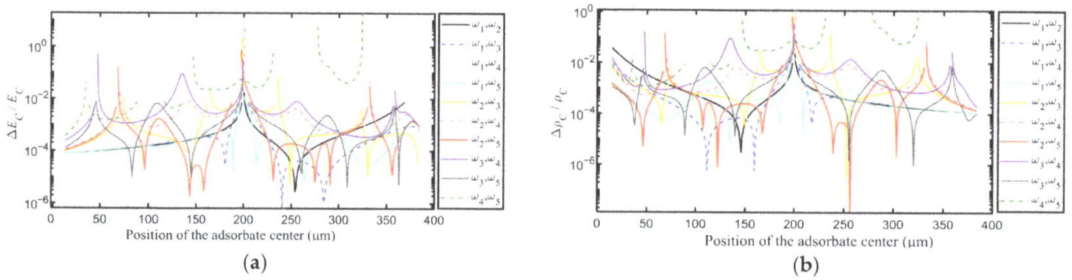

Figure 4. The relative physical parameter errors of the polymer adsorbate when 0.01% frequency error were added to the first five order natural frequencies. (**a**) Young's modulus errors, (**b**) Density errors.

It can be clearly seen that higher frequency errors will lead to higher Young's modulus and density errors. Furthermore, by comparing Figure 3a with Figures 4a and 3b with Figure 4b, it is obvious that the curves drawn from different frequency errors show the same trend along the longitude direction. All the peaks in the two relative figures are perfectly located at the same positions. Hence, Figures 3 and 4 proved that the value of the natural frequency error will not change the most appropriate location and frequency pairs for both Young's modulus and density measurement. It should be noted that some curves appear to be discontinuous in Figures 3 and 4. This is caused by the fact that for specific frequency pairs, their E_C-ρ_C curves did not intersect at the adsorbate positions so there were no available values.

Another simulation with different geometry dimensions of the adsorbate and the resonator was carried out. The material of the adsorbate was set as platinum in this simulation. The parameters of the adsorbate and the resonator are shown in Table 3.

Table 3. The parameters of the silicon cantilever resonator and platinum adsorbate.

Resonator	Material	L	W_R	T_R	E_R	ρ_R
	Silicon	800 μm	40 μm	1 μm	168 GPa	2329 kg/m^3
Adsorbate	Material	L_C	W_C	T_C	E_C	ρ_C
	Platinum	20 μm	20 μm	1 μm	168 GPa	21,450 kg/m^3

By comparing Figure 5a with Figures 4a and 5b with Figure 4b, it can be noticed that the curves drawn from different materials or geometry sizes of the adsorbate and resonator show almost the same trend along the longitude direction. For example, the highest peaks all appear at 50% of the length of the cantilever. All the peaks in the two relative figures are perfectly located at the same relative positions. Hence, Figures 4 and 5 proved that the parameters of the adsorbate and resonator do not make a difference to the most appropriate location or frequency pairs for either Young's modulus or the density measurement.

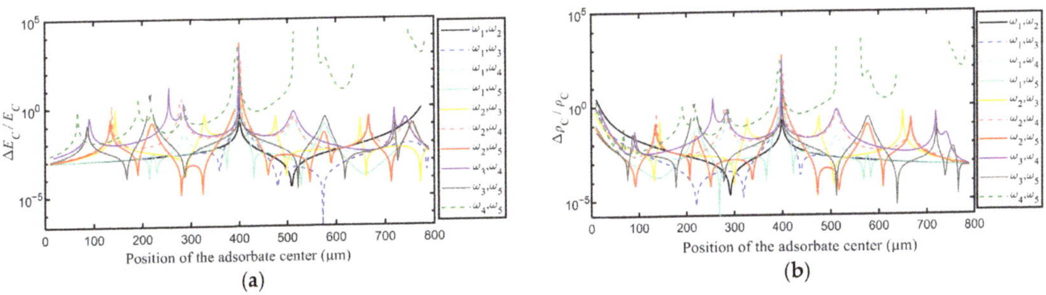

Figure 5. The relative physical parameter errors of the platinum adsorbate when 0.01% frequency error were added to the first five order natural frequencies. (**a**) Young's modulus errors, (**b**) Density errors.

The numerical simulation results revealed that the measurement accuracy and reliability are affected by the relative position of the adsorbate on the cantilever and the frequency pairs applied. The best frequency pairs for the measurement can be obtained by the following principles. (1) The $\Delta E_C / E_C$ and $\Delta\rho_C / \rho_C$ are smallest at the corresponding adsorbate location. This means the calculated Young's modulus and density are close to the ideal values. (2) The curves change smoothly along the longitudinal axis. As frequency errors are always random, small slopes make the results less sensitive to position errors, as the position is highly related to frequency error.

Based on the principles, the suggestions for Young's modulus and density measurement are as follows, when the adsorbate is located at a specific position on the resonator, the corresponding best frequency pair for the Young's modulus measurement can be found in Figure 4a. For example, when the distance between the adsorbate center and the fixed end of the resonator is less than 44% L, the results from (ω_1, ω_2), (ω_1, ω_3), and (ω_1, ω_5) change smoothly within a low level. From 56% to 79% L, the result from (ω_1, ω_3) appears to be smooth with low error. From 79% L to the free end, (ω_1, ω_5) can provide more accurate results.

As for the measurement of density, the best frequency pair can be found in Figure 4b. From the fixed end to about 23% L, (ω_1, ω_5) could provide a better result. From 23% to 35%L, (ω_1, ω_3) will be the best choice. Beyond 54% L, (ω_1, ω_2), (ω_1, ω_3), and (ω_1, ω_5) can provide good results and their results will get closer as the adsorbate move to the free end of the resonator.

By combining Figure 4a,b, several recommended positions on the resonator and relative natural frequency pairs can be found to measure the density and the Young's modulus of the adsorbate with good results. The recommended choices are given in Table 4.

Table 4. Recommended adsorbate center positions and relative natural frequency pairs to measure Young's modulus and density.

Position	Young's Modulus	Density
Fixed end-23% L	(ω_1, ω_2), (ω_1, ω_3), (ω_1, ω_5)	(ω_1, ω_5)
23–35% L	(ω_1, ω_2), (ω_1, ω_3), (ω_1, ω_5)	(ω_1, ω_3)
56–79% L	(ω_1, ω_3)	(ω_1, ω_2), (ω_1, ω_3), (ω_1, ω_5)
79% L-Free end	(ω_1, ω_5)	(ω_1, ω_2), (ω_1, ω_3), (ω_1, ω_5)

3.3. The Relationship between Error Peaks and the Density to Young's Modulus Frequency Shift Ratio

To find out what affects the error of density and Young's modulus measurement of the adsorbate, a simulation was performed and revealed the effects of the density and Young's modulus on the natural frequencies of the resonator. The result of the polymer adsorbate is shown in Figure 6a. Another simulation result with a platinum adsorbate is shown in Figure 6b.

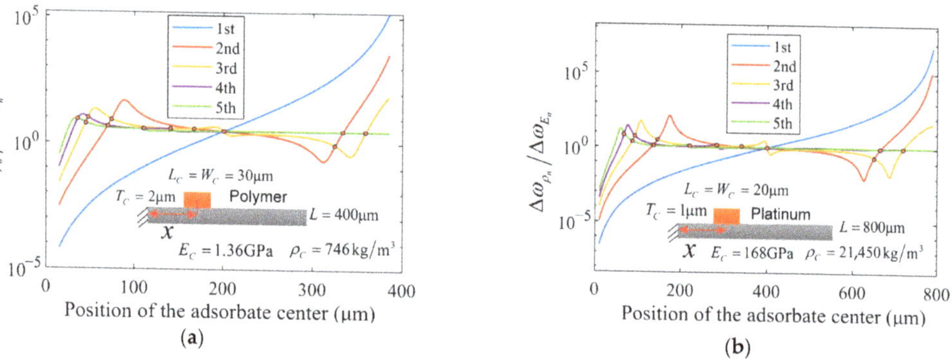

Figure 6. The ratio of the frequency shift caused by added density to the frequency shift caused by added Young's modulus. (**a**) The curves are simulated with parameters in Table 1, the material of the adsorbate is polymer (**b**) The curves are simulated with parameters in Table 3, the material of the adsorbate is platinum.

ω_{ρ_n} represents the nth order natural frequency of the resonator only affected by the density of the adsorbate, while ω_{E_n} means the nth order natural frequency only affected by the Young's modulus of the adsorbate. ω_{Bare_n} is the nth-order natural frequency of the cantilever without the adsorbate. $\Delta\omega_{E_n}/\Delta\omega_{\rho_n}$ is the ratio of the frequency shift caused by the added Young's modulus to the frequency shift caused by added density. ω_{ρ_n},ω_{E_n}, and ω_{Bare_n} can be acquired with Equation (2) by using relative parameters.

$$\Delta\omega_{\rho_n}/\Delta\omega_{E_n} = |\omega_{\rho_n} - \omega_{Bare_n}|/|\omega_{E_n} - \omega_{Bare_n}| \tag{12}$$

In Figure 6a, red points are the intersections of different order natural frequency curves. Curves of the fourth and fifth order natural frequencies coincided with each other as the adsorbate moved toward the free end of the resonator.

Interestingly, the horizontal coordinate values of the interactions in Figure 6a matched perfectly with those of the peaks in Figure 4a,b, which means when the adsorbate is located at these positions, the calculated Young's modulus and density will be inaccurate. Especially for the central point on the resonator, or the middle of the cantilever, all the five order natural frequency curves were almost crossing at the same point in Figure 6a, causing the peaks in Figures 3 and 4 for all different order frequency pairs. Hence, it will be better to locate the adsorbate away from the center of the resonator in the longitudinal direction.

Furthermore, comparing Figure 6a with Figure 6b, the geometry parameters and mechanical properties of the adsorbate and resonator mainly influence the amplitude of the plots in Figure 6a but hardly affect the trend of the plots and the positions of the interactions. As a result, the suggestions provided for the choice of adsorbate position and frequency pair are capable of measuring adsorbates with different materials and different dimensions.

4. Measurement Procedures and Finite Element Analysis Simulation Validation

The detailed procedures to measure Young's modulus and density of the adsorbate on a cantilever in a real-world application are listed as follows.

4.1. The Procedures of Young's Modulus and Density Measurement

1. Firstly, prepare a large length-to-thickness ratio rectangle cantilever resonator with known Young's modulus and density values.
2. Place the adsorbate that needs to be measured on one surface of the cantilever resonator. The adsorbate should be securely fixed to the cantilever so it will not separate or change its location during frequency measurement. Commonly, the fixed end of the cantilever is better for Young's modulus measurement, while the free end of the resonator performs better for density measurement. Furthermore, avoid putting the adsorbate in the center of the longitudinal direction of the cantilever.
3. Measure the geometry parameters of the resonator and the adsorbate, including the length, width, thickness, and relative position in a scanning electron microscope.
4. Measure the vertical bending mode natural frequencies of the cantilever and the adsorbate with a contactless method. The measurement device can be atomic force microscopy or a laser doppler vibrometer.
5. Find the best frequency pairs for the Young's modulus and density measurement from Table 4 according to the relative position of the adsorbate measured in step 3.
6. Input the length, width, and thickness of the adsorbate and the resonator, the location of the adsorbate center on the resonator, the first five order natural frequencies, and the Young's modulus and density of the cantilever into Equation (11) to plot the Young's modulus and density curves of the adsorbate for different order natural frequencies. The interaction of the frequency pairs chosen in step 5 will be the determined Young's modulus and density.

In real-world applications, air damping will decrease the natural frequencies of the cantilever. The best way is to carry out the experiment in a vacuum. If not, it is better to measure the natural frequency of the cantilever without the adsorbate in the air to evaluate the effect of the air damping. The residual stress in the microcantilever may make the mode shape of the cantilever different from the theoretical result and affect the measurement accuracy. It is recommended to use the resonator after its residual stress has been released. Temperature may affect the mechanical properties of the cantilever and the adsorbate. It may cause residual stress between the adsorbate and resonator, and change the density, and the stiffness. Therefore, keeping the temperature stable during the experiment would help to improve the measurement accuracy and reliability.

To validate the proposed method, an FEA model was applied in this research.

4.2. Finite Element Analysis Validation

The parameters of the adsorbate and resonator were set according to Table 1. The length, width, and thickness of the cantilever were 400 μm, 40 μm, and 2 μm, respectively. The length, width, and thickness of the adsorbate were 30 μm, 30 μm, and 2 μm, respectively. The Poisson's ratio of them was set as 0. The interface between the adsorbate and the resonator was constrained as a Tie type in Abaqus software. The fixed-end surface of the cantilever was constrained with no displacement in any direction. The center of the adsorbate was located at 50 μm from the fixed end of the cantilever. The mesh element shape was Hexahedron. The FEA model is shown in Figure 7.

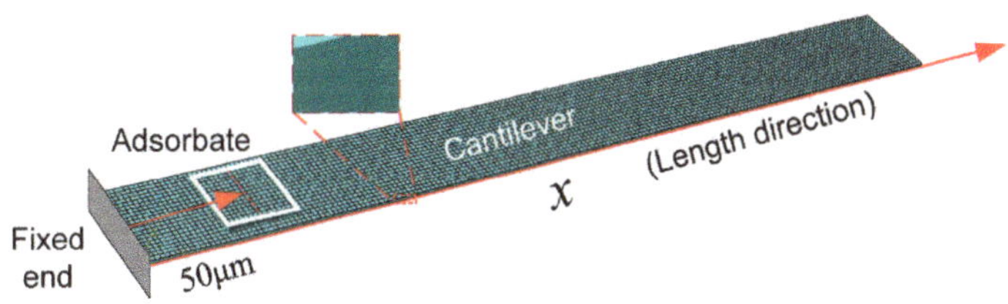

Figure 7. The FEA model of the cantilever resonator and adsorbate.

The first five order vertical bending angular natural frequencies and relative mode shapes simulated with the FEA method are shown in Figure 8.

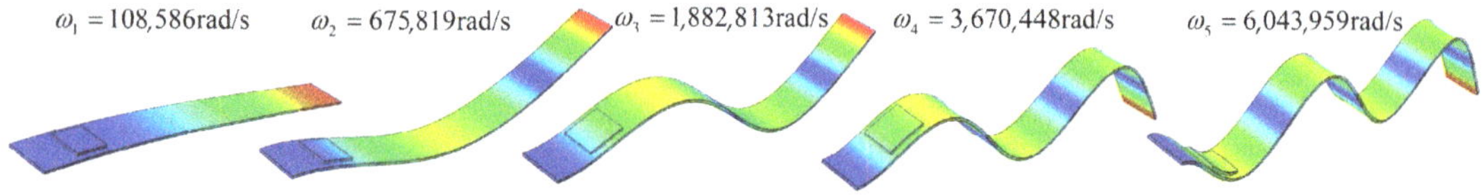

Figure 8. The first five order bending angular natural frequencies and relative mode shapes.

Putting these bending natural frequencies into Equation (11), the Young's modulus and density of the adsorbate were calculated from each frequency pair, as shown in Table 5.

Table 5. The Young's modulus and density calculated from different natural frequency pairs.

Parameters	(ω_1,ω_2)	(ω_1,ω_3)	(ω_1,ω_4)	(ω_1,ω_5)	(ω_2,ω_3)
E_C (GPa)	1.309	1.037	1.306	1.306	1.087
ρ_C (kg/m^3)	1180	862.2	775.0	760.5	853.4
Parameters	(ω_2,ω_4)	(ω_2,ω_5)	(ω_3,ω_4)	(ω_3,ω_5)	(ω_4,ω_5)
E_C (GPa)	1.022	1.003	5.502	2.918	1.650
ρ_C (kg/m^3)	757.7	729.0	1027	926.0	795.9

Based on the curves in Figure 9a, the $\Delta E_C/E_C$ values calculated from (ω_1,ω_2), (ω_1,ω_3), (ω_1,ω_4), and (ω_1,ω_5) are relatively lower than other frequency pairs at the 50 µm position. As for the $\Delta\rho_C/\rho_C$ curves of different frequency pairs shown in Figure 9b, (ω_1,ω_5) and (ω_2,ω_5) can provide better results at this location. However, the curves of (ω_1,ω_5) changed more gently without sudden change, making the calculated result less sensitive, so the result from this frequency pair will be more stable and reliable. Additionally, these frequency pairs coincided with the recommended frequency pairs mentioned in Table 3.

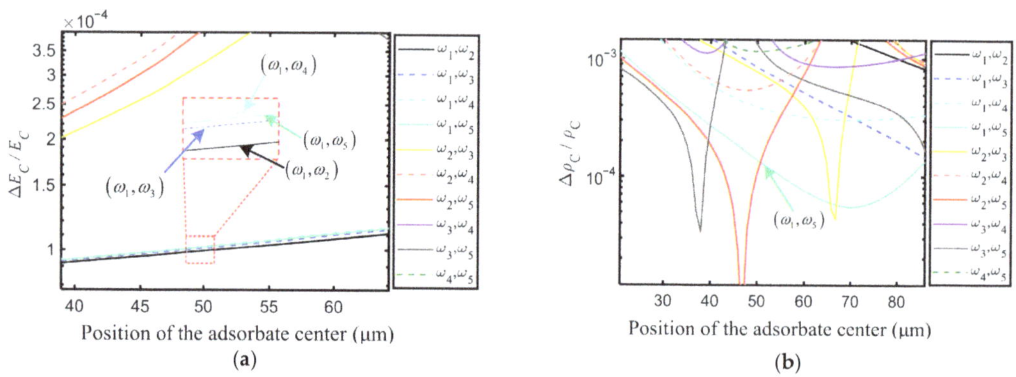

Figure 9. The partial zoom of Figure 4 contains the curves of different frequency pairs of the adsorbate locating at 50 µm from the fixed end of the resonator. (**a**) Young's modulus errors, (**b**) Density errors.

Finally, (ω_1,ω_2) was picked out for the Young's modulus measurement and (ω_1,ω_5) was picked out for the density measurement of the adsorbate. The E_C was calculated as 1.309 GPa, with an error of 3.75% compared with the ideal value of 1.36 GPa. Furthermore, the ρ_C was calculated as 760.5 kg/m^3, with an error of -1.94% compared with the ideal density value of 746 kg/m^3. The selected frequency pairs and the relative Young's modulus and density result are shown in Figure 10.

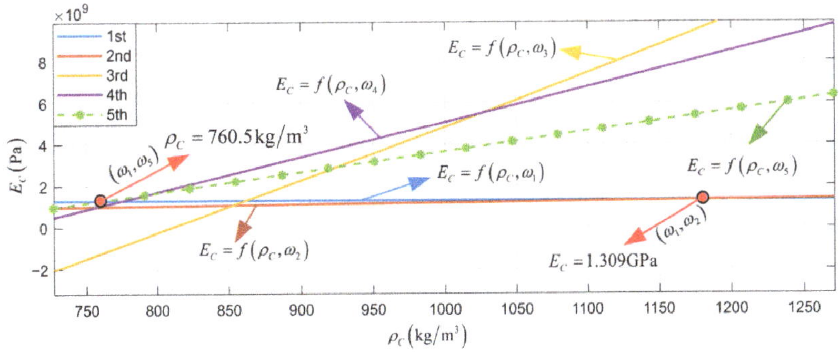

Figure 10. E_C-ρ_C curves of the first five order natural frequencies and the recommended frequency pairs to calculate the Young's modulus and density values of the adsorbate.

5. Conclusions

In summary, a mathematical model based on the Rayleigh–Ritz method was built to determine the Young's modulus and density of an adsorbate by measuring several vertical vibration natural frequencies of the resonator and extracting the intersections of their nonlinear Young's modulus and density curves. Ideally, the intersections of different natural frequency curves would coincide, which revealed the Young's modulus and density of the adsorbate simultaneously. This research revealed a vital conclusion, that when there were errors in the measurement of natural frequencies, the position of the adsorbate had a complicated and nonlinear effect on the measurement accuracy of Young's modulus and density. Furthermore, this study revealed that frequency pairs and the relative position of the adsorbate on the cantilever are two important factors that affect the accuracy and reliability of the measurement. Using appropriate frequency pairs can improve the accuracy and reliability of measurement. This study provided a method to set the adsorbate at specific positions and choose suitable natural frequency pairs to obtain more accurate and stable results. Finally, the method was validated by the finite element analysis method.

The results from this study show the potential application of combining several high-order natural frequencies to improve measurement accuracy. It can be applied in real-time micro/nano-scale research and provide researchers with intensive data to monitor the mechanical properties variation in the experimental process. It is a non-invasive and label-free scheme that works to monitor the biological reaction or to distinguish the cell by its mechanical properties without destroying the sample.

Author Contributions: Conceptualization, Y.Y. and Y.T.; methodology, Y.Y. and Y.S.; software, Y.Y. and Y.T.; validation, Y.T., Y.S., X.L. and H.T.; formal analysis, Y.Y., Y.S. and Y.T.; investigation, Y.T.; resources, Y.T.; data curation, Y.Y. and Y.S.; writing—original draft preparation, Y.Y.; writing—review and editing, Y.T. and Y.S.; visualization, Y.Y.; supervision, X.L. and H.T.; project administration, Y.T. All authors have read and agreed to the published version of the manuscript.

Funding: This work was supported by program of Horizon Europe (MSCA SE ENSIGN 101086226).

Data Availability Statement: Not applicable.

Acknowledgments: The authors would like to express their sincere gratitude to Lee A. J. Davis, for many valuable suggestions that helped to improve the paper.

Conflicts of Interest: The authors declare no conflict of interest.

References

1. Johnson, B.N.; Mutharasan, R. Biosensing using dynamic-mode cantilever sensors: A review. *Biosens. Bioelectron.* **2012**, *32*, 1–18. [CrossRef] [PubMed]
2. Wang, K.; Wang, B. Vibration modeling of carbon-nanotube-based biosensors incorporating thermal and nonlocal effects. *J. Vib. Control.* **2014**, *22*, 1405–1414. [CrossRef]
3. Dohn, S.; Schmid, S.; Amiot, F.; Boisen, A. Position and mass determination of multiple particles using cantilever based mass sensors. *Appl. Phys. Lett.* **2010**, *97*, 044103. [CrossRef]
4. Ekinci, K.L.; Huang, X.M.H.; Roukes, M.L. Ultrasensitive nanoelectromechanical mass detection. *Appl. Phys. Lett.* **2004**, *84*, 4469–4471. [CrossRef]
5. Dohn, S.; Svendsen, W.; Boisen, A.; Hansen, O. Mass and position determination of attached particles on cantilever based mass sensors. *Rev. Sci. Instrum.* **2007**, *78*, 103303. [CrossRef]
6. Sader, J.E.; Hanay, M.S.; Neumann, A.P.; Roukes, M.L. Mass Spectrometry Using Nanomechanical Systems: Beyond the Point-Mass Approximation. *Nano Lett.* **2018**, *18*, 1608–1614. [CrossRef]
7. Yuksel, M.; Orhan, E.; Yanik, C.; Ari, A.B.; Demir, A.; Hanay, M.S. Nonlinear Nanomechanical Mass Spectrometry at the Single-Nanoparticle Level. *Nano Lett.* **2019**, *19*, 3583–3589. [CrossRef]
8. Tsai, T.-T.; Fang, Y.-C.; Wu, S.-K.; Chang, R.-H.; Chien, W.-C.; Chang, Y.-H.; Wu, C.-S.; Kuo, W. High throughput and label-free particle sensor based on microwave resonators. *Sens. Actuators A Phys.* **2019**, *285*, 652–658. [CrossRef]
9. Fritz, J.; Baller, M.K.; Lang, H.P.; Rothuizen, H.; Vettiger, P.; Meyer, E.; Guntherodt, H.; Gerber, C.; Gimzewski, J.K. Translating biomolecular recognition into nanomechanics. *Science* **2000**, *288*, 316–318. [CrossRef]
10. Majstrzyk, W.; Mognaschi, M.E.; Orłowska, K.; di Barba, P.; Sierakowski, A.; Dobrowolski, R.; Grabiec, P.; Gotszalk, T. Electromagnetic cantilever reference for the calibration of optical nanodisplacement systems. *Sens. Actuators A Phys.* **2018**, *2018*, 149–156. [CrossRef]
11. Chen, F.; Gao, J.; Tian, W. Force-frequency characteristics of multi-electrode quartz crystal resonator cluster. *Sens. Actuators A Phys.* **2018**, *269*, 427–434. [CrossRef]
12. Ekmekci, E.; Kose, U.; Cinar, A.; Ertan, O.; Ekmekci, Z. The use of metamaterial type double-sided resonator structures in humidity and concentration sensing applications. *Sens. Actuators A Phys.* **2019**, *297*, 111559. [CrossRef]
13. Laidoudi, F.; Kanouni, F.; Assali, A.; Caliendo, C.; Amara, S.; Nezzari, H.; Boubenider, F. Thickness shear SMR resonator based on Yttrium-doped AlN for high sensitive liquid sensors. *Sens. Actuators A Phys.* **2021**, *333*, 113238. [CrossRef]
14. Eom, K.; Park, H.S.; Yoon, D.S.; Kwon, T. Nanomechanical resonators and their applications in biological/chemical detection: Nanomechanics principles. *Phys. Rep.* **2011**, *503*, 115–163. [CrossRef]
15. Parmar, J.; Patel, S.K. Tunable and highly sensitive graphene-based biosensor with circle/split ring resonator metasurface for sensing hemoglobin/urine biomolecules. *Phys. B Condens. Matter* **2022**, *624*, 413399. [CrossRef]
16. Suresh, S. Biomechanics and biophysics of cancer cells. *Acta Biomater.* **2007**, *3*, 413–438. [CrossRef] [PubMed]
17. Garcia, R.; Herruzo, E.T. The emergence of multifrequency force microscopy. *Nat. Nanotechnol.* **2012**, *7*, 217–226. [CrossRef]
18. Ramos, D.; Tamayo, J.; Mertens, J.; Calleja, M.; Villanueva, L.G.; Zaballos, A. Detection of bacteria based on the thermomechanical noise of a nanomechanical resonator: Origin of the response and detection limits. *Nanotechnology* **2008**, *19*, 035503. [CrossRef]
19. Calleja, M.; Kosaka, P.M.; San Paulo, A.; Tamayo, J. Challenges for nanomechanical sensors in biological detection. *Nanoscale* **2012**, *4*, 4925–4938. [CrossRef]
20. Teva, J.; Abadal, G.; Torres, F.; Verd, J.; Perez-Murano, F.; Barniol, N. A femtogram resolution mass sensor platform based on SOI electrostatically driven resonant cantilever. Part II: Sensor calibration and glycerine evaporation rate measurement. *Ultramicroscopy* **2006**, *106*, 808–914. [CrossRef]
21. Jensen, K.; Kim, K.; Zettl, A. An atomic-resolution nanomechanical mass sensor. *Nat. Nanotechnol.* **2008**, *3*, 533–537. [CrossRef] [PubMed]
22. Lassagne, B.; Garcia-Sanchez, D.; Aguasca, A.; Bachtold, A. Ultrasensitive mass sensing with a nanotube electromechanical resonator. *Nano Lett.* **2008**, *8*, 3735–3738. [CrossRef] [PubMed]
23. Gil-Santos, E.; Ramos, D.; Martinez, J.; Fernandez-Regulez, M.; Garcia, R.; San Paulo, A.; Calleja, M.; Tamayo, J. Nanomechanical mass sensing and stiffness spectrometry based on two-dimensional vibrations of resonant nanowires. *Nat. Nanotechnol.* **2010**, *5*, 641–645. [CrossRef] [PubMed]
24. Belardinelli, P.; Hauzer, L.M.F.R.; Šiškins, M.; Ghatkesar, M.K.; Alijani, F. Modal analysis for density and anisotropic elasticity identification of adsorbates on microcantilevers. *Appl. Phys. Lett.* **2018**, *113*, 143102. [CrossRef]
25. Lavrik, N.V.; Sepaniak, M.J.; Datskos, P.G. Cantilever transducers as a platform for chemical and biological sensors. *Rev. Sci. Instrum.* **2004**, *75*, 2229–2253. [CrossRef]
26. Zhang, Y. Detecting the stiffness and mass of biochemical adsorbates by a resonator sensor. *Sens. Actuators B Chem.* **2014**, *202*, 286–293. [CrossRef]
27. Keeler, E.G.; Jing, P.; Wu, J.; Zou, C.; Lin, L.Y. MEMS Resonant Mass Sensor With Integrated Optical Manipulation. *IEEE Trans. Nanotechnol.* **2018**, *17*, 714–718. [CrossRef]
28. Goodno, B.J.; Gere, J.M. *Mechanics of Materials*, 9th ed.; Cengage Learning: South Melbourne, VIC, Australia, 2018; pp. 553–637.

actuators

MDPI

Article

Design and Simulation of a Single Piezoelectric-Driven Rotary Actuator with Double-Layer Flexible Mechanism

Zhiyong Guo [1,*], Pengchao Zhao [2], Wenchao Zhang [2], Yanling Tian [3] and Gaofeng Hu [4]

1 College of Aeronautical Engineering, Civil Aviation University of China, Tianjin 300300, China
2 China Automotive Technology and Research Center Company Limited, Tianjin 300162, China
3 School of Engineering, University of Warwick, Coventry CV4 7AL, UK
4 School of Mechanical Engineering, Tianjin University of Technology and Education, Tianjin 300222, China
* Correspondence: zyguo@cauc.edu.cn

Abstract: A novel pure rotary actuator with a double-layer flexible mechanism (RA-DFM), which is driven by a single piezoelectric ceramic in the lower mechanism and generates rotational motion in the upper mechanism, is proposed in this paper. The output of piezoelectric ceramic is successively amplified using an enhanced double Scott–Russell mechanism and two lever-type mechanisms to obtain a large rotation range. The static, kinematic and dynamic properties of the RA-DFM are numerically analyzed, and the actual output of the piezoelectric is analyzed considering the input stiffness. The geometric parameters of the RA-DFM are optimized based on the constructed numerical models. Finite element analysis has been implemented to validate the correctness of the theoretical models and further evaluate the output property. The simulation results show the maximal rotation angle of the RA-DFM is 15.14 mrad with 0.44% center drift.

Keywords: flexible mechanism; nano-rotation; amplification mechanism

Citation: Guo, Z.; Zhao, P.; Zhang, W.; Tian, Y.; Hu, G. Design and Simulation of a Single Piezoelectric-Driven Rotary Actuator with Double-Layer Flexible Mechanism. *Actuators* **2023**, *12*, 231. https://doi.org/10.3390/act12060231

Academic Editor: Micky Rakotondrabe

Received: 4 May 2023
Revised: 30 May 2023
Accepted: 31 May 2023
Published: 2 June 2023

1. Introduction

The precise positioning stage with nanoscale accuracy has been widely used in surface topography, biological medicine and optics [1–3]. In the actuation of the nano-positioning stage, voice coil motor (VCM) and piezoelectric ceramic are the most popular candidates. The voice coil motor (VCM) is generally used to realize large strokes, but the small output force limits the working frequency of the positioning stage [4,5]. Piezoelectric ceramic has also been widely applied due to the high resolution, fast response and large output force [6–8], but the micron-level stroke impedes more extensive applications.

In recent years, the nano-rotational motion was usually included in the multi degrees of freedom (DOF) positioning stage, especially as the $XY\theta$-type stage generates the planar translation and the rotation around the vertical direction of the plane. For instance, Wang designed a 3-DOF monolithic manipulator based on the 3-revolute-revolute-revolute (3RRR) parallel mechanism; it is used to generate translations in the X and Y axes and rotation around the Z axis [9]. A similar kinematic mechanism has been applied by many researchers to generate rotational motion in the micro positioning stage [10,11]. In the developed 3-DOF mechanism of Kim [12], the rotational motion was generated through controlling the two piezoelectric ceramics in one direction. The 6-DOF positioning system is also generally designed with three rotational motions about the X, Y and Z axes [13]. However, the rotational motion in these nano-propositioning stages is usually coupled with other motions, which leads to the drift of the rotation center, and the rotational motion is realized through the actuation of multiple actuators, bringing in the issue of actuation redundancy.

In order to avoid motion coupling, a pure rotation stage has been developed. Lee proposed a pure nano-rotation scanner based on circular hinges and leaf springs, but it was still driven by two piezoelectric ceramics [14]. The rotary micro-positioning stage of Zhu employed a single piezoelectric ceramic driving the stage to avoid actuation redundancy, but the rotation range of 1.58 mrad is not enough for some applications [15]. To facilitate the machinability of the flexure-based mechanism, Clark presented a single piezoelectric-driven multi-level rotation stage using emerging additive manufacturing techniques; the working range reached 2.54 mrad, but the stage output drifted seriously over time [16].

For the requirements of large range, the stick-slip mechanism has been applied to drive the mobile slider, which could generate both linear movement and rotation with a large stroke [17,18]. The piezoelectric inchworm rotary actuator is also an effective method to obtain a large workspace [19,20]. However, the backward motion of the stick-slip mechanism is difficult to eliminate. The inchworm driving method requires simultaneous control of multiple piezoelectric ceramics, leading to control complexity and actuation redundancy.

Aiming at the requirements of large stoke, compact structure and without actuation redundancy, this paper proposes a single piezoelectric-driven rotary actuator with a double-layer flexible mechanism, which integrates the lever type and enhances the double Scott–Russell amplification mechanism. In this paper, mechanical design of the rotary stage is firstly introduced in Section 2. Then, the theoretical models, including the actual output model of piezoelectric ceramic, the maximum stress, kinematic and dynamic model of the rotary stage are established in Section 3. The main geometrical parameters are optimized in Section 4, and then the finite element analysis (FEA) is implemented in Section 5.

2. Mechanical Design of the RA-DFM

The proposed pure rotary actuator is shown in Figure 1a, which includes double-layer flexible mechanisms with a thread-connected base. The lower mechanism shown in Figure 1b displays a symmetrical layout. The piezoelectric ceramic is utilized as the actuator, one side of which is contacted with the input end and the other side is preloaded with a screw. A half column is designed on the input end to avoid the piezoelectric ceramic suffering a bending moment. A dual leaf parallelogram hinge (DLPH) is connected on two symmetric traditional Scott–Russell mechanisms to organize the enhanced double Scott–Russell mechanism (EDSRM), which has been demonstrated to possess excellent dynamic properties [21]. The EDSRM is utilized to amplify the input displacement and transfer the single translation to the symmetric and contrary translation. Afterwards, the displacements are secondly amplified via the lever type mechanism of the lower mechanism (LTM_{low}).

As shown in Figure 1c, the upper mechanism is designed with a rotary symmetric structure. The symmetric outputs from the lower mechanism are thirdly amplified and transferred to parallel and opposite translations via the lever-type mechanisms in the upper mechanism (LTM_{up}). The amplified displacements act on the top and bottom sides of the output platform to form the moment of couple and further generate rotational motion on the output platform. In addition, the output platform is over-constrained by a pair of flexible support mechanisms (FSM) to enlarge the stiffness in the out-off-plane direction, but it has little effect on the rotational motion due to the compliance in this direction, which could restrain the unwanted resonant vibration.

The motion transferred from the lower mechanism to the upper mechanism seriously affects the output efficiency. As shown in Figure 1a, the lower connecting block is connected to the upper connecting block with a horizontally arranged screw in the structure design. This connecting method generates normal pressure between the upper and lower connecting block, which could transfer a larger force compared with the friction-type connection.

Furthermore, the velocity diagrams of the connecting block, which are shown in Figure 2, are analyzed using the pseudo rigid body theory. The right circular hinges are simplified as a revolute joint and marked as A–K, and the FSM is regarded as the revolute joint L due to the actual rotary motion of the output platform. The linkage GH is the connection bar between the lower and upper mechanism. If the rotation centers of output hinge F and fulcrum hinge G of LTM_{low} as well as the input hinge I and fulcrum hinge H of LTM_{up} are respectively designed in a strictly straight line, the velocity of flexure hinge

G will be same as flexure hinge H, and the linkage GH will generate translational motion as an integral. Otherwise, there will be a rotary motion on the linkage GH, as shown in Figure 3, which may result in the separation of the two connecting blocks. Therefore, the flexure hinge positions of LTM_{low} and LTM_{up} are designed as in Figure 2.

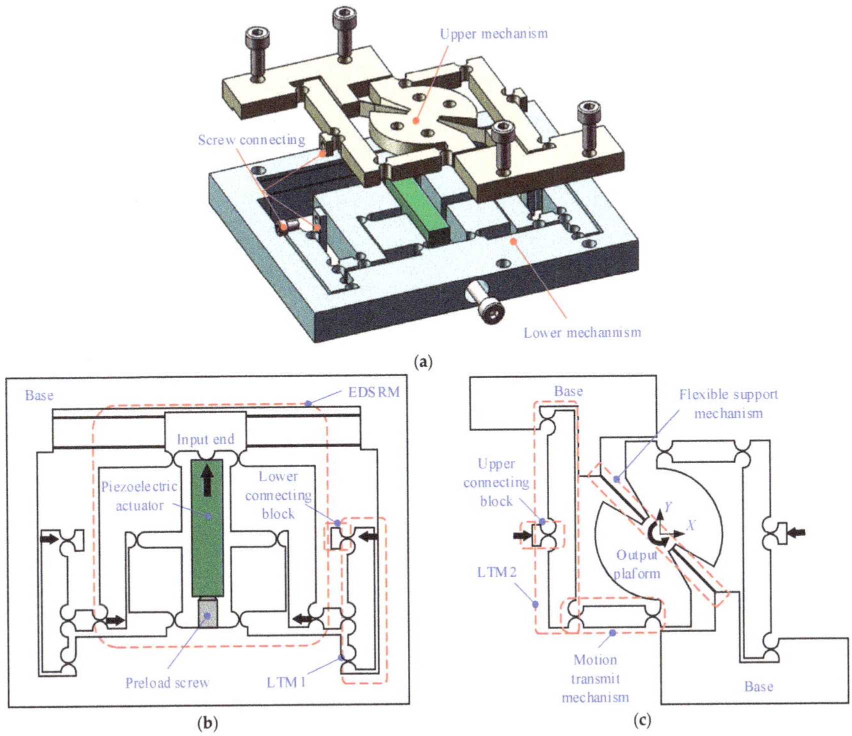

Figure 1. Mechanical design of the RA-DFM: (**a**) the assembly diagram; (**b**) the lower mechanism; (**c**) the upper mechanism.

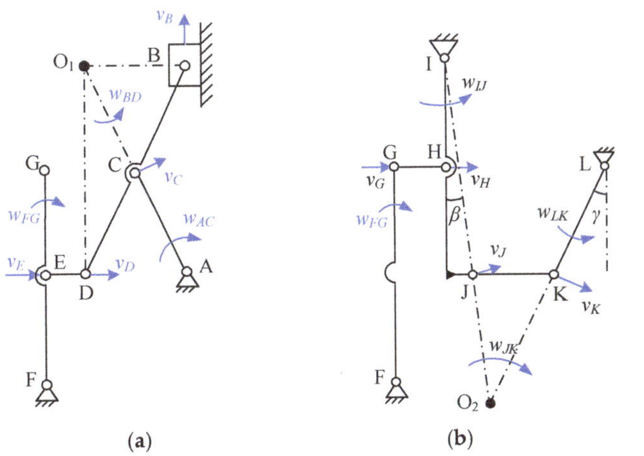

Figure 2. The velocity diagram of half RA-DFM: (**a**) the lower stage; (**b**) the upper stage.

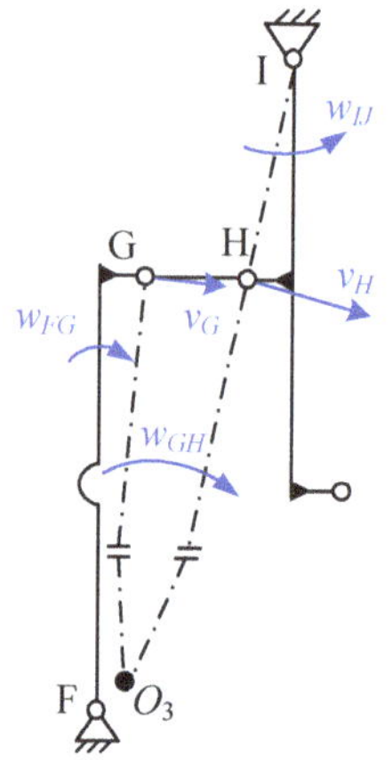

Figure 3. The velocity diagram of the connecting block with unaligned rotation centers in the LTM_{low} and LTM_{up}.

3. Theoretical Modeling and Analysis of the RA-DFM

3.1. The Kinematic Modeling

Based on the velocity diagram of the half RA-DFM in Figure 2, the velocity response of each flexure hinge is analyzed under the action of the input velocity v_B. It is obvious that the linkages AC, FG and IJ rotate around their revolute joints A, F and I; the linkage LK rotates around the assumed center L. According to the input and output velocity directions, the linkages BD and JK rotate with O_1 and O_2, respectively. Considering the small deformation of the whole system, the velocity directions of the linkages are seen as constant, so the linkage DE generates pure translational movement similar to linkage GH. The velocities of revolute joints B–K can be obtained as follows:

$$v_B = w_{BD} l_{O_1B} \tag{1}$$

$$v_C = w_{BD} l_{O_1C} = w_{AC} l_{AC} \tag{2}$$

$$v_D = v_E = w_{BD} l_{O_1D} = w_{FG} l_{EF} \tag{3}$$

$$v_G = v_H = w_{FG} l_{FG} = w_{IJ} l_{IH} \tag{4}$$

$$v_J = w_{IJ} l_{IJ} = w_{JK} l_{O_2J} \tag{5}$$

$$v_K = w_{JK} l_{O_2K} = w_{LK} l_{LK} \tag{6}$$

where $l_{AC}, l_{EF}, l_{FG}, l_{IH}, l_{IJ}, l_{JK}$ and $w_{BD}, w_{AC}, w_{FG}, w_{IJ}, w_{JK}, w_{LK}$ are the length and angular velocity of the corresponding linkages, respectively. $l_{O_1B}, l_{O_1C}, l_{O_1D}, l_{O_2J}, l_{O_2K}$ and l_{LK} are the rotation radii of the corresponding revolute joints, which could be easily calculated using the sine theorem of triangles.

The rotational angles of flexure hinges are analyzed with the input displacement y_B. In order to clearly describe the rotational deformation, the amplification ratios of the EDSR and LTMs are calculated using a simplified analytical model [21].

$$amp_{EDSR} = \frac{l_{AB}}{l_{AD}} \tag{7}$$

$$amp_{LTM-low} = \frac{l_{FG}}{l_{EF}} \tag{8}$$

$$amp_{LTM-up} = \frac{l_{IJ}}{l_{IH}} \tag{9}$$

where l_{AB} and l_{AD} are the distance of revolute joins A, B and A, D, respectively.

As shown in Figure 4, the linkages AC, BD, FG, IJ, JK and KL generate rotation angles Δ_1, Δ_2, Δ_3, Δ_4, Δ_5 and Δ_6, respectively, and the rotation of linkages DE and GH are ignored for the approximate translational motion. Based on the formula of arc length, the rotation angles of revolute joints A–L are calculated using the following equations:

$$\varphi_A = \Delta_1 = \frac{y_B}{l_{AD}} \tag{10}$$

$$\varphi_C = \Delta_1 + \Delta_2 = \frac{2y_B}{l_{AD}} \tag{11}$$

$$\varphi_B = \varphi_D = \Delta_2 = \frac{y_B}{l_{AD}} \tag{12}$$

$$\varphi_E = \varphi_F = \varphi_G = \Delta_3 = \frac{amp_{EDSR} y_B}{l_{EF}} \tag{13}$$

$$\varphi_H = \varphi_I = \Delta_4 = \frac{amp_{EDSR} amp_{LTM-low} y_B}{l_{IH}} \tag{14}$$

$$\varphi_J = \Delta_4 + \Delta_5 = \frac{amp_{EDSR} amp_{LTM-low} y_B}{l_{IH}} + \frac{amp_{EDSR} amp_{LTM-low} amp_{LTM-up} y_B}{l_{O_2 J}} \tag{15}$$

$$\varphi_K = \Delta_5 + \Delta_6 = \frac{amp_{EDSR} amp_{LTM-low} amp_{LTM-up} y_B}{l_{O_2 J}} + \frac{amp_{EDSR} amp_{LTM-low} amp_{LTM-up} y_B}{l_{LK} \cos\gamma} \tag{16}$$

$$\varphi_L = \Delta_6 = \frac{amp_{EDSR} amp_{LTM-low} amp_{LTM-up} y_B}{l_{LK} \cos\gamma} \tag{17}$$

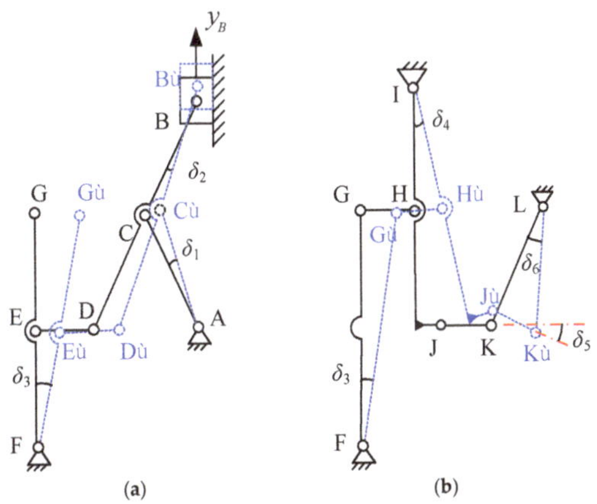

Figure 4. The deformation diagram of half RA-DFM: (**a**) the lower stage; (**b**) the upper stage.

Based on the above analysis, the rotational deformation of the assumed revolute joint L is the final output angle of the RA-DFM, which is expressed in Equation (17). The output angle is affected by the total amplification ratio, the arm length of the output platform and the input displacement.

3.2. The Output of Piezoelectric Ceramic

The actual output displacement of piezoelectric ceramic is related to the external stiffness; thus, the input stiffness of RA-DFM is firstly modeled. Assuming an input displacement y_B is applied on the input end, the generated force of the DLPH and torques of the right circular flexure hinges A~L are:

$$F_{DLPHy} = -k_{DLPHy}y_B \tag{18}$$

$$M_{ri} = -K_{ri}\varphi_i \quad i = A \sim L \tag{19}$$

where k_{DLPHy} is the translational stiffness of DLPH in the Y direction, and $K_{DLPHy} = \frac{4Eb_{low}t_{DLPH}^3}{l_{DLPH}^3}$ [21]. b_{low}, t_{DLPH} and l_{DLPH} are the width, thickness and length of the leaf spring hinge in the DLPH, respectively. K_{ri} is the rational stiffness of flexure hinges A~L, and $K_{ri} = \left(\frac{9\pi r_{Ri}^{1/2}}{2Eb_i t_{Ri}^{5/2}}\right)^{-1}$ for the right circular hinges A–K [21]. b, r_R and t_R are the width, radius and thickness of the right circular hinge, respectively. It is noted $b = b_{low}$ for the right circular hinge in the lower mechanism, and $b = b_{up}$ for the right circular hinge in the upper mechanism.

The revolute joint L is simplified from the FSM, which includes two leaf spring hinges. The compliance matrix of flexure hinges has been widely applied to analyze the stiffness of flexible mechanisms [12,21–23]; thus, it is used to model the rotational stiffness of the revolute joint L in this paper. As shown in Figure 5, the local coordinates of leaf spring hinges are expressed as o_i-$x_iy_iz_i$, and the global coordinate O_0-$X_0Y_0Z_0$ is defined in the center of the output platform. The compliance matrix of the leaf spring hinge in the local coordinate is expressed as

$$\mathbf{C}_0 = \begin{bmatrix} \frac{l_{FSM}}{Eb_{up}t_{FSM}} & 0 & 0 & 0 & 0 & 0 \\ 0 & \frac{4l_{FSM}^3}{Eb_{up}t_{FSM}^3} & 0 & 0 & 0 & \frac{6l_{FSM}^2}{Eb_{up}t_{FSM}^3} \\ 0 & 0 & \frac{4l_{FSM}^3}{Eb_{up}^3t_{FSM}} & 0 & -\frac{6l_{FSM}^2}{Eb_{up}^3t_{FSM}} & 0 \\ 0 & 0 & 0 & \frac{l_{FSM}}{Gk_2b_{up}t_{FSM}^3} & 0 & 0 \\ 0 & 0 & -\frac{6l_{FSM}^2}{Eb_{up}^3t_{FSM}} & 0 & \frac{12l_{FSM}}{Eb_{up}^3t_{FSM}} & 0 \\ 0 & \frac{6l_{FSM}^2}{Eb_{up}t_{FSM}^3} & 0 & 0 & 0 & \frac{12l_{FSM}}{Eb_{up}t_{FSM}^3} \end{bmatrix} \tag{20}$$

where l_{FSM}, t_{FSM} and b_{up} are the length, thickness and width of the leaf spring hinge in the FSM, respectively. E and G are the Young's modulus and shear modulus of the material, respectively. k_2 is a geometric parameter, and $k_2 = b_{up}/t_{FSM}$.

In order to obtain the compliance model of the FSM in the global coordinate O_0-$X_0Y_0Z_0$, the local compliance matrices of the two leaf spring hinges are transferred to the global coordinate.

$$C_{FSM} = \sum_{i=1}^{2} \begin{bmatrix} \boldsymbol{R}_i & 0 \\ 0 & \boldsymbol{R}_i \end{bmatrix} \begin{bmatrix} \boldsymbol{I} & \boldsymbol{P}_i^T \\ 0 & \boldsymbol{I} \end{bmatrix} \mathbf{C}_0 \begin{bmatrix} \boldsymbol{I} & 0 \\ \boldsymbol{P}_i & \boldsymbol{I} \end{bmatrix} \begin{bmatrix} \boldsymbol{R}_i^T & 0 \\ 0 & \boldsymbol{R}_i^T \end{bmatrix} \tag{21}$$

where $\boldsymbol{P}_i$ and $\boldsymbol{R}_i$ are the translational and rotational matrices, respectively, from the local coordinate to the global coordinate. For the ith flexure hinge, $\boldsymbol{P}_i = \begin{bmatrix} 0 & -r_{iz} & r_{iy} \\ r_{iz} & 0 & -r_{ix} \\ -r_{iy} & r_{ix} & 0 \end{bmatrix}$, $r_i = \overrightarrow{o_iO_0} = \begin{bmatrix} r_{ix} & r_{iy} & r_{iz} \end{bmatrix}$ is the vector from the origin o_i to the global origin O_0 in the local coordinate o_i-$x_iy_iz_i$. R_i is the rotation matrix of the coordinate o_i-$x_iy_iz_i$ with respect to O_0-$X_0Y_0Z_0$, and the rotation matrices about the X, Y and Z axis are $\boldsymbol{R}_x(\theta_x) = \begin{bmatrix} 1 & 0 & 0 \\ 0 & \cos\theta_x & -\sin\theta_x \\ 0 & \sin\theta_x & \cos\theta_x \end{bmatrix}$, $\boldsymbol{R}_y(\theta_y) = \begin{bmatrix} \cos\theta_y & 0 & \sin\theta_y \\ 0 & 1 & 0 \\ -\sin\theta_y & 0 & \cos\theta_y \end{bmatrix}$ and $\boldsymbol{R}_z(\theta_z) = \begin{bmatrix} \cos\theta_z & -\sin\theta_z & 0 \\ \sin\theta_z & \cos\theta_z & 0 \\ 0 & 0 & 1 \end{bmatrix}$, respectively.

The relationship between the deformation and the applied force on the FSM is

$$\boldsymbol{F} = \boldsymbol{C}_{FSM}^{-1}\boldsymbol{S} \tag{22}$$

where $S = [X_0, Y_0, Z_0, \theta_{X0}, \theta_{Y0}, \theta_{Z0}]$ and $F = [F_{X0}, F_{Y0}, F_{Z0}, M_{X0}, M_{Y0}, M_{Z0}]$.

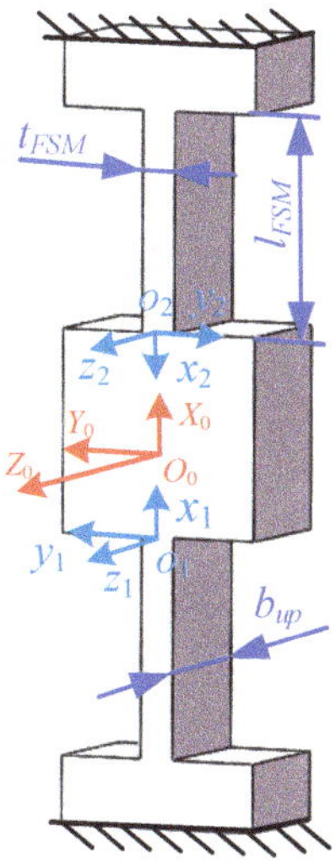

Figure 5. The miniature output platform and FSM.

According to the simplification rule of the revolute join L, the rotation direction of the revolute join L is the θ_{Z0} direction of the FSM, and considering the symmetry of the FSM, the rotational stiffness of L is half of the rotational stiffness of FSM in the θ_{Z0} direction, which could be calculated via M_{Z0}/θ_{Z0}; it is also the $6_{th} \times 6_{th}$ element of $\boldsymbol{C}_{FSM}^{-1}$ based on Equation (22).

$$K_{rL} = \left(\boldsymbol{C}_{FSM}^{-1}\right)_{6,6}/2 \tag{23}$$

The flexible mechanism is a stable system after an input force F_{yB} from the piezoelectric actuator is applied on the input end; the generated displacement on the input end is y_B and the angular deformation of each joint is listed in Equations (10)–(17). Based on the principle of virtual work, the total virtual work of the system is zero if a virtual displacement Δ_{yB} is applied on the system.

$$\Delta W = F_{yB}\Delta_{yB} + F_{DLPHy}\Delta_{yB} + 2\sum_{i=A}^{L} M_{ri}\Delta_{\varphi i} = 0 \quad i = A \sim L \tag{24}$$

where the first term denotes the virtual work caused by the input force F_{yB} and the virtual displacement Δ_{yB}. The second term denotes the virtual work of the DLPH due to the deformation of DLPH and elastic force F_{DLPHy}. As shown in Figure 6, the third term is the virtual work caused by torsional springs with rotary torque M_{ri} and virtual angular deformation $\Delta_{\varphi i}$. The relationships between $\Delta_{\varphi i}$ and Δ_{yB} are same as those between φ_i and y_B.

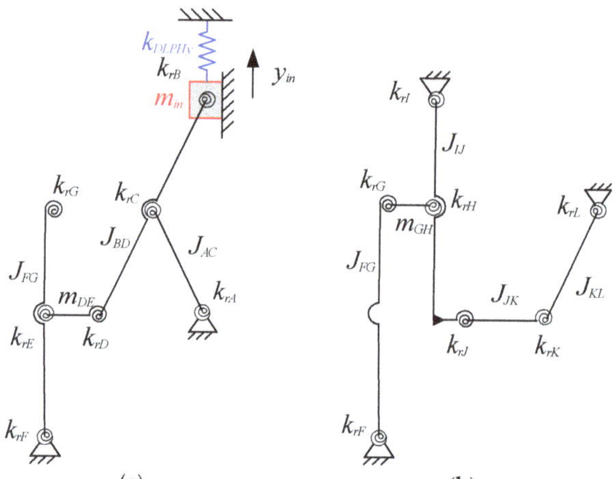

Figure 6. The dynamic model of half RA-DAFM: (**a**) the lower stage; (**b**) the upper stage.

Substitute Equations (18) and (19) into Equation (24), and the input stiffness of the RA-DFM in the Y direction is obtained.

$$K_{in-Y} = F_{yB}/y_B \tag{25}$$

Generally, the piezoelectric ceramic could be seen as a spring with constant stiffness, and the flexible mechanism acts as the load for the piezoelectric ceramic. Based on the Hooke's law, the actual output of piezoelectric ceramic is reduced according to

$$y_{apzt} = \frac{K_{pzt}}{K_{pzt} + K_{in-Y}} y_{npzt} \tag{26}$$

where K_{pzt} and $Snpzt$ are the stiffness and nominal output of piezoelectric ceramic, respectively.

3.3. The Stress Analysis

In order to guarantee the repeatability and durability of the RA-DFM, the maximal stress of the flexure hinges must be less than the allowable stress of the material. According to the kinematic analysis, the leaf spring hinges and right circular hinges mainly generate translational deformation and rotational deformation, respectively. Based on the stress analysis in Ref. [24], the maximum stress of each can be expressed as follows:

$$\sigma_{leaf-max} = \frac{3k_l \Delta E t_{leaf}}{l_{leaf}^2} \tag{27}$$

$$\sigma_{rch-max} = \frac{6k_c K_r \varphi}{bt_R^2} \tag{28}$$

where k_c and k_l are the stress concentration factor of the right circular hinge and leaf spring hinge, respectively. Δ and φ are the translational displacement of a leaf spring hinge and the angular deformation of a right circular hinge, respectively.

3.4. The Dynamic Modeling

In order to maximize the working frequency of a rotary actuator, the first resonant mode is better along the working direction because far less working frequency is usually required compared to the first resonant frequency to avoid activating the resonant vibration.

Therefore, the first mode shape of the RA-DFM is designed to be the rotation of the output platform, which is also the translation in the Y direction for the input end. The Lagrange equation is applied for the modeling of the resonant frequency.

$$\frac{d}{d_t}\left(\frac{\partial T}{\partial \dot{y}_{in}}\right) - \frac{\partial U}{\partial y_{in}} = 0 \tag{29}$$

where T and U are the total kinetic energy and elastic potential energy, respectively. y_{in} is the assumed input displacement.

The dynamic model of the half RA-DFM is shown in Figure 6; the flexure hinges are simplified as torsional springs with constant stiffness. It is noted that only planar motion is considered in the dynamic model due to the high rigidity of the flexure hinges in the out-of-plane configuration. Based on the velocity and deformation analysis in Section 3.1, when an input displacement y_{in} is enacted on the input end, the total kinetic energy of the system is expressed as

$$T = \frac{1}{2}m_{in}\dot{y}_{in}^2 + m_{DE}v_D^2 + m_{GH}v_G^2 + +\sum J_i w_i^2 \qquad i = AC, BD, FG, IJ, JK, KL \tag{30}$$

where m_{in}, m_{DE} and m_{GH} are the masses of the input end and linkages DE and GH, respectively. J_i is the rotational inertia of the linkage i. w_i is the angular velocity of the linkage i, which is also the first-order derivative of φ_i.

Potential energy mainly results from the deformation of the flexure hinges. The total potential energy of the system can be expressed as

$$U = \frac{1}{2}K_{DLPHy}y_{in}^2 + \sum_{i=A}^{L} K_{ri}\varphi_i^2 \qquad i = A \cdots L \tag{31}$$

Substitute Equations (30) and (31) into (29) and transfer it to the standard dynamic equation $M_{in-Y}\ddot{y}_{in} + K_{in-Y}y_{in} = 0$. Based on the theory of vibrations and the dynamic equation, the first resonant frequency of RA-DFM can be obtained using the characteristic equation:

$$f_1 = \frac{\sqrt{K_{in-Y}/M_{in-Y}}}{2\pi} \tag{32}$$

where $M_{in\text{-}Y}$ is the equivalent mass of the RA-DFM in the input end.

4. The Optimization

According to the above analyses, the performance of the RA-DFM is significantly affected by the geometric parameters; thus, it is necessary to conduct mechanical optimization to improve the static and dynamic characteristics. A large output range is able to extend the application, which is related to both the geometric parameter and the input displacement according to Equation (17). This section focuses on the maximization of the output range through optimizing the geometric parameters. The objective function is expressed as:

$$F_{\max} = \frac{amp_{EDSR}amp_{LTM-low}amp_{LTM-up}}{l_{LK}\cos\gamma} \tag{33}$$

Considering the stress, output of piezoelectric ceramic, resonant frequency, structure compactness and machining accuracy of the RA-DFM, the constraint conditions for the optimization are given as follows:

(1) Maximum stress: As mentioned previously, the maximum stresses occur at the surface of the leaf spring hinge and/or the minimal thickness of the fight circular hinge. Both of the maximum stresses should be lower than the allowable stress of the material.

$$\begin{cases} \sigma_{\text{leaf}-\max} \le \sigma_{\text{m}}/s_f \\ \sigma_{rch-max} \le \sigma_{\text{m}}/s_f \end{cases} \tag{34}$$

where σ_m is yield strength, and it is about 400 MPa for the aluminum alloy 7075-T6, which has been widely applied as the material of the flexible mechanism. s_f is the safety factor and has been chosen as 1.2.

(2) Input stiffness: According to Equation (26), a smaller input stiffness could improve the actual output of the piezoelectric actuator; thus, the input stiffness of the RA-DFM is limited in the following condition based on the stiffness of the ordinary commercial piezoelectric actuator:

$$K_{in-Y} \leq 2\ \text{N}/\mu\text{m} \tag{35}$$

(3) First resonant frequency and mode shape: The first resonant frequency of the developed mechanism decides the working frequency; thus, the low-limit frequency of the RA-DFM is restrained to guarantee the working bands.

$$f_1 \geq 200\ \text{Hz} \tag{36}$$

The resonant mode shape of the FSM in the vertical direction (Y_0 direction) is possible to activate due to compliance. In order to guarantee the rotational motion is the first mode shape, the resonant frequency of the FSM in the Y_0 direction is required to be larger than that in the rotational direction.

$$\frac{\sqrt{\left(C_{FSM}^{-1}\right)^{2,2}/m_{FSM}}}{2\pi} \geq \frac{\sqrt{\left(C_{FSM}^{-1}\right)^{6,6}/J_{FSM}}}{2\pi} \tag{37}$$

where m_{FSM} and J_{FSM} are the mass and rotational inertia of the moving platform in the FSM, respectively.

(4) Machining accuracy and structure compactness: To guarantee the machining accuracy and the material buckling of the leaf spring hinge and right circular hinge, it is feasible to choose $0.05 \leq t_R/r_R \leq 0.65$, $l_{DLPH} \geq 15t_{DLPH}$ and $l_{FSM} \geq 15t_{FSM}$. Considering the assembly of the upper and lower mechanism, it is better to satisfy $l_{EG} = l_{HJ} = l_{AB}/2$. Further, taking the theoretical modeling precision and structural compactness into account, the ranges of the variables are chosen as the third column of Table 1.

Table 1. The key sizes of RA-DFM.

Variables	Original Values (mm)	Ranges (mm)	Optimization Results (mm)
b_{low}	12.00	[10, 15]	10.00
b_{up}	8.00	[5, 10]	5.00
t_{DLPH}	0.5	[0.3, 1]	0.30
l_{DLPH}	15	[10, 30]	30.00
l_{AB}	40	[30, 50]	41.11
l_{AD}	20	[15, 30]	20.00
l_{FG}	40	[30, 50]	31.45
l_{IJ}	40	[30, 50]	49.54
t_{FSM}	0.5	[0.3, 1]	0.30
l_{FSM}	15	[10, 30]	10.98
l_{JK}	15	[10, 20]	20.00
r_R	1	[0.5, 2]	2.00
t_R	0.5	[0.3, 1]	0.30

As listed in the first column of Table 1, the variables mainly include geometrical sizes of the right circular hinge, leaf spring hinge and the linkage lengths of the amplification mechanisms. The optimization process is conducted using the "fmincon" function in Matlab software, which is used to find local minima. Therefore, the optimization objective adopts the negative of Equation (33), and several different groups of original values in the defined ranges are selected to break the local optimization. After many attempts, for the smallest value of the objective function, the corresponding original values and approximative optimization results are obtained and shown in Table 1.

5. The Finite Element Analysis of RA-DFM

In the FEA modeling of the RA-DFM, the lower and the upper mechanisms are firstly modeled separately based on the optimized parameters, which are then formed into a new part in Workbench software to couple the nodes between the related bases and connecting blocks, and the threaded connections are simplified as rigid connections to reduce the structure complexity. Aluminum alloy is selected as the material, with Young's modules $E = 71$ GPa, density $\rho = 2770$ kg/m^3 and Poisson's ratio $\nu = 0.33$. The RA-DFM is meshed using the adaptive method with 0.5 mm element size; the whole mechanism is meshed in 347,947 elements. The surrounding surfaces are completely fixed in the following simulation, as shown in Figure 7.

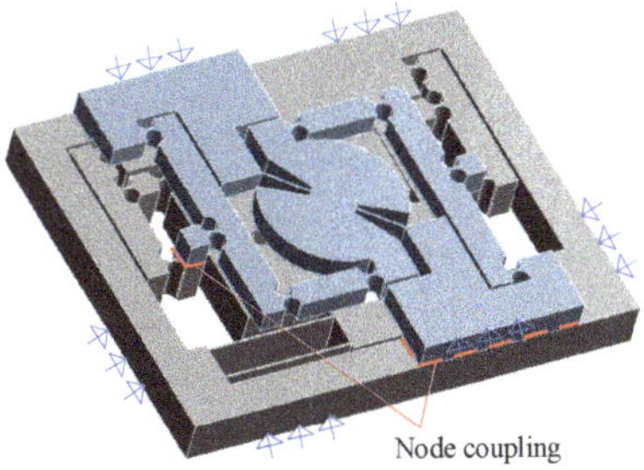

Figure 7. The FEA model of the RA-DFM.

An input force $F_{in} = 10$ N is applied on the input end in the Y direction to examine the input stiffness of the RA-DFM; the total deformation is shown in Figure 8. The input end translates in the Y direction by 5.83 μm, and the output platform generates a rotary angle as in the design. The input stiffness of RA-DFM is calculated as 1.71 N/μm, and the modeling error is 2.34% compared with the theoretical result of 1.75 N/μm, indicating the correctness of the theoretical model. In order to guarantee the output of the piezoelectric ceramic, the stiffness of the piezoelectric ceramic is preferred to exceed 10 times that of the external mechanism.

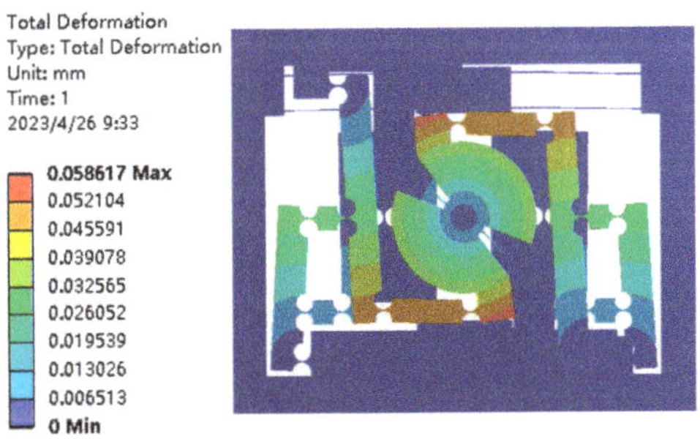

Figure 8. The deformation of the RA-DFM with input force 10 N.

The space reserved for the piezoelectric ceramic is 13 × 10 × 43.11 mm, which is suitable for commercial piezoelectric ceramic (MTP150/7 × 7/36, Coremorrow, Harbin, China). The nominal output displacement and stiffness are 38 μm and 50 N/μm, respectively. Based on Equation (24), the actual output of the selected piezoelectric ceramic is about 37 μm. A series of inputs y_{in} = [5, 10, 15, 20, 25, 30, 35, 37] μm is used to assess the output property. As shown in Figure 9 and Table 2, the output rotary angle is almost linear to the input displacement, and the total simulated amplification ratio of the RA-DFM is about 8.42, which is obviously less than the theoretical calculation of 10.10. The separate amplification ratios of the three amplification mechanisms are simulated as 1.78, 2.82 and

1.67, respectively. The largest modeling error occurs at the EDSR mechanism; it is because the rest of the flexible mechanisms generate a large load on the output terminal of the EDSR mechanism, which leads to serious deformation of the right circular hinges A and C in the X direction as well as the rotation. In addition, the linkages of the EDSR generate much more deformation compared to the lever-type mechanism.

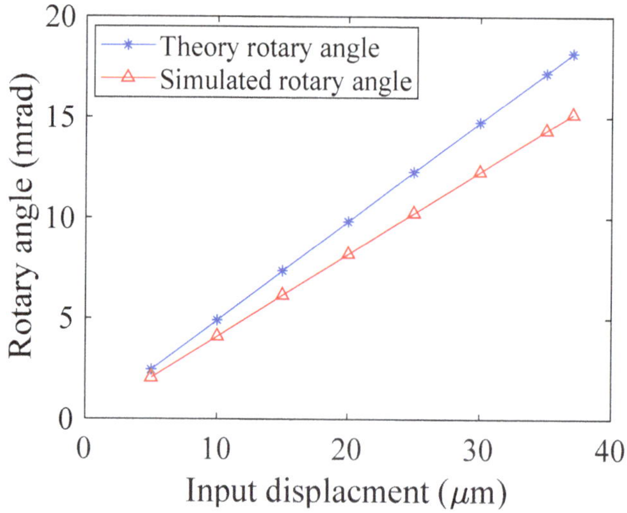

Figure 9. The output comparison of the theoretical and simulation results with different input displacement.

Table 2. The comparison of the theoretical and simulation results.

	amp_{EDSR}	$amp_{LTM\text{-}low}$	$amp_{LTM\text{-}up}$	$K_{in\text{-}Y}$	f_1
Theoretical calculation	2.05	2.88	1.71	1.75 N/μm	189.35 Hz
Simulation results	1.78	2.82	1.68	1.71 N/μm	192.45 Hz
Modeling error	15.16%	2.13%	1.79%	2.34%	1.61%

For the input placement y_{in} = 37 μm, the horizontal displacement generated on flexure hinge K is 296.46 μm, leading to a rotary angle of 15.14 mrad. Furthermore, the position of the output platform center is extracted to assess the drift of the rotational center. The total displacement change of the center is about 1.29 μm, which only accounts for 0.44% of the rotary displacement, showing the excellent rotational accuracy. In addition, the maximum stress of 194.08 MPa occurred at the circular hinge K in the FEA; it is because the stress of the circular hinge is positively correlated to the angle deformation. Flexure hinge K generates the largest deformation according to Equations (10)–(17). According to Equation (34), the simulated maximum stress is far less than the allowable yield stress of 333.33 Mpa for aluminum alloy 7075-T6, which indicates that the designed RA-DFM could work normally in this situation.

The modal simulation of the RA-DFM is also implemented, and the first three mode shapes are shown in Figure 10. It shows that the first mode shape is the rotation of the output platform in the θ_Z direction with the resonant frequency 192.45 Hz, which coincides well with the design objective. The modeling error is 1.61% compared with the theoretical calculation. The second mode shape is the translation in the Y_0 direction of the FSM with the resonant frequency 240.47 Hz, which is 24.95% larger than the first resonant frequency; this is mainly caused by the stiffness difference of the FSM in the θ_{Z0} and Y_0 directions.

The third mode shape is out of the plane with the resonant frequency 642.43 Hz, which is about 3.34 times the first resonant frequency; it means the third mode shape is hard to activate in low-frequency operation.

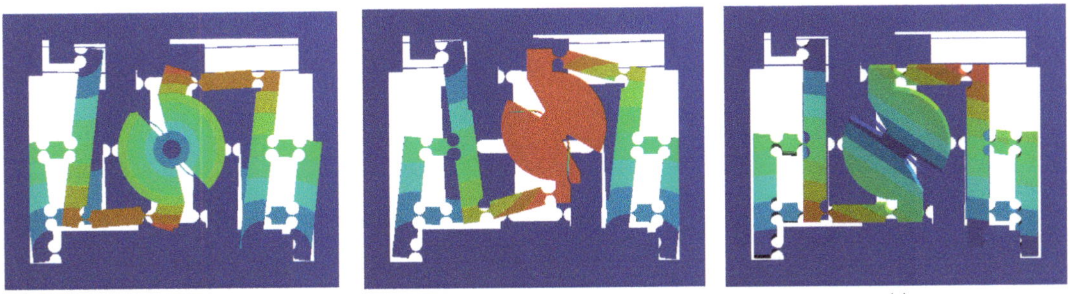

(**a**) (**b**) (**c**)

Figure 10. The first three mode shapes of RA-DFM: (**a**) the first mode shape; (**b**) the second mode shape; (**c**) the third mode shape.

6. Conclusions

A single piezoelectric-driven rotary actuator including a double-layer flexible mechanism was proposed in this paper. The actuator integrated an EDSRM and an LTM to amplify the output displacement of the piezoelectric ceramic in the lower mechanism. The rotational motion was generated on the upper mechanism as well as the third displacement amplification. The rotation angle of the mobile platform, actual output of the piezoelectric ceramic, maximal stress and first resonant frequency of the RA-DFM were theoretically modeled, and they were utilized to optimize the geometrical parameters.

The numerical models were validated to be efficient at predicting properties using the FEA method. According to the simulation results, the proposed RA-DFM possesses an amplification ratio of 8.42 and a first resonant frequency of 192.45 Hz. With 37 μm input displacement, the maximum stress of the RA-DFM is 194.08 MPa, and the maximal rotational angle of the output platform is 15.14 mrad with 0.44% center drift, showing excellent output properties.

Author Contributions: Conceptualization, Z.G. and Y.T.; methodology, Z.G. and P.Z.; software, P.Z. and W.Z.; validation, P.Z. and W.Z.; formal analysis, W.Z. and G.H.; investigation, Z.G., P.Z. and W.Z.; resources, Z.G. and Y.T.; data curation, G.H.; writing—original draft preparation, Z.G., P.Z. and W.Z.; writing—review and editing, G.H. and Y.T.; project administration, Z.G. and Y.T.; funding acquisition, Z.G. and Y.T. All authors have read and agreed to the published version of the manuscript.

Funding: This research was supported by the "National Natural Science Foundation of China (No. 52205138)", "Guangdong International Cooperation Program of Science and Technology (No. 2022A0505050078)", "Open Fund of State Key Laboratory of Precision Electronic Manufacturing Technology and Equipment, Guangdong University of Technology (No. JMDZ2021001)" and "Fundamental Research Funds for the Central Universities, CAUC (No. 3122019095)".

Data Availability Statement: Not applicable.

Conflicts of Interest: The authors declare no conflict of interest.

References

1. Habibullah, H. 30 Years of atomic force microscopy: Creep, hysteresis, cross-coupling, and vibration problems of piezoelectric tube scanners. *Measurement* **2020**, *159*, 107776. [CrossRef]
2. Wei, Y.; Xu, Q. Design and Testing of a New Force-Sensing Cell Microinjector Based on Soft Flexure Mechanism. *IEEE Sens. J.* **2019**, *19*, 6012–6019. [CrossRef]
3. Qin, Y.; Soundararajan, R.; Jia, R.; Huang, S. Direct inverse linearization of piezoelectric ceramic's initial loading curve and its applications in full-field optical coherence tomography (FF-OCT). *Mech. Syst. Signal Process.* **2021**, *148*, 107147. [CrossRef]

4. Zhang, Z.; Yang, X.; Yan, P. Large dynamic range tracking of an XY compliant nanomanipulator with cross-axis coupling reduction. *Mech. Syst. Signal Process.* **2019**, *117*, 757–770. [CrossRef]
5. Wang, T.; Li, Y.; Zhang, Y.; Lin, R.; Qain, J.; Dou, Z. Design of a flexure-based parallel XY micropositioning stage with millimeter workspace and high bandwidth. *Sens. Actuators A Phys.* **2021**, *331*, 112899. [CrossRef]
6. Bazaei, A.; Ettefagh, M.; Chen, Z. Displacement amplification and differential actuation in piezo driven nanopositioners. *Mech. Syst. Signal Process.* **2021**, *151*, 107356. [CrossRef]
7. Lyu, Z.; Wu, Z.; Xu, Q. Design and development of a novel piezoelectrically actuated asymmetrical flexible microgripper. *Mech. Mach. Theory* **2022**, *171*, 104736. [CrossRef]
8. Baek, S.; Cho, N.; Lee, D. Fabrication of a piezoelectrically driven micropositioning 3-DOF stage with elastic body using a multi-material 3D printer. *Rapid Prototyp. J.* **2020**, *26*, 1579–1591. [CrossRef]
9. Wang, R.; Zhang, X. Optimal design of a planar parallel 3-DOF nanopositioner with multi-objective. *Mech. Mach. Theory* **2017**, *112*, 61–83. [CrossRef]
10. Zhan, W.; He, X.; Yang, J.; Lai, J.; Zhu, D. Optimal design method for 3-DOF planar compliant mechanisms based on mapping matrix constraints. *Structures* **2020**, *26*, 1–5. [CrossRef]
11. Al-Jodah, A.; Shirinzadeh, B.; Ghafarian, M.; Das, T.; Pinskier, J. Design, modeling, and control of a large range 3-DOF micropositioning stage. *Mech. Mach. Theory* **2021**, *156*, 104159. [CrossRef]
12. Kim, H.; Gweon, D. Development of a compact and long range $XY\theta_z$ nano-positioning stage. *Rev. Sci. Instrum.* **2012**, *83*, 085102. [CrossRef] [PubMed]
13. Kang, S.; Lee, M.; Choi, Y. Six Degrees-of-Freedom Direct-Driven Nanopositioning Stage Using Crab-Leg Flexures. *IEEE/ASME Trans. Mechatron.* **2020**, *25*, 513–525. [CrossRef]
14. Lee, M.; Park, E.; Yeom, J.; Hong, D.; Lee, D. Pure Nano-Rotation Scanner. *Adv. Mech. Eng.* **2012**, *4*, 962439. [CrossRef]
15. Zhu, X.; Wen, Z.; Chen, G.; Liang, J.; Liu, P. A decoupled flexure-based rotary micropositioning stage with compact size. *Proc. Inst. Mech. Eng. Part C J. Mech. Eng. Sci.* **2018**, *232*, 4167–4179. [CrossRef]
16. Clark, L.; Shirinzadeh, B.; Zhong, Y.; Tian, Y.; Zhang, D. Design and analysis of a compact flexure-based precision pure rotation stage without actuator redundancy. *Mech. Mach. Theory* **2016**, *105*, 129–144. [CrossRef]
17. Yang, Z.; Zhou, X.; Huang, H.; Dong, J.; Fan, Z.; Zhao, H. On the Suppression of the Backward Motion of a Piezo-Driven Precision Positioning Platform Designed by the Parasitic Motion Principle. *IEEE Trans. Ind. Electron.* **2020**, *67*, 3870–3878. [CrossRef]
18. Chang, Q.; Liu, Y.; Deng, J.; Zhang, S.; Chen, W. Design of a precise linear-rotary positioning stage for optical focusing based on the stick-slip mechanism. *Mech. Syst. Signal Process.* **2022**, *165*, 108398. [CrossRef]
19. Sun, X.; Chen, W.; Zhang, J.; Zhou, R.; Chen, W. A novel piezo-driven linear-rotary inchworm actuator. *Sens. Actuators A Phys.* **2015**, *224*, 78–86. [CrossRef]
20. Song, S.; Shao, S.; Xu, M.; Shao, Y.; Tian, Z.; Feng, B. Piezoelectric inchworm rotary actuator with high driving torque and self-locking ability. *Sens. Actuators A Phys.* **2018**, *282*, 174–182. [CrossRef]
21. Guo, Z.; Zhang, W.; Tian, Y.; Zhang, Z.; Cao, Y.; Lu, X.; Zhang, T. Design, modeling, and testing of a one degree of freedom manipulator with three stage amplification mechanism. *Rev. Sci. Instrum.* **2022**, *93*, 123705. [CrossRef] [PubMed]
22. Liu, Y.; Zhang, Z. A large range compliant XY nano-manipulator with active parasitic rotation rejection. *Precis. Eng.* **2021**, *72*, 640–652. [CrossRef]
23. Yuan, L.; Wang, L.; Qi, R.; Zhao, Z.; Jin, J.; Zhao, C. A novel hollow-type XY piezoelectric positioning platform. *Int. J. Mech. Sci.* **2023**, *255*, 108496. [CrossRef]
24. Duan, Y.; Ou, K.; Sun, Y.; Tian, Y. Stress Analysis of the Circular Arc Flexure Hinge. *J. Mech. Transm.* **2016**, *40*, 148–170.

MDPI
St. Alban-Anlage 66
4052 Basel
Switzerland
www.mdpi.com

Actuators Editorial Office
E-mail: actuators@mdpi.com
www.mdpi.com/journal/actuators

www.ingramcontent.com/pod-product-compliance
Lightning Source LLC
Chambersburg PA
CBHW061446210926
51850CB00008B/1633